FOREST RESOURCE ECONOMICS
AND POLICY RESEARCH

FOREST RESOURCE ECONOMICS AND POLICY RESEARCH

Strategic Directions for the Future

EDITED BY

Paul V. Ellefson

Westview Press

BOULDER, SAN FRANCISCO, & LONDON

Copyright © 1989 by Westview Press, Inc., except for chapters 3, 11, 13, 15, 16, 19, 21, and 25, which are works of the U.S. government

Published in 1989 in the United States of America by Westview Press, Inc., 5500 Central Avenue, Boulder, Colorado 80301, and in the United Kingdom by Westview Press, Inc., 13 Brunswick Centre, London WC1N 1AF, England

Library of Congress Cataloging-in-Publication Data
Forest resource economics and policy research : strategic directions
   for the future / edited by Paul V. Ellefson.
      p.   cm.
   Includes bibliographies and index.
   ISBN 0-8133-7780-3
   1. Forest products industry—Government policy—United States.
2. Forest policy—United States.   3. Forests and forestry—Economic
aspects—United States.   I. Ellefson, Paul V.
HD9756.F66   1989
333.75′0973—dc20   89-32434
                                                                                    CIP

Printed and bound in the United States of America

The paper used in this publication meets the requirements of the American National Standard for Permanence of Paper for Printed Library Materials Z39.48-1984.

10   9   8   7   6   5   4   3   2   1

CONTENTS

v

FOREWORD

Forest economics emerged as a field of inquiry during the early decades of this century when a significant portion of the nation's conservation sentiment crystallized into concern for the allocation and use of forest resources for the greatest well-being of men and women (William A. Duerr and Henry J. Vaux, *Research in the Economics of Forestry*, 1953). A number of forestry issues with significant economic consequences attracted public attention during the 1920s and 1930s. Among them were growing concerns over scarcity of timber in the South, rising timber prices in the Northeast, social problems arising from the use of forest land for agriculture in the Lake States, and public timber sales, labor relations, and community stability in the West. Comprehensive analyses of such issues and the development of solutions to them required an understanding of forestry's technical and institutional framework as well as the economic environment in which forestry is practiced (forest economics).

Recognition and respect for forest economics grew as researchers in the field completed several studies of national significance, including the 1933 *National Plan for American Forestry* (Copeland Report), the 1935 *Forest Taxation Inquiry*, and the 1941 *Forests and National Prosperity: A Reappraisal of the Forest Situation in the United States*. These were followed by a number of national timber and renewable resource assessments, including, most recently, *The 1990 RPA Assessment of the Forest and Range Land Situation in the United States*. Forest economics thrived in an environment of concern over scarcity of forest resources. The work of forest economists helped other professionals as well as lay citizens to understand important forestry issues and the strategies available for their resolution.

The field of forest economics has continued to grow in importance. The number of forest economists increased as the usefulness of their contributions became more widely recognized. By 1985, more than 500 forest economists were directing investments of $30 million annually in forest economics research. Focused on a wide array of economic problems concerning the management and use of forest resources, forest economics research now encompasses virtually all of the benefits produced by forests and associated rangelands (e.g., timber, water, recreation, range, forage, wildlife, fish, wilderness). Forest economists also address a complex variety of markets that affect the production and distribution of these benefits and have stimulated relevant insights and contributions by sociologists, lawyers, political scientists, and other social scientists.

Forest economics of the future will be determined in large measure by the importance that society places on issues concerning the use and management of forests and by the effectiveness with which forest economists assist society in its search for means of resolving such issues. By the year 2020, the U.S. population will increase by nearly 30 million, and disposable income will triple, greatly increasing the demand for goods and services produced by the nation's forests. Compared to the late 1970s, the year 2030 will see a doubling of demand for timber and participation in dispersed camping and a 60 percent increase in recreational hiking and consumption of water flowing from forested watersheds. The demands for other forest benefits is also expected to increase—all to be met from a shrinking forest land base. In a worldwide context, the *Global 2000 Report to the President* projects that world population will increase from 4.5 billion in 1980 to more than 6 billion people by the year 2000. In such a setting, demands on the world's forested resources will become unprecedented. How can forest lands best be used to meet the expectations of growing populations? Which forestry benefits will be produced, in what quantities, and using what management practices? How well will existing markets allocate production resources to achieve social goals for the goods and services to be provided by forests, and what market interventions offer the greatest promise for achieving desired adjustments? These are but some of an extensive array of important forest resource issues that will confront society in this and other countries.

Clearly the issues will be important—what then of the effectiveness of forest economists in helping society resolve such issues? In large measure, this will be determined by the ability of forest economists to identify and focus limited research resources on the most important of many emerging issues. Marion Clawson's 1977 *Research in Forest Economics and Forest Policy* and the pioneering 1953 work of William A. Duerr and Henry J. Vaux, *Research in the Economics of Forestry*, sought to assess the status of forest economics and related policy research and to identify forest resource issues having economic implications deserving of the field's attention. This book represents a concerted effort to again assess the status of forest economics and policy research. A number of the nation's leading forest economics and policy researchers were asked to set forth their perspectives on future strategic research directions within their fields of expertise. More than 100 strategic directions are identified; more than 300 research topics are set forth.

As we begin further transitions to eras of increasing economic scarcity, the community of researchers applying economic and policy sciences to national and worldwide forest resource issues will become even more important to societal interest in high levels of social and economic well-being. Forest economics and policy researchers have a responsibility to rise to the challenge and to boldly address the innumerable forestry issues that will ultimately flow from social efforts to achieve such interests. This compendium of strategic research directions in the field of forest economics and policy is designed to facilitate these actions.

Carl H. Stoltenberg
Dean, College of Forestry
Oregon State University

PREFACE

The application of economic and policy sciences to issues emanating from the use and management of renewable forest resources has a substantial and respectable history. Forest resource professionals in various employment categories have applied the tools of economic and policy evaluation to issues ranging from economic potentials for producing timber on public lands to evaluation of recreation and wildlife outputs in a multiple production context. In recent years, however, technical and institutional advances have changed the manner in which economic and policy analyses are carried out and have influenced the effectiveness of products produced by such analyses. For example, issues demanding the attention of forest economics have grown by leaps and bounds and at the same time have become increasingly complex. Tools and techniques available for undertaking analyses have improved considerably, often leading to insatiable demands for costly information. And clients for the products of economic and policy analysis have grown in number and become more sophisticated in their expectations. Such changes, however, are the products of more deep-seated structural changes in the forestry community in general. For example:

- Ever-growing multiplicity of interests concerned with the outputs that forests are capable of producing.
- Major scientific and technical advances that have had an impact on the production of forest outputs and the efficiency of forestry programs in general.
- Increasingly complex policy-making processes, and incorporation therein of new, often noneconomic, decision rules.
- Mandated assessment and planning activities requiring significant economic information and analyses.
- Significant focus of forestry on economic development goals, especially in rural areas.
- Continuing and significant long-term economic instability within certain sectors of the wood-based industry.
- Growing focus on forestry in a global context, especially international trade and economic development of lesser-developed countries.

Long-term structural changes in technical, economic, and social systems will continue to have a major impact on the manner in which forest economics and policy sciences are applied. Within such a context, there is significant virtue in reflecting on the current status of forest economics and policy research and in characterizing a vision of future research in the field. Reviews can stimulate

ix

discussion of where the field is tending, what informational and intellectual products are likely to result, and the effects such outcomes are likely to have on forestry and related subject areas. Reviews can also encourage healthy debate and discussion, which can facilitate the identification of neglected research topics and highlight opportunities for the application of newly developed research techniques. This collection is intended to foster such reflection. Specifically, the intent of this collection is to

- Review the status of discipline-wide activity in forest economics and policy research, especially investment levels, past and current program emphasis, program planning, and organizational involvement.
- Define strategic directions for forest economics and policy research—directions that will respond to important economic and social conditions occurring now or likely to occur in the near future.

Discipline-wide reviews of forest economics and policy research are not new to the forestry community. In 1936, the Social Science Research Council undertook an "investigation of work being done in forest economics within the United States and of the institutions engaged therein," the product of which was *A Survey of Research in Forest Economics* (Social Science Research Council, New York, NY). In 1953, William A. Duerr and Henry J. Vaux completed a definitive assessment entitled *Research in the Economics of Forestry* (Charles Lathrop Pack Forestry Foundation, Washington, DC). And in 1977, Marion C. Clawson compiled *Research in Forest Economics and Forest Policy* (Resources for the Future, Inc., Washington, DC). Canadian regional interests in forest economics research were identified in "Forest Economics Research Needs for West-Central Canada" (W. E. Phillips, J. A. Beck and G. W. Lamble, University of Alberta, Edmonton, Alberta, 1986). Related research reviews important to forest economics and policy research include *Land Economics Research* (J. Ackerman, M. Clawson, and M. Harris, The Johns Hopkins University Press, 1962) and "Economics Research in Transition" (USDA–Economic Research Service, Washington, DC, 1985).

A systematic review of forest economics and policy research is a major undertaking that involves a number of important steps. Of special concern here was the identification of economic and social conditions that are likely to have a major influence on the forest economics and policy research agenda during the next eight to ten years. And within such a setting, identification of issue areas most worthy of attention by forest economics and policy researchers is important. The setting for economics and policy research and the issue areas emanating from such a setting were identified after careful review of more than seventy-five comprehensive international, national, and regional planning and issue documents prepared by a variety of public and private organizations ("Forestry Issues as a Focus for Forest Economics and Policy Research: A Review of Selected Issue Defining Documents," Paul V. Ellefson, Department of Forest Resources, University of Minnesota, St. Paul, MN). In addition, the objectives of nearly 150 existing forest economics and policy research projects were identified (via Current Research Information System [CRIS]) and assessed. After considerable debate and discussion, eight "settings" were selected, within which twenty issues were identified (e.g., international trade in forest and related products, management of fire in forested environments). The settings are as follows:

- Forestry will experience growing linkages to the world economy.
- Forest resource policy-making and program development will increase in complexity.
- Forest-based industries will continue significant restructuring within domestic and worldwide economies.
- Rural needs for social and economic development will increase as a focus of forestry interests.
- Technological developments will increasingly influence resource use and productivity.
- Management and production processes will grow increasingly complex.
- Information requirements and information management will increase as a concern of forestry interests.
- Forest use and management decisions will increasingly reflect environmental quality interests.

For each issue area, knowledgeable researchers, administrators, and policy-makers were commissioned to review the status of economic and policy information pertinent to an area; to define broad strategic research directions to be emphasized within an area over the next eight to ten years; and to specify sample research topics to be undertaken within each strategic direction. Authors placed considerable emphasis on reviews of existing literature and on definition of strategic directions. The directions were broadly defined so as to be suggestive of the scope of research that might be undertaken. Although an attempt was made to cover a broad range of topics important to forest economics and policy research, omission of subject matter important to some readers will surely have occurred. In view of the breadth and detail of the subject matter involved, no attempt was made to summarize the commissioned papers.

In addition to assessing strategic research directions, other topics concerning forest economics and policy research were addressed. Included in this volume is a review of research program development and accomplishments since the mid-1940s, an assessment of trends in research investments and program directions, an assessment of program evaluation and planning activities, and a review of organizational involvement in forest economics and policy research. Of special note is the volume's concluding chapter, which sets forth a number of professional and managerial challenges to those involved in forest economics and policy research.

The undertaking of an effort of this magnitude would not have been possible without the support of a number of organizations. In this respect, attention must be directed to the University of Minnesota's Department of Forest Resources, College of Natural Resources, and Agricultural Experiment Station. The generous support of the USDA–Forest Service is also acknowledged, especially the Policy Analysis Staff and the Forest Inventory, Economics and Recreation Research Staff of the agency's Washington, DC, office.

Paul V. Ellefson
Richard A. Skok

PART ONE

Historical, Institutional, and Investment Context

1

DEVELOPMENT AND ACCOMPLISHMENTS OF RESEARCH PROGRAMS

Henry J. Vaux and H. R. Josephson

Forest resource economics and policy research has evolved strikingly over the past forty years in a number of fundamental respects. For example, the size of the effort devoted to such research has grown substantially. The research methods available for use have become much more sophisticated. And the array of problems needing and receiving attention has multiplied in number and become much more complex. But what can be said of major program developments and accomplishments in the field of forest economics and policy research? And how has forest economics and policy research related to major contemporary trends in national affairs, of which forestry research is a part? Of necessity, answers to such questions must be broad in scope. A fully detailed account of program developments and accomplishments would far transcend the limits of space and time that apply here. Moreover, excessive detail could overwhelm the highlighting of fundamental societal forces that have strongly influenced research activity in the field of forest economics. In what follows, major emphasis will be given to the period 1947-1987. As appropriate, research begun before 1947 will be recognized in order to provide essential historical perspective.

Basic Trends Affecting the Forestry Sector

The four decades following World War II were, for the most part, periods of great dynamism in the whole of American society. The population of the country increased by more than seventy percent, real national product almost tripled, per capita real income doubled, and use of natural resources and energy rose substantially. Improved technology and better developed infrastructure expanded both the leisure time and the mobility of the steadily growing population. This almost unprecedented growth in numbers of people and their ability to exert effective demand for goods and services would, under any circumstances, have had strong repercussions within the forestry sector. But such effects were intensified by another fact. Prior to World

War II, the inventory of timber in the contiguous forty-eight states was so vast that both virgin timber and good site commercial forest land could be readily purchased at nominal prices in many Pacific Coast localities. As such, the primary determinants of both market and long run supplies of timber were inventory volume and its accessibility to the market. Under such circumstances, stumpage prices bore little relation to the cost of growing timber—market incentives to grow timber were generally lacking. Because transport costs for wood products were low in relation to product prices, the economic effect of large virgin timber inventories in the West also depressed timber prices in the South and East. About the time of World War II, these fundamental conditions of timber supply changed greatly. Old-growth timber inventories were finally reduced to levels where they were perceived as no longer adequate to sustain timber supply in the long run. Real prices for stumpage and forest land began to rise. For the first time, timber growing became a feasible economic activity and the costs of growing timber became a major determinant in long-run timber supply.

A number of events signalled this fundamental and permanent change in the economics of timber. For example, the first of many industrial tree farms was dedicated in 1941. At the same time, the pulp and paper industry, especially in the South, acquired extensive areas of cutover land to assure long-term wood supply for its rapidly expanding markets. After World War II, western industries brought pressure to greatly increase extension of roads and the annual harvest from the National Forests. And new enterprises were established in the virgin forests of Canada to supply growing U.S. markets that could no longer be fully served by opening up previously undeveloped domestic forests.

In contrast to earlier times, control over additional long-term supplies of timber was no longer to be had by moving to a little settled geographic area. New reserves for forest parks, wilderness, and other forms of dispersed forest recreation could no longer be found without reducing the apparent timber supply of existing industries and significantly modifying opportunities for other forest uses. These forces and events had, by the 1950s, wrought fundamental changes in the nature of American forestry—changes that greatly stimulated development of forest economics as a field and whose consequences continue to preoccupy forest economics and policy research. The more important and immediate impacts on forestry included the following.

- Steady and widespread increase in the economic value of forests and forest products. In real terms, stumpage prices increased two to five-fold (depending on region and species) during the period 1947-1987. The real value of commercial forest land used for timber-growing showed comparable increases in market price. Lumber prices increased roughly 50 percent in real terms.

Rising values of timber and timberland provided rapidly expanding incentives for analysis of timber supply, wood products demand, forest industry and market organization, wood product prices, and related economic topics. Increasing competition for timber and the rising value of wood commodities increased pressures for greater efficiency in the growing, manufacturing, and marketing of such materials.

Achieving increased efficiency required more and better economic data and more refined analysis across all wood-producing sectors. Expanded needs for housing, packaging materials, and numerous other products required to support national growth inspired continued attention to the nature and adequacy of raw material supplies and called for an emphasis on economic and geographic, rather than merely physical, dimensions of timber supplies.

Similarly, rising land and timber values intensified the need for more information and improved methods of stumpage valuation, land and timber assessment and taxation, efficiency of alternative contractual arrangements, and knowledge about the nature of markets in the forestry sector.

- Expanding population and income increased the demand for conversion of forest land to nonforestry uses for agriculture, home subdivisions, and attendant industrial and commercial purposes. Effects of these forces on the area of land available for forestry and on forest land values were more localized than those of rising wood prices. But, whether actually or potentially at work, they resulted in substantial land value increases and thus created incentives for additional research.

Ongoing land use adjustments affecting the forest base required research to determine their effects, project their future impacts, and provide factual and analytical bases. Such information was important for development of zoning controls and taxation, for forest planning within the remaining forest area, and for analysis of other problems associated with the geographic interface between the forest and adjoining areas.

- Growth in population, income, and leisure time led to dramatic increases in people's demands for outdoor recreation of all kinds, including forest-based recreation. The result was strong and sustained political pressure for the dedication of additional public forest land for wilderness, parks, and wildlife and recreation areas and increased incentives for protection and development of recreational values on private forest land.

These pressures led forest economics and policy research into areas of study and methodology that previously had been virtually unexplored. Because most forest recreation services are unmarketed, the solving of acute economic and policy problems required the development of data and methods appropriate for establishing reasonable equivalents to market values for wilderness use, hunting and fishing, camping sites, and a variety of other activities within the spectrum of outdoor recreation activities. Transferring land from what had been considered by many as part of the timber growing economy was often vigorously opposed by industry groups and others because of perceived adverse economic impacts. Hence, the new methods were quickly applied in an effort to make reasonably valid comparisons of the economic and social values at stake under various combinations of forest uses. Research approaches to such comparisons were further complicated by the geographically specific requirements of many recreational uses and by the need to consider cumulative impacts of changes in resource allocation.

• Public concern and specialists' attention broadened from almost exclusive preoccupation with timber-related problems to considering the entire spectrum of multiple-use values. This occurred partly as a result of the growing body of research information available about forests and partly because of growing public interest in forests for recreational use and related concerns.

This shift in emphasis, combined with the fact that the major short-run supply response to increased demand for forest recreation was accommodated on public land, was reflected in research. More attention was directed to the nature and content of decision making in the public sector, to economic implications of budgeting for forestry agencies, and to interactions between economic and political forces in determining policy. On the one hand, forest economists had to devote more attention to factors outside the traditional market framework. On the other, specialists in social science disciplines other than economics were drawn into the analysis of forestry problems.

• Increasing public concern with environmental considerations such as air and water quality, control of pesticides and toxins, habitat requirements for rare and endangered species, and the adverse effects of congestion in certain critical areas also pushed research and analysis into new terrain.

The biological foundations on which economic and policy analysis rested had to be broadened to include not only the vegetative cover but the entire forest ecosystem. Off-site and ambient effects and influences had to be incorporated in analysis to a greatly increased degree. Because policy approaches to such issues as air and water quality often involved legally mandated constraints, the interaction between such constraints and established ownership rights in forest property received increasing attention from economists and lawyers. Application of environmental quality constraints required the establishment of standards (such as those for water quality and for tolerable use of toxic material) that raised significant economic issues and provided incentives for their analysis.

• Advanced technologies became available at reasonable cost, especially recon-naissance satellites with ability to provide earth-surface imagery of great scientific sophistication and computers with ability to store, recall, and analyze relatively vast amounts of data. These and other technological advances provided new tools for resource managers and gave forest economics research and policy analysis increased capabilities.

Using such technologies, resource inventories could be made in greater detail and could be kept more current; market behaviors could be monitored, recorded, and recalled more comprehensively; and models for the analysis of economic and planning problems could be designed with a level of comprehensiveness well in advance of what had been previously possible. Problems that had been insoluble because of requirements for data or computational time became the objects of applied studies. An example is the detailed projection of timber supply, demand, and price over several decades—a problem in which intertemporal relations between variables must be recognized and accounted for. Such technological advances permitted analysts to take account of many more variables than had previously been practical. And the biological production functions on which much analysis in forest economics must rest could be built on much improved quantitative and qualitative bases.

Technological advances, particularly those in transportation and communications, also affected the management of forests and their societal context. They permitted forests to be organized and managed in units of increasing size and encouraged consolidation of many land ownerships. Consumers, whether of wood or recreation, were brought effectively closer to the woods. Development of improved equipment resulted in greater efficiency in timber harvesting and fire control. Developments in applied biochemistry led to much broader use of herbicides and pesticides in forests, although there was increased public concern over potential problems of toxicity. Such changes in resource management not only posed new questions for both production-oriented and environmentally oriented research but also intensified the need for information to help resolve administrative, legal, and judicial conflicts within a benefit-cost framework.

By 1987 more people were demanding a greater quantity and diversity of products and services from forests than in previous American experience. As a result, there was widespread public concern for the protection and production of forest values and much more intense political conflict over what balance among forest products and services was appropriate and how best to distribute them among competing interests. The geographic focus of forest problems became more localized as members of the public became intimately familiar with specific resource areas. Conflicts between traditional, frontier-oriented perceptions of forest values and the perceptions of urban publics emerged in most sections of the country, whether the issue was more wilderness reservation, additional protection for rare species of wildlife, the role of fire and clearcutting in forest management, or the use of rationing devices such as user fees or reservation systems for previously freely available forest benefits.

Greater attention to local forest use issues served to intensify apparent conflicts between national, regional, and local forest community points of view. Economists and policy analysts were pushed to develop methods for analyzing tradeoffs in resource management from the standpoint of their national versus regional versus local impacts. Such problems required much more attention to questions about the distribution of income than earlier required of production-focused research. "Who gains and who loses?" became a common question in analyses of forestry program alternatives.

All aforementioned developments tended to stimulate forest economics research and the analysis of policy problems. The growing importance of forest resources in the minds of many members of the public—and persistent demands for changes in forest use by some of them—provided more incentives to conduct such research.

More powerful methodologies and more and better-trained research workers provided the means to broaden and intensify research and analysis efforts. As a result, substantial advances in methods for evaluating extra-market benefits were achieved— some by forest economists, others by the growing body of economists who specialized in environmental problems or in natural resources other than forests. Collection of data relevant for such problems accelerated. As a result, useful comparative evaluations of multiple use became feasible and were more widely applied. Computer capabilities permitted much more robust analyses of the consequences of various alternatives in resource use over the long-time horizons inherent in timber production. Such advances, along with the development of tools such as modeling, intensified the collection and publication of basic resource information and related economic data, the appearance of continually updated bibliographies, and the proliferation of national, regional, and local associations of forest economists along with conferences and workshops. All such activities evidenced both growth in the field of forest economics and a clearer definition of the subject as a valuable area of specialization.

Major Problem Areas Addressed: 1947-1987

Collection of Economics Statistics

The collection and publication of a wide variety of economic statistics provides information essential for most research purposes. The U.S. Bureau of the Census, for example, has long compiled annual or periodic data on industrial operations, industry production, employment, expenditures, and related matters for the various forest industries. The USDA–Forest Service for many years collected statistics on production of lumber, pulpwood, naval stores and other products, wood preservation, and stumpage and log prices. Some data on production of wood products are still collected by the USDA–Forest Service and by various state agencies. The U.S. Department of Labor has maintained statistical series on employment and wages in various occupations, together with prices of many products. Import and export data have been compiled by the U.S. Department of Commerce and the U.S. Tariff Commission. The Census of Agriculture has included data on forest lands in farms and income from forest products. And many state agencies issue localized price and market reports for forest products, generally with the aim of aiding owners of small forest properties to market their timber more effectively. Over the 1947-1987 period, the richness of statistics available for forest economics research increased. New statistical time series became available; the quality of data improved; definitions were sharpened; and additional, more localized information became available.

Economics of Timber Growing and Utilization

The fact that more than seventy percent of the commercial forest in the United States is owned by private individuals or corporations has served to direct much research attention to the economics of timber growing by private owners. Fostering early interest in timber-growing economics was taxation of private forest land and timber, which was often perceived as a serious economic obstacle to long-term profitable timber growing. The federal Forest Taxation Inquiry, initiated in 1924 and headed by distinguished Yale University economist Fred R. Fairchild, studied the effects of tax law and practice on forest perpetuation and devised alternative forms of taxation designed to encourage the conservation and growing of timber (Fairchild 1935). These proposals led to the adoption of numerous state laws providing special treatment of property taxes on private forests. During the last four decades, additional research on alternatives to the general property tax, usually by state level agencies, has contributed to the adoption of yield or severance tax alternatives to the general property tax in a number of states.

Federal capital gains treatment of timber income and expensing of the costs of forestry operations also emerged as an important focus of economic analysis. Legislation for both federal and state income taxes was influenced by these efforts. In addition, studies of federal contributions in lieu of state and local taxes on federally owned resources have aided in analysis of recurring legislative proposals for change in existing policies.

Many investigations have provided evaluations of the financial aspects of private forestry. Early studies of the costs and returns of timber growing undertaken at Crossett, Arkansas, and other experimental forests demonstrated the practicality of regenerating and managing young-growth forests. Information on costs and benefits of forest protection and management practices has been made available to many parts of the country through studies by the USDA–Forest Service, universities, and other organizations.

Selective timber management has been guided in part by concepts of financial maturity. Capital budgeting concepts and guidelines have aided in the allocation of funds available for forestry purposes. And studies of investment opportunities in reforestation, stand management, and other forestry practices have been of major importance to the evaluation of timber-growing opportunities.

Research aimed at improving the efficiency of marketing, harvesting, and manufacturing timber products has involved forest resource economists for many years. Logging and manufacturing costs for different species, sizes, and grades of timber have been related to the value of the several products that could be produced by such timber. Research of this nature aided in increasing operating efficiency by simplifying the identification of marginal logs, trees, and stands and provided economic guides for partial cuttings. They contributed to more accurate appraisals of timber values. Today, less research effort is devoted to such topics, largely because the methods developed by research have been incorporated into operating systems. Market analyses have dealt with such matters as the impacts of tariffs or other trade restrictions on prices and shipments of wood products and the impacts of log exports or wood product imports on forest industries and other domestic interests. Following the rapid increase in timber sales on federal land in the West after 1950, significant research was done on the structure of such stumpage markets, the influence of alternative timber sale practices, and the implications of marketing practices for timber sale policy.

Because of their diverse character, the response of private forest owners to market stimuli or to public forestry programs have differed sharply depending on the type of private owner. Hence, research of private forest ownerships have evaluated differences in ownership intent, size of holdings, relationship to markets, and other factors. Research of this nature has revealed certain owner characteristics (e.g., residence, age, managerial abilities and background) as reasons for ownership of forests, availability of capital, and attitude towards public and private forestry assistance programs. Determinations of the costs and responses to publicly provided financial incentives, technical assistance, regulation of timber harvesting, and other measures have provided an indication of the effectiveness of such programs in enhancing timber supplies and other benefits from such ownerships. In the case of industrial forestry, studies have revealed the economic structure of forest land owning companies, the comparative economic performance of such companies, and financial aspects of major ownership changes.

Economics of Resource Protection

Protection of forest resources from fire, insects, and disease was the central objective of some of the earliest and most extensive governmental forestry programs. For such programs, economic research has aided in the development of economic criteria for blister rust control programs in California, the northern Rocky Mountains, and the Northeast; the concepts developed have been applied at least in part in control programs aimed at other forest pests. Development of the least-cost-plus-loss notion permitted significantly improved efficiencies in the planning of fire control programs. Both examples illustrate cases in which the application of economic principles (once economists had identified them and suggested methodologies) quickly became incorporated in routine resource management and administrative planning.

Community Dependency and Land Use Changes

Research that focused on community dependency on forest resources and forest product industries has quantified the relationships between forest production and employment, income, tax base, and other regionally significant economic variables. Development of methodologies and data for input-output analysis has permitted more widespread study and a more effective analysis of regional policy issues. Usually such research has been aimed at identifying trends in dependency or at assessing effects of alternative forest policies. Examples of the latter include evaluation of proposals for major land use changes, such as creation of a national park, alternative timber harvest and investment levels on public land management units, limiting log exports, or comparing the advantages of recreation versus timber development in particular forest areas. They have also been employed in research addressing policies for economic development of stagnant rural economies.

Nontimber Uses of Forests

As the use of forests diversified, the need for information on the economic aspects of forest recreation, wildlife, water yields and quality, flood control, and other forest-derived goods and services rapidly expanded. In virtually all such areas the contributions of forest economists have been closely interwoven with those of other resource economists who have approached the problems from the perspective of general recreation, flood control, water quality control, or some other nonforest vantage point. Methods and measurement of demand for forest recreation (including social and economic characterization of the recreationists), evaluation of flood control costs and benefits, estimation of benefits from fish and wildlife, and costs and benefits of accommodating water and air quality standards have been among the topics with which forest resource economists have been concerned.

Considerable information on nontimber resources and uses has been brought together in a number of general appraisals conducted by federal agencies, commissions, and private organizations. Examples of comprehensive investigations include reports of the Outdoor Recreation Resources Review Commission (1962), the National Water Commission (1973), the President's Materials Policy Commission (1952), and the Public Land Law Review Commission (1970) as well as recent national assessments of the forest rangeland situation called for by the Forest and Rangeland Renewable Resources Planning Act of 1974 (RPA). As in the case of appraisals relating to timber, reports of this nature have been designed to provide the factual and analytical basis for strengthening resource policies and programs.

Multiple-Use Planning

Adoption of the National Forest Management Act of 1976 and the Federal Land Policy and Management Act of 1976 led to major and comprehensive administrative programs for multiple-use planning for USDA–Forest Service and USDI–Bureau of Land Management lands. Pursuant to these statutes, the role of economic analysis in designing and choosing among alternatives for the management of specific public forest areas was greatly broadened. Supply-demand relationships for all resources on each planning unit and the economic implications of adopting alternative combinations of multiple uses in land management programs were developed and given explicit consideration in planning decisions. The extent to which such planning is to be considered research is, if not a sterile question, largely a matter of definition. However, there is no doubt that several considerations highly significant to the development of forest economics research have emerged from these very extensive multiple-use planning projects. Forest economists have been involved with multidisciplinary teams in planning forest operations on public land to a degree not previously experienced. They have had to deal with problems in obtaining data, specifying variables, and integrating information from a variety of specialties on a scale not previously undertaken. Statutory requirements for public input into the forest resource planning process have raised questions about the nature of decision making for public land that may redefine the role of economics in such decision making. Regardless of whether land use and management planning activities are embraced within the definition of the field of forest economics research, such activities are fostering important interactions with the theories, conceptualizations, and research of forest economists.

Attention in federal forest planning has clearly been broadened from analysis of single uses, such as timber or forage production, to consideration of the costs and benefits of integrated resource management for various combinations of timber, recreation, wilderness, forage, water yield, wildlife, and other forest uses, each at various levels of management intensity. This has required development of new techniques for multi-resource inventories, determination of interproduct production relationships, and new technologies (planning models such as FORPLAN) to analyze various management alternatives. Substantial and largely unsolved problems persist in achieving fully comparable methods for evaluating market and nonmarket values, identifying and quantifying joint product relationships, and means for properly considering intergenerational comparisons of both supply and demand.

International Trade and Development

The 1947-1987 period was marked by a substantial increase in U.S. foreign trade in wood products, particularly for lumber and related imports. Large integrated wood-products corporations also became increasingly active in exporting wood products, especially logs, and (for a time) in developing additional supplies of wood products from off-shore sources. Research on international trade in forest products was responsive to this growth in activity as information was required to guide marketing strategies, international trade policies (e.g., tariffs), and estimates of demand and prices for domestic timber supplies. In addition to monitoring timber trade on essentially a world-wide basis, specific studies were made of the economic effects of log export restrictions, and of the comparative advantage for timber products of domestic and foreign competitors in U.S. and foreign markets.

U.S. forest economists have increased the scope of their international activities beyond that of international trade in wood. They have widened their participation in scholarly exchanges with other countries, in technical assistance programs in the Third World, and in direct employment or consultation with international or foreign forestry agencies. Such contacts resulted in U.S. forest economists incorporating into their work some analyses of economic and policy problems of forestry in other countries. Comparisons and contrasts emerging from such studies served to enrich the field.

Also originating from the international scene was the rise in domestic energy prices experienced during the 1970s. In most regions, such increases resulted in rapid increases in fuelwood consumption, keen interest and some investment in plants to produce energy from wood (biomass either from utilization of residues from existing wood products plants or from fuelwood plantations), and in some reevaluation of alternative production methods in light of a changed energy price structure. Although the short-term incentives for economic research on the effects of high cost energy on the forestry sector largely disappeared with the ending of the energy crises in the 1980s, this area of research is likely to see renewed activity when longer-run forces once again dominate international petroleum markets.

Goals for National Forest Policy

Forest economics research has long been identified with appraisals of the nation's timber situation and outlook and related policy implications. These appraisals have dealt primarily with the economic subject of supply and demand for timber and timber products. They have also illustrated the close relationship between economic considerations and formulation of forest policy that has generally been characteristic of forest economics research. Among the early appraisals of timber resources and uses in the United States was the 1911-1914 investigation by the Bureau of Corporations (1911-1914) in response to congressional resolutions concerned with possible trusts and restraint of trade in the lumber industry. Reports from this investigation focused on concentration of timber ownership, impacts of concentration on timber prices and lumber supplies, and recommendations aimed at retention of public lands and timber in federal ownership.

Beginning in 1920, the USDA–Forest Service periodically studied issues concerning appropriate goals for U.S. forest policy and progress toward their achievement. After World War II, these studies were published under various titles in 1948, 1958, 1965, 1974, and 1982 (Forest Service 1948, 1958, 1965, 1974, 1982). All were focused on timber production goals and prospects for meeting them; leadership for their accomplishment was lodged in the Washington office of the USDA–Forest Service. Each study summarized and analyzed a large amount of inventory and research information drawn from many sources around the nation. Even a cursory review of these reports demonstrates the evolution of increasingly sophisticated economic terminology and methodology. In large measure, these advances were due to the increasing influence of well-trained economists in the USDA–Forest Service's Division of Forest Economics Research and cooperating agencies.

Following adoption of the Forest and Rangeland Renewable Resources Planning Act of 1974 (RPA), the timber supply and demand analyses characteristic of earlier USDA–Forest Service reports were subsumed within congressionally mandated periodic assessments called for by the RPA. They were broadened to embrace several multiple forest uses, rather than limiting attention to timber as the earlier reports had done. Both the RPA and the subsequent National Forest Management Act of 1976 expressly provided for the use of economic methodologies and criteria as components of planning and policy analysis. As a result, the role of economic analysis in resource assessments was significantly enlarged to include supply-demand relationships for all resources on forests and related lands.

The statistical basis of periodic timber appraisals has been steadily strengthened with resource data from a nationwide forest survey and with related studies of trends in markets for timber products as authorized by the McSweeney-McNary Forest Research Act of 1928. The forest survey has furnished increasingly accurate data on timber inventories, growth, mortality and removals, ownership of forest land and timber, and opportunities for enhancing timber yields on different types and sites. About half the country was surveyed by World War II and the remainder by the 1960s. Since the late 1940s, surveys have provided updated information at ten- to fifteen-year intervals. Analytical reports for individual states as well as the national appraisals have provided a basis for judging the effectiveness of existing forestry policies and programs and have indicated measures that would help increase future timber supplies. Survey data have also been of major importance in industrial programs of land acquisition, forest management, and plant construction—witness the spectacular expansion of the southern pulp and paper industry beginning in the 1930s and the establishment of a southern plywood industry in the 1960s.

The efficiency and accuracy of timber surveys have been significantly enhanced by development of new and improved techniques for tree and stand measurements, projections of timber growth and available harvests, and use of a variety of computer programs. The usefulness of survey findings has been enhanced by improved evaluations of forest investment opportunities and supply-demand-price relationships for timber and timber products. Many scientists in universities and other agencies as well as the USDA–Forest Service have contributed to these developments.

Periodic appraisals of the forest situation and the emergent analyses pursuant to the RPA have all been strongly oriented toward policy analysis and prescription, each of which resulted in the summarizing of much research. In addition to these purposes, however, such appraisals and analyses have had two other very important consequences for forest economics research. First, the studies embraced the entire forestry sector of the U.S. economy. They included, for example, material varying from the biological characteristics of wood production on the land to demographic relationships and projections that influence the demand for housing and other wood products. Thus, they required integration of economic and other information across all of the complex interrelationships that characterize a major sector of the national economy. They served to put together into a single cohesive framework the results of a large body of forestry research—ranging from silviculture to wood technology—and fostered its interpretation in terms of broad economic significance. Given the extreme difficulty of this integration task, the degree of success achieved by periodic timber appraisals and by analyses pursuant to RPA is far more remarkable than the fact that they fell short of perfection in some respects.

A second and very important aspect of the integrative effort involved in national appraisals and analyses is their effect on research per se. Because the timber appraisal and the RPA studies were policy oriented, they received particularly critical scrutiny both inside and outside the USDA–Forest Service. At each stage, this critical scrutiny highlighted significant gaps in information or method that hampered or undermined the integrative effort. Over the decades, forest economics research was able to respond positively to much of this criticism. Significant gaps in information were filled, data were made more comparable, and conceptualization of the structure of the forest economy made more consistent. As a consequence, integrated analysis of broad forestry problems was a much more refined art at the end of the period than at the beginning.

Broad Dimensions of Accomplishments

Of these accomplishments of forest economics research and policy analysis over the past four decades, both the variety of issues addressed and the degree to which economic research has been embodied in policy analysis appear impressive, although they defy quantification or other form of summary. However, certain common denominators run through virtually all of the work surveyed. These, too, suggest something about both the magnitude and quality of the accomplishments.

Extent and Quality of Information Base

Because it is an applied field and responds directly to real world problems, a primary constraint on progress in forest economics is the availability of data that accurately reflect real world economic phenomena. The amount and quality of such data appear to have expanded dramatically over the past forty years. Today, much more economic information is available on timber and other resource inventories, on demands for goods and services provided by those resources, and on the prices of commodities, services, and factors of production. Much more is known about how major features of the forest production system can be expected to perform, including the production capabilities of forest lands, the managerial capacities and goals of forest owners, or the biological or societal interactions between competing forms of forest use. Much more powerful models of forest economic systems relevant for public policy have been built. Because they incorporate a variety of conceptual advances and have the capacity to deal with more variables and a much bigger data base, these models have great capability to develop valid comparisons between economic alternatives and to project alternative scenarios into the murky future. The Timber Assessment Market Model and FORPLAN are but two examples of such accomplishments. Substantial progress is evident in the ability to focus such improved tools of analysis on local and regional problems as well as national problems.

Conceptual Foundations

Forest resource economists and researchers in related fields have significantly improved the definition, rigor, and applicability of concepts used to analyze economic and policy problems. Examples include the replacement of simple growth-drain or requirements concepts for timber supply assessment by concepts based on rigorous application of supply and demand models, the development of several conceptual approaches to estimating market value equivalents of extra market benefits, the application of benefit-cost and cost-effectiveness concepts in the design and selection of major projects for resource investment, and the availability of more sophisticated models to aid managers faced with decisions about resource management and utilization. In recent years, significant progress has been made in the analysis of multiple use problems by modeling the underlying resource production system in a way that accounts for interactions between localized geographic units.

Improved data base and strengthened conceptualizations permit today's forest economists to attack important analytical and policy problems that were simply beyond reach forty years ago. Adding to such capabilities has been the success with which individuals with strong foundations in social sciences other than economics (e.g., law, sociology, and public administration) have been drawn into forestry research. Contributions from such fields have helped to more sharply define both the limits of effective economic analysis and how it may be used effectively in some untraditional areas.

Training and Employment of Forest Economists

Beginning in the 1930s, a growing number of foresters began to acquire advanced training in economics and, along with a few general and agricultural economists, helped staff the expanding programs of economics research and new teaching positions at universities. In the period 1940-1950, 452 researchers were reported to be leaders of economic studies pertaining to forestry, of which thirty-three could be regarded "as the active corps of forest economists" (Duerr and

Vaux 1953). About half the leaders of economics research had advanced degrees, of which about two-thirds had undergraduate degrees in forestry and one-third in general economics or another specialty. For some years, the training of forest economists was largely in departments of agricultural economics; by 1950, some forestry schools were granting advanced degrees in forest economics. During ensuing decades, such programs became much more numerous and their graduates became the dominant source of recruits for research organizations.

Forestry research at universities expanded significantly following enactment of the McIntire-Stennis Forest Research Act of 1962. This provided for federal matching funds for a broad program of forest and related range and watershed research at state-supported colleges and universities. The requirement for matching funds led to greater financing from state sources; the buildup of research capabilities at many universities was a factor in obtaining additional funding from other sources. University research in forest economics expanded along with other research, and the increased training of graduate students associated with McIntire-Stennis funding also added to the supply of forest economists and other forest scientists. The USDA–Forest Service's Division of Forest Economics Research played a role in the training program by identifying individuals with appropriate interests and potentialities and encouraging them to undertake and complete Ph.D. programs.

A major contribution to the training of forest economists (and of others) was the publication of a number of college-level textbooks on forest economics and forest policy. Those by Worrell (1959, 1970), Duerr (1960), and Gregory (1972, 1987) have been widely used and have added much to the definition of the field of forest economics.

As larger numbers of trained forest economists became available and as demands for forest economics and policy analyses grew and diversified, the numbers and kinds of agencies supporting forest economics research expanded. In the late 1940s, over ninety percent of forest economics researchers were employed in federal agencies or at forestry schools and state experiment stations. Forty years later, a large number of forest economists are employed by forest industries, consulting forestry firms, independent nonprofit foundations, staffs of legislative bodies, and lobbying organizations. Some of the work of these economists involves activities other than research, but their efforts also testify to the widening influence of accomplishments attributable to forest economics.

A great enrichment of the research environment in which forest economists work has also occurred. In 1950, with aid from the Charles Lathrop Pack Foundation and the USDA Library, Professor Duerr arranged for compilation and publication of a comprehensive research bibliography for the field. Subsequently, with additional help from the USDA–Forest Service and a number of universities, this and other bibliographies have been updated at regular intervals (now available from the University of Minnesota). The availability of research tools such as bibliographies has facilitated the work of many forest economists.

Communications among researchers in the field have been fostered also by the emergence of a number of informal but very active regional associations of forest economists and by the Society of American Foresters' Division of Forest Economics and Policy and the more recent Working Group on Forest Economics, Policy and

Law. These groups, through regular meetings, newsletters, and a variety of projects, have provided both stimulus and identity to individuals in the field and have facilitated the personal interactions among researchers that help invigorate any scientific endeavor.

Scope of the Specialized Field

In 1902, Bernard E. Fernow published *Economics of Forestry* in which he wrote that the field of forest economics included "such information as will enable [one] to form an intelligent view and a true estimate . . . of the position which the community and governments should take with reference to their forest resources" (Fernow 1902). During the period between the two world wars, the central focus of U.S. forest economics and forest policy was the adequacy of future timber supply. After 1950, it became apparent that prospective timber supply-demand relationships were only part of a broader issue in need of resolution if an "intelligent view" of forest resources was to be reflected in community and government policy. The increasing scarcity of forest land had made competition from nontimber forest uses an increasingly important factor in influencing future timber supplies. As the competition between timber and nontimber uses intensified, it became necessary to place timber, recreation, high quality water yield, wilderness, and other forest uses in a framework of analysis that could balance the competition among all such uses.

Providing information of such comprehensiveness was a formidable challenge. By 1987, forest economics research had provided much information to guide the management of timber resources. Some methods and data for valuing nontimber goods and services had been pioneered, and significant applications had been made. Much less has been accomplished in providing information on costs of producing extra-market values. In the case of most environmental services, the need to quantify with meaningful precision the real costs of such benefits presents a challenge not only to forest economists but also to biologists and other scientists. Although progress is being made on such tasks, there will be a continuing need for analysis of various market sectors of the forest economy. The societal context that surrounds the forest and influences forest uses undoubtedly will remain highly dynamic. Answers to questions about the impact of these societal dynamics on the market uses of forests will have to be continually reevaluated.

Literature Cited

Bureau of Corporations. 1911-14. Report of the Commissioner of Corporations on the lumber industry. U.S. Department of Commerce and Labor. Washington, D.C.

Duerr, W. A. and H. J. Vaux. 1953. Research in the economics of forestry. Charles Lathrop Pack Foundation. Washington, D.C.

Duerr, W. A. 1960. Fundamentals of forestry economics. McGraw-Hill Publisher. New York, NY.

Fairchild, F. R. and Associates. 1935. Forest taxation in the United States. Misc. Pub. 218. U.S. Department of Agriculture. Washington, DC.

Fernow, B. E. 1902. Economics of forestry. T. Y. Crowell Publishers. New York, N.Y.

Forest Service. 1948. Forests and national prosperity. Ag. Misc. Publ. 668. U.S. Department of Agriculture. Washington, D.C.

Forest Service. 1958. Timber resources for America's future. Forest Resource Report 14. U.S. Department of Agriculture. Washington, D.C.

Forest Service. 1965. Timber trends in the United States. Forest Resource Report 17. U.S. Department of Agriculture. Washington, D.C.

Forest Service. 1974. The outlook for timber in the United States. Forest Resource Report 20. U.S. Department of Agriculture. Washington, D.C.

Forest Service. 1982. An analysis of the timber situation in the U.S. 1952-2030. Forest Resource Report 23. U.S. Department of Agriculture. Washington, DC.

Gregory, G. R. 1972. Forest resource economics. Ronald Press. New York, NY.

Gregory, G. R. 1987. Resource economics for foresters. John Wiley Publishers. New York, NY.

National Water Commission. 1973. New directions in U.S. water policy: Summary, conclusions and recommendations. Arlington, VA.

Outdoor Recreation Resource Review Commission. 1962. Outdoor recreation for America: A report to the President and to the Congress. Washington, D.C.

President's Materials Policy Commission. 1952. Resources for freedom: A report to the President. Washington, D.C.

Public Land Law Review Commission. 1970. One-third of the nation's land: A report to the President and to the Congress. Washington, D.C.

Worrell, A. C. 1959. Economics of American forestry. Wiley Publishers. New York, NY.

Worrell, A. C. 1970. Principles of forest policy. McGraw-Hill Publishers. New York, NY.

2

PROBLEM ORIENTATION AND INVESTMENTS IN RESEARCH PROGRAMS

Paul V. Ellefson

Forest economics as a field of interest is not new to the American forestry scene, even though specific conditions leading to its origin as a distinct field of interest (apart from a more general concern over forest conservation) are not clear. If a period of time for its origin were to be identified, focus would have to be on the early decades of this century, when conservation sentiments began to crystallize into a concern for the allocation and use of forest resources. As might be expected, forest economics was born in a shower of data; it developed amidst the gathering and compiling of statistics that sought to define the magnitude of forests in terms of area, timber volumes, growth, and harvests. Statistics of an economic nature soon followed, especially information on prices, trade, consumption, shipments, and timber requirements. Most of the statistical evaluations were performed by federal agencies, especially the USDA–Forest Service and the U.S. Department of Commerce.

A major surge in interest in forest economics occurred in the late 1940s and early 1950s. This was the result of leadership exercised by newly trained "forest economists" as well as a perception that timber supplies of the future must be founded on the production of wood fiber in a specific locale. The wood-based industry's migratory era had come to an end. Implied was the concept of scarcity and the need for the application of an allocation science—economics. Personalities and issues involving forest economics of the 1950s and 1960s form an especially interesting segment of the nation's forest history (Lage 1986). Brought to attention by significant reports such as *Timber Resources for America's Future* in 1958; *Timber Trends in the United States* in 1965; *The Outlook for Timber in the United States* in 1974; and *An Analysis of the Timber Situation in the United States* in 1982, forest economics and research involving forest economics became refined and focused as a discipline within forest conservation. Today it has attained prominence as a field providing much information about the use and management of forests.

Forest economics and forest economics research gained clarity and definition as discussion of its nature and application became more widespread. A major effort to define the economics of forestry (forest economics) and to explore forest economics as a field of research was the 1953 *Research in the Economics of Forestry* (Duerr and Vaux 1953). The topical areas of interest included the forest economy at large (location and flow of economic activity, planning and evaluation of programs); agents of production (labor, capital, land, entrepreneurship); forest management (forest valuation, nontimber management, enterprise activities); timber harvesting and processing; supply and marketing (marketing, marketing agencies, pricing, requirements); and demand for forest products. These attempts to define forest economics research did much to crystallize the nature of the topic and focus the efforts of those involved.

In 1977, forest economics research was the subject for further review by Resources for the Future; a review that led to *Research in Forest Economics and Forest Policy* (Clawson 1977). The breadth of research topics had expanded considerably since 1953 to include projecting demand and supply of forest products; pricing timber products; evaluating nontimber forest outputs; modeling multiple forest land-use interactions; evaluating the social role of forests; organizing information resources; international forestry; evaluating agents of forest production; and policy and administration, including agenda setting, policy processes, administrative-political decision rules and alternative institutional arrangements.

Definition, Scope, and Intent of
Forest Resource Economics

As with any new discipline, forest economics and forest economics research was beset by definitional problems in its early years; in some respects, these problems continue today. Attempts to define the subject in a definitive sense might be viewed as futile; at a minimum, however, such efforts serve to focus attention on the broadening scope of forest economics and the transitory nature of its boundaries. In the early 1950s, Duerr and Vaux (1953) did much to clarify forest economics. However, they still found it "immature" in three aspects, namely, unclear in content and scope, imperfect in terms of purpose and function, and unsystematic in the use of research methods. Although to a lesser degree than in the past, concern over such shortcomings continues to this date.

The content of forest economics can be clarified somewhat by focusing on the component terms "economics" and "forest." This approach led Vaux (Duerr and Vaux 1953, p. 15-16) to define forest economics as

> concerned with problems of [economically] allocating productive resources so as to maximize the returns from them, wherever the use of forest resources is involved. It is comparable to the parallel fields of labor economics, agricultural economics and land economics. . . . Since forest economics enters into the solution of all of the forester's practical problems, the boundaries of the field are coextensive with those of the whole of forestry. . . . [It] is parallel to each of the technical [forestry] subdivisions, from forest soils to wood chemistry, and from animal ecology to aerial photogrammetry.

The significant definitional constraints are "economics" and "forest resources." As these terms are defined and redefined, so is the field of forest economics and forest economics research. In recent times, "forest economists" have seen fit to apply noneconomic disciplines (e.g., organizational-administrative theories; statistical-biometrics theories, mathematical theories; sociological theories; and political science theories) to forest resource problems, and in doing so have made the label "forest economics" less useful. The latter also occurs as forest economists move to apply talent to a broader scope of natural resources, both renewable and nonrenewable (e.g., minerals, wildlife, water). In such circumstances, the term "forest economist" fails to communicate an accurate picture of the subject material involved; alternative labels are then assumed (e.g., resource economist, natural resource economist, renewable natural resource economist).

Many organizations undertake forest economics research, not all of which have their interests focused primarily on forest resources. This does not facilitate a clear understanding of the bounds of the subject area. For example, significant forest economics research is conducted by the USDA–Economic Research Service, the USDA–Business Innovative Research Grant Program, and the USDA–Agricultural Cooperative Service—all of which have missions focused primarily on agriculture. Likewise, research on the economics of using wood fiber as an energy source is undertaken by a number of university research centers that have the broader mission of identifying economically efficient sources of energy—wood fiber being one possibility. Forest economics research has the involvement of many institutions; most certainly it is not solely the domain of the forestry organizations that fostered its original growth, namely, the USDA–Forest Service or the state agricultural experiment stations.

Intent can also be a source of difficulty when attempting to define forest economics. If agreement is attained on subject matter, definitional questions of research versus application can surface. Rooted in fundamental questions of what "research" is, the problem becomes important as forest economists move from a research role to an analyst role. In the latter, the tools of economics are applied to current forest resources problems of concern to administrators interested in information for decisionmaking—persons not necessarily interested in a solution to a fundamental problem involving scarcity (e.g., valuation of nonmarket forest outputs) (Ellefson 1986). The research versus analyst roles were surfaced by Bromley (Economic Research Service 1977, p. 105):

> [USDA–Economic Research Service] has been much too involved as a "Congressional Research Service for Agriculture," and is also at the beck and call of the executive branch. Of course, only the politically naive would expect otherwise, but the capacity of ERS to do truly imaginative anticipatory thinking and research on important issues has been sacrificed for the job of providing numbers on demand. While I appreciate the political problems of saying "go away and leave us alone for a spell," it would seem that there must be several places in the agency where more of this sort of activity could take place.

Assessment of forest economics research will also invariably confront the "basic" versus "applied" and the "research" versus "applied" continua. Time spent at-

22 *Paul V. Ellefson*

tempting to define these extremes may be unwise; better that it be devoted to developing an understanding of which direction forest economics is tending.

Magnitude of Research Investments

The magnitude of forest economics research in the United States can be viewed as substantial or virtually nonexistent, depending on the scope of activities included within the domain of forest economics. For discussion purposes, six research problem areas as defined by the USDA–Cooperative State Research Service seem appropriate (Cooperative State Research Service 1978): (1) alternative uses of land; (2) economic and legal problems in the management of water and watersheds; (3) economics of timber production; (4) development of markets and efficient marketing of timber and related products; (5) supply, demand, and price analysis of forest products; and (6) multiple-use potential of forest land and evaluation of forestry programs. Investments in these problem areas include those made by the U.S. Department of Agriculture, the state agricultural experiment stations, and a variety of other organizations such as forestry schools and certain private sector entities acting as project cooperators with public agencies (Cooperative State Research Service 1988). Not specifically included is research carried out by private organizations (e.g., foundations and industry) or by specialized public institutions.

Research Projects

Research projects involving forest economics and related research numbered 543 in 1987, seven more than the previous high that occurred in 1983 (see Table 2.1). Since 1966, the trend in project numbers has—for the most part—been steadily upward. The vast majority of the projects (seven out of ten) are administered by state agricultural experiment stations. Although increasing significantly in 1986 and 1987, projects administered by the USDA–Forest Service averaged twenty seven per year over the period 1966 through 1985. The number of agricultural experiment station projects involving forest economics research increased nearly 64 percent between 1966 and 1987.

The average 1987 forest economics research project involved an investment of slightly more than $54,000, which supported 0.38 scientist years of effort. The latter is down considerably from the approximate one-half scientist year of effort that occurred annually between 1966 and 1985. As measured by scientist years, state agricultural experiment station projects involving forest economics research are substantially smaller than USDA–Forest Service projects. Between 1980 and 1985, the former averaged 0.3 scientist years while the latter ranged from 1.9 to 2.5 scientist years per year. For the eight-year period beginning in 1980, USDA–Forest Service investments per project (averaging $228,000) were approximately seven times that invested per project by state agricultural experiment stations.

Scientist Years

Scientist years devoted to forest economics research totaled 206 in 1987—down from the 262 annual average that occurred from 1981 through 1985 but up from the scientist years of effort that occurred in the mid-1970s. State agricultural

Table 2.1 Forest Economics and Related Research, by Organization, Number of Projects, Scientist Years, and Funding, 1966-1987

Year	U.S. Department of Agriculture Forest Service	Other Agencies	State Agricultural Experiment Stations	Other Organizations	Total
Number of Projects					
1987	59	26	396	62	543
1986	56	30	380	64	530
1985	30	28	392	53	503
1984	31	33	387	54	505
1983	33	25	413	65	536
1982	31	29	397	60	517
1981	34	31	377	70	512
1980	30	24	355	72	481
1979	30	31	350	66	477
1978	28	22	340	57	447
1977	29	21	344	56	450
1976	23	22	346	57	448
1975	20	18	302	40	380
1974	22	17	274	14	327
1973	20	17	259	24	320
1972	20	16	220	27	283
1971	20	18	216	32	286
1970	19	45	247	31	342
1969	20	35	248	34	337
1968	21	38	230	18	307
1967[a]	28	47	236	30	341
1966	35	57	242	43	377
Scientist Years					
1987	52.0	46.2	93.4	14.6	206.2
1986	48.4	43.4	91.9	13.2	196.9
1985	68.4	77.4	102.6	12.2	260.6
1984	59.7	73.4	104.7	16.6	254.4
1983	68.1	51.0	122.5	13.0	254.6
1982	74.6	59.0	116.4	11.6	261.6
1981	83.2	66.5	112.6	18.1	280.4
1980	76.6	26.5	105.1	20.3	228.5
1979	74.0	32.4	99.7	15.1	221.2
1978	84.9	35.1	103.7	13.7	237.4
1977	74.9	28.2	104.8	11.0	218.7
1976	59.6	32.6	94.8	16.4	203.4
1975	54.9	27.8	76.3	6.6	165.6
1974	61.8	28.8	70.6	2.5	163.7
1973	60.2	14.1	70.1	5.2	149.6
1972	62.4	14.7	57.0	5.6	139.7
1971	63.7	20.6	51.0	7.0	142.3
1970	60.4	21.7	57.6	3.8	143.5
1969	51.7	24.8	48.6	4.0	129.1
1968	56.3	29.0	47.1	0.3	132.7
1967[a]	61.8	31.3	61.8	3.5	158.4
1966	69.1	30.2	57.5	6.8	163.6

(continues)

Table 2.1 *(continued)*

Year	U.S. Department of Agriculture Forest Service	Other Agencies	State Agricultural Experiment Stations	Other Organizations	Total
		Investment ($-million)			
1987	8.07	4.85	14.67	1.93	29.52
1986	6.04	5.81	14.06	1.43	27.34
1985	7.85	8.22	12.89	1.50	30.46
1984	7.84	8.38	12.95	0.60	29.77
1983	8.84	5.76	12.80	0.01	27.41
1982	9.62	4.61	13.04	1.43	28.70
1981	10.20	4.55	11.49	1.51	27.75
1980	8.72	2.53	10.26	1.53	23.04
1979	7.38	2.10	9.28	0.91	19.67
1978	7.79	4.48	8.48	0.88	21.63
1977	5.48	1.55	7.42	0.74	15.19
1976	3.64	1.64	6.33	0.72	12.33
1975	3.20	1.31	4.82	0.34	9.67
1974	3.25	1.11	3.95	0.09	8.40
1973	3.06	0.53	3.20	0.23	7.02
1972	2.79	0.60	2.34	0.20	5.93
1971	2.75	0.64	2.32	0.23	5.94
1970	2.48	0.47	1.98	0.16	5.09
1969	2.23	0.66	1.70	0.15	4.74
1968	2.15	0.70	1.56	0.01	4.42
1967[a]	2.52	0.80	1.51	0.11	4.94
1966	2.52	0.81	1.50	0.15	4.98

[a]Estimate by Cooperative State Research Service.

Note: Forest economics and related research includes the following problem areas as defined by USDA–Cooperative State Research Service: Alternative uses of land (104); economic and legal problems in management of water and watersheds (108); economics of timber production (303); development of markets and efficient marketing of timber and related products (502); supply, demand, and price analysis—forest products (513); and multiple-use potential of forest land and evaluation of forestry programs (903). Other U.S. Department of Agriculture agencies include Economic Research Service, Agricultural Research Service, Agricultural Cooperative Service, Cooperative State Research Service, Competitive Research Grant Office, and Small Business Innovative Research Grant Program. Other organizations include forestry schools, 1890/Tuskegee institutions, and cooperating institutions (e.g., industry). Data reflect conditions during federal fiscal years.

Source: Inventory of Agricultural Research. USDA–Cooperative State Research Service, Washington, D.C., 1988 (and earlier years).

experiment stations were responsible for more than four out of ten scientist years (45 percent) in 1987, whereas the USDA–Forest Service and other agencies (especially the Economic Research Service) accounted for nearly half (48 percent) in the same year. Significant increases in effort were made by the agricultural experiment stations between 1966 and 1983. Investments in the latter year were twice the former. Substantial increases have also occurred in the "other organization" category (e.g., forestry schools, cooperative ventures with private organizations), and in other agencies within the U.S. Department of Agriculture. Except for the late 1970s and early 1980s, scientist years devoted to forest economics research by the USDA–Forest Service has consistently remained in the fifties or low sixties. As a proportion of total scientist years invested in forest economics research, USDA–Forest Service efforts have declined significantly. In 1966, 43 percent of total scientist years invested in forest economics research was made by the agency; by 1987 it had declined to twenty five percent with a consistent proportional erosion of scientist years over the intervening years.

The number of research forest economists per se in the United States is not easily determined. Information that is available is offered only for select organizations. Confounding attempts to obtain such information is the elusiveness of a definition of "forest economist." In 1984-1985, 84 individuals at state agricultural experiment stations were identified as forest economists involved in research activities (Cooperative State Research Service 1984). Forty percent were located in the North, 32 percent in the West and 27 percent in the South. An additional 143 persons were identified as resource economists (Cooperative State Research Service 1984) and another 140 persons were identified as being involved in research in a closely related field such as range economics, land economics, resource development, resource planning, forest products marketing, and natural resources policy and administration. In total, persons involved in forest economics or closely related research at state agricultural experiment stations totaled a conservative 367 in 1984-1985. Research forest economists in the USDA–Forest Service totaled sixty two in 1985; eight of whom were located in the South, seventeen in the West, thirty seven in the North (including six in Washington, D.C.) (Forest Service 1985). In addition to forest economists assigned to USDA–Forest Service research work units dealing with forest economics and related problems, professionals with training in the following fields may also be so assigned: forest products technology, mathematics, statistics, wildlife biology, engineering, and operations research. In addition to this professional staff, USDA–Forest Service professional staff assigned to forest economics research units totaled seventy seven in 1985. The number of "forest or closely related" economists in other public agencies or in private employment has yet to be accurately determined. In sum, the total number of individuals working as researchers in the field of forest economics in all employments probably exceeds 500.

Research Investments

Investments in forest economics and related research in 1987 totaled $29.5 million. Fifty percent of the total originated with state agricultural experiment stations; 27 percent with the USDA–Forest Service; and 16 percent with other U.S. Department of Agriculture agencies, especially the USDA–Economic Research Service. In 1972 dollars, the 1987 national total is approximately five times the 1972 level. Especially dramatic since 1980 has been the rise in research investments made by U.S. Department of Agriculture agencies other than the Forest Service (over six times 1972 levels, current dollars). This is in sharp contrast to the decline in funds allocated to forest economics research by the USDA–Forest Service over the period 1981 through 1986. The rise in non-Forest Service investments in forest economics research is consistent with the rise in scientist years that has occurred in non–Forest Service U.S. Department of Agriculture agencies during similar periods.

Problem Orientation and
Geographic Focus of Research

Forest economics and related research addresses an extremely wide range of forestry problems of concern to every region of the nation and to all sectors within the domain of public and private forestry. This research reaches international proportions via investigations that focus on problems concerning economic development and trade. Consider two approaches to understanding the scope of forest economics research, namely a problem orientation (Cooperative State Research Service 1988) and a project objective orientation (DIALOG Information Services 1985).

Problem Orientation

Employing the six problem areas previously mentioned and defined by the Cooperative State Research Service, 33 percent of the $29.5 million invested in forest economics research in 1987 was devoted to problems involving alternative uses of land (e.g., appraisal of current and potential uses of land, economics of conservation programs, models for evaluating economic benefits) (see Table 2.2). Following was research focused on the economics of timber production ($6.4 million, 22 percent) and multiple-use potential of forest land and evaluation of forestry programs (e.g., efficiency of various combinations of programs and land uses, economic response of landowners to assistance programs) ($5.1 million or 17 percent). As might be expected, these problem areas also led in terms of number of research projects and scientist years. Investments in research (current dollars) that focused on alternative land use problems experienced significant gains in 1984 and 1985, with slight declines occurring in 1986 and 1987. Other problem areas experiencing gradual upward trends in funding over the years (current dollars) are the economics of timber production (e.g., capital requirements, investment returns, income and local tax impacts) and the supply, demand, and price analysis of forest products (e.g., new technology assessment, price consequences of supply and demand changes, macro influences on timber demand).

In terms of scientist years invested by problem area by agency in 1987, state agricultural experiment stations lead in three of the six problem areas (i.e., economic and legal problems in the management of water, economics of timber production, market development and marketing efficiency) and the USDA–Forest Service led in two (i.e., supply, demand, and price analysis and multiple use potential of forest land) (see Table 2.3). The USDA–Forest Service has made especially significant increases in investments (scientist years) in research concerning supply, demand, and price analysis of forest products. From a low of 3.8 in 1982, scientist years in this problem area have risen to 18.2—five times the 1982 level. This is overshadowed by the agency's declining investments in market development research which, from a high of 40.3 scientist years in 1978, fell to a low of 0.6 scientist years in 1986 and 1987. Similar declines, though less dramatic, have occurred in the agency's research concerning the management of water and watersheds (e.g., benefits and costs of water projects, institutional arrangements for allocating water,

Table 2.2   Forest Economics and Related Research, by Problem Area, Number of Projects, Scientist Years, and Funding, 1975-1987

Year	Number of Projects	Scientist Years	Funding ($-000)
	Alternative Use of Land		
1987	178	76.8	9,766.5
1986	177	78.1	10,797.4
1985	191	115.4	12,733.1
1984	193	96.9	11,359.4
1983	206	73.6	7,434.8
1982	210	80.3	8,020.4
1981	214	76.0	7,176.9
1980	206	74.0	7,067.6
1979	201	77.5	6,381.6
1978	201	84.1	8,216.1
1977	204	81.3	5,682.7
1976	197	72.6	4,429.1
1975	160	51.9	3,240.6
	Economic and Legal Problems in the Management of Water and Watersheds		
1987	77	27.3	3,284.1
1986	65	20.9	2,849.8
1985	56	24.2	2,599.6
1984	52	39.7	3,611.6
1983	62	50.6	4,649.3
1982	55	56.7	5,101.5
1981	49	68.1	5,011.7
1980	45	25.7	2,472.6
1979	55	24.5	1,887.9
1978	46	24.8	2,141.8
1977	47	19.4	1,276.1
1976	55	24.0	1,254.4
1975	55	17.7	954.7
	Economics of Timber Production		
1987	98	37.2	6,428.6
1986	99	36.7	5,169.7
1985	78	37.4	4,882.2
1984	76	38.5	4,970.9
1983	72	34.7	4,059.1
1982	64	30.5	3,857.2
1981	67	32.6	3,534.5
1980	59	31.7	2,950.7
1979	57	23.5	2,383.2
1978	54	29.6	2,251.0
1977	55	29.1	1,958.0
1976	43	28.2	1,886.5
1975	39	34.6	1,859.2

(continues)

Table 2.2 *(Continued)*

Year	Number of Projects	Scientist Years	Funding ($-000)
Market Development and Efficient Marketing of Timber and Related Products			
1987	43	9.5	1,427.8
1986	41	7.5	1,164.8
1985	45	10.0	1,396.0
1984	43	12.3	1,733.0
1983	52	21.7	2,608.5
1982	49	29.7	3,500.3
1981	43	34.0	3,182.3
1980	38	35.8	3,195.1
1979	37	37.1	3,325.4
1978	34	46.5	3,822.1
1977	39	45.0	4,230.4
1976	38	37.2	2,194.4
1975	33	31.7	1,855.8
Supply, Demand, and Price Analysis of Forest Products			
1987	34	23.3	3,540.8
1986	34	22.0	2,719.7
1985	31	21.9	2,429.8
1984	30	14.2	1,608.1
1983	31	13.0	1,458.0
1982	28	9.6	1,364.9
1981	28	12.8	1,347.5
1980	28	12.4	1,111.5
1979	28	12.3	903.9
1978	27	15.7	1,004.5
1977	20	13.5	923.5
1976	25	10.9	754.8
1975	19	6.8	347.9
Multiple-use Potential of Forest Land and Evaluation of Forestry Programs			
1987	113	32.1	5,075.8
1986	114	31.7	4,631.8
1985	108	51.7	6,423.2
1984	111	52.8	6,490.3
1983	113	61.0	7,194.5
1982	111	54.8	6,849.1
1981	111	56.9	7,504.8
1980	105	48.9	6,242.1
1979	99	46.3	4,803.8
1978	85	36.7	4,201.4
1977	85	31.7	2,563.1
1976	90	30.5	1,820.8
1975	74	22.9	1,405.3

Note: Data reflect conditions during federal fiscal years.

Source: Inventory of Agricultural Research. USDA–Cooperative State Research Service, Washington, D.C., 1988 (and earlier years).

and allocation of water among competing uses) (3.6 scientist years in the late 1970s to 0 in 1986).

Objective Orientation

An inventory of 146 forest economics and related research projects located throughout the United States in 1984-1985 provides additional insight to the focus of forest economics research (Table 2.4). The projects' funding-agency sponsors were distributed as follows: McIntire-Stennis (49 percent), State (31 percent), USDA–Forest Service (12 percent), Hatch (7 percent), and special grants (1 percent). As to geographic distribution, 145 of the projects were distributed as follows: New England (13 projects), Middle Atlantic (12 projects), Lake States (9 projects), Central States (6 projects), South Atlantic (9 projects), East-Central (12 projects), West Gulf (8 projects), Pacific Northwest (17 projects), Pacific Southwest (4 projects), Northern Rocky Mountain (8 projects), and Southern Rocky Mountain (6 projects). The projects involved a variety of federal, state, and some private organizations.

A total of 267 research objectives were specified by forest economics researchers associated with the 146 projects. Of these objectives, 38 percent involved assessment of demand, supply, and prices of timber or closely related products. Next most common were objectives involving specific issue and administrative evaluations and development of techniques to assist decision-making (e.g., investment models for computer use)—20 percent. Research addressing the economics of production agents important for timber production involved a like proportion, 20 percent. The remaining 13 percent of the objective statements (excluding "other") dealt with evaluations of nontimber, nonforest, and environmental problems.

Geographic and Sector Orientation

Additional insight into the interests of forest economics and related research can be gained by reviewing the geographic and sector orientation of specific research projects. Again, consider the 146 projects previously referred to; specifically, the frequency of researchers' statements about the focus of research in terms of geography and sector (Table 2.5). Of 150 statements of geographic orientation, most frequent were statements indicating a research focus on a specific state (47 percent); followed by statements indicating focus on a specific region (23 percent). Very few researchers specified a national or foreign orientation for research (less than 5 percent). Likewise, a specific timber type was infrequently specified as a geographic boundary for research (6 percent of geographic statements). Interpretations of geographic orientation must be made cautiously because a significant portion (18 percent) of the research project descriptions did not specify geographic boundaries. There is no reason, however, to believe that if known, the "unspecified" statements would be distributed more heavily to one location than another.

As for sector orientation of forest economics research, over two-thirds of the 83 sector statements specified in project descriptions were focused on private sector problems, especially the timber industry, the nonindustrial private forests, and the nontimber industries (e.g., tourism). The remaining one-third of the 83 statements were publicly oriented, especially toward federal agencies and federal programs.

TABLE 2.3 Forest Economics and Related Research, by Research Problem, Organization, Number of Projects, Scientist Years and Funding, 1975-1987[a]

Year	U.S. Department of Agriculture						State Agricultural Experiment Stations			Other Organizations			Total		
	Forest Service			Other Agencies											
	1	2	3	1	2	3	1	2	3	1	2	3	1	2	3
Alternative Uses of Land															
1987	3	1.0	270.4	23	41.9	4,497.2	143	32.0	4,679.6	9	1.9	319.3	178	76.8	9,766.5
1986	3	1.0	177.0	23	42.3	5,242.5	145	34.2	5,372.1	16	0.6	5.8	177	78.1	10,797.4
1985	0		0.0	22	71.0	7,526.6	162	43.5	5,322.0	7	0.9	105.9	191	115.4	12,733.1
1984	1	0.6	62.6	24	53.2	5,996.7	160	42.6	5,237.5	8	0.5	62.6	193	96.9	11,359.4
1983	0	0.0	0.0	18	20.1	1,829.5	180	51.3	5,460.5	8	2.2	144.8	206	73.6	7,434.8
1982	0	0.0	0.0	25	21.2	1,643.4	176	57.5	6,255.1	9	1.6	121.9	210	80.3	8,020.4
1981	0	0.0	0.0	23	18.8	1,914.8	178	55.1	5,221.1	13	2.1	41.0	214	76.0	7,176.9
1980	0	0.0	0.0	19	19.3	1,843.5	173	52.5	5,075.0	14	2.2	149.1	206	74.0	7,067.6
1979	0	0.0	0.0	19	25.5	1,526.8	173	51.3	4,844.0	9	0.7	10.8	201	77.5	6,381.6
1978	0	0.0	0.0	17	29.4	3,685.7	174	52.9	4,477.6	10	1.8	52.8	201	84.1	8,216.1
1977	0	0.0	0.0	17	23.7	1,218.1	176	57.1	4,324.1	11	5.2	140.5	204	81.3	5,682.7
1976	0	0.0	0.0	15	18.8	994.1	174	48.6	3,342.2	5	1.1	92.8	197	72.6	4,429.1
1975	0	0.0	0.0	13	20.3	988.9	174	30.5	2,200.0	0	0.0	51.7	160	51.9	3,240.6
Economic and Legal Problems in Management of Water and Watersheds															
1987	3	1.9	204.0	3	4.3	351.5	70	20.4	2,641.5	1	0.7	86.8	77	27.3	3,284.1
1986	0	0.0	0.0	3	1.1	289.0	61	19.8	2,512.8	1	0.0	48.0	65	20.9	2,849.8
1985	1	0.7	62.8	2	6.4	477.0	53	17.1	2,059.8	0	0.0	0.0	56	24.2	2,599.6
1984	1	0.6	62.6	3	20.2	961.1	48	18.9	2,173.5	0	0.0	414.4	52	39.7	3,611.6
1983	2	2.2	271.8	3	30.8	2,524.4	57	17.6	1,853.1	0	0.0	0.0	62	50.6	4,649.3
1982	2	2.7	468.8	3	37.7	2,688.7	50	16.3	1,944.0	0	0.0	0.0	55	56.7	5,101.5
1981	1	2.8	398.0	6	46.4	2,520.0	42	18.9	2,093.7	0	0.0	0.0	49	68.1	5,011.7
1980	1	2.8	391.4	4	7.2	606.4	40	15.7	1,474.8	0	0.0	0.0	45	25.7	2,472.6
1979	1	3.6	370.0	10	6.8	391.4	44	14.1	1,126.5	0	0.0	0.0	55	24.5	1,887.9
1978	1	3.6	403.4	4	5.7	734.0	40	15.5	1,004.3	1	0.0	0.1	46	24.8	2,141.8
1977	1	1.1	80.0	4	4.5	332.4	41	13.8	863.6	1	0.0	0.1	47	19.4	1,276.1
1976	0	0.0	0.0	6	13.8	562.8	48	10.2	691.1	1	0.0	0.5	55	24.0	1,254.4
1975	0	0.0	0.0	4	7.5	303.6	50	10.1	650.8	1	0.0	0.3	55	17.7	954.7
Economics of Timber Production															
1987	21	14.8	2,245.1	0	0.0	0.0	59	17.6	3,483.3	18	4.8	700.2	98	37.2	6,428.6
1986	21	14.4	1,739.0	3	0.0	188.2	54	15.7	2,818.4	21	6.6	424.1	99	36.7	5,169.7

Year														
1985	7	17.7	1,741.5	1	0.0	96.0	50	16.0	2,520.2	20	3.7	524.5	78	37.4	4,882.2
1984	7	18.0	2,039.0	0	0.0	0.0	49	15.5	2,306.0	20	5.0	625.9	76	38.5	4,970.9
1983	6	15.2	1,608.2	0	0.0	0.0	47	15.8	1,972.7	19	3.7	478.2	72	34.7	4,059.1
1982	6	12.9	1,519.0	0	0.0	0.0	44	13.8	1,825.9	14	3.8	512.3	64	30.5	3,857.2
1981	6	13.1	1,315.1	0	0.0	0.0	46	13.6	1,658.2	15	5.9	561.2	67	32.6	3,534.5
1980	5	11.2	778.3	1	0.0	34.7	38	13.3	1,653.7	15	7.2	484.0	59	31.7	2,950.7
1979	4	7.1	555.5	1	0.0	74.3	33	11.8	1,506.8	19	4.6	246.6	57	23.5	2,383.2
1978	4	13.2	819.8	1	0.0	49.6	30	12.4	1,131.1	19	4.0	250.5	54	29.6	2,251.0
1977	5	10.9	742.5	0	0.0	0.0	29	14.9	969.1	21	3.3	246.4	55	29.1	1,958.0
1976	4	14.0	778.7	1	0.0	57.4	22	10.8	874.6	16	3.4	175.8	43	28.2	1,886.5
1975	5	18.8	961.5	1	0.0	14.4	22	4.5	182.4	11	11.3	700.9	39	34.6	1,859.2

Market Development and Efficient Marketing of Timber and Related Products

Year														
1987	2	0.6	221.7	0	0.0	0.0	34	6.2	856.5	7	2.7	349.6	43	9.5	1,427.8
1986	2	0.6	120.7	1	0.0	90.1	29	5.2	787.4	9	1.7	166.6	41	7.5	1,164.8
1985	3	2.3	260.0	2	0.0	369.2	29	5.8	730.5	11	1.9	36.3	45	10.0	1,396.0
1984	5	5.0	826.0	1	0.0	4.3	28	5.5	603.3	9	1.8	299.4	43	12.3	1,733.0
1983	6	12.0	1,415.4	1	0.1	10.7	32	7.7	919.8	13	1.9	262.6	52	21.7	2,608.5
1982	6	20.8	2,406.8	1	0.1	11.2	32	7.1	758.1	10	1.8	324.2	49	29.7	3,500.3
1981	7	24.7	2,396.1	1	0.1	16.1	26	5.3	506.9	9	3.9	263.2	43	34.0	3,182.3
1980	7	28.1	2,606.6	0	0.0	0.0	23	4.9	400.6	8	2.8	187.9	38	35.8	3,195.1
1979	10	30.4	2,851.6	1	0.0	62.3	22	4.8	339.4	4	1.9	72.1	37	37.1	3,325.4
1978	11	40.3	3,151.8	0	0.0	0.0	19	5.1	548.5	4	1.1	121.8	34	46.5	3,822.1
1977	11	35.6	2,440.5	0	0.0	0.0	23	4.9	264.3	5	4.5	1,525.6	39	45.0	4,230.4
1976	11	31.2	1,864.0	0	0.0	0.0	22	5.0	260.6	5	1.0	69.8	38	37.2	2,194.4
1975	10	27.7	1,653.9	0	0.0	8.0	20	3.3	168.1	3	0.7	25.8	33	31.7	1,855.8

Supply, Demand, and Price Analysis of Forest Products

Year														
1987	13	18.2	2,803.3	0	0.0	0.0	20	3.8	634.7	1	1.3	102.8	34	23.3	3,540.8
1986	13	17.4	2,124.8	0	0.0	0.0	19	4.0	539.7	2	0.6	55.2	34	22.0	2,719.7
1985	4	15.5	1,633.9	0	0.0	0.0	22	5.4	535.1	5	1.0	260.8	31	21.9	2,429.8
1984	3	7.5	844.9	0	0.0	0.0	22	5.5	535.3	5	1.2	227.9	30	14.2	1,608.1
1983	3	5.0	751.3	0	0.0	0.0	20	5.6	448.4	8	2.4	258.3	31	13.0	1,458.0
1982	2	3.8	652.0	0	0.0	0.0	19	4.4	377.8	7	1.4	355.1	28	9.6	1,364.9
1981	3	5.2	727.8	1	1.1	59.9	16	4.2	364.4	8	2.3	195.4	28	12.8	1,347.5
1980	3	6.2	603.5	0	0.0	0.0	15	3.1	270.2	10	3.1	237.8	28	12.4	1,111.5
1979	2	5.5	410.4	0	0.0	0.0	15	3.4	283.4	11	3.4	210.1	28	12.3	903.9
1978	3	9.8	549.2	0	0.0	0.0	16	3.6	263.6	8	2.3	191.7	27	15.7	1,004.5
1977	3	8.6	593.2	0	0.0	0.0	13	2.6	198.3	4	2.3	132.0	20	13.5	923.5
1976	3	5.7	415.1	0	0.0	0.0	14	2.8	167.9	8	2.4	171.8	25	10.9	754.8
1975	2	3.2	162.4	0	0.0	0.0	12	3.0	157.5	5	0.6	28.0	19	6.8	347.9

(continues)

TABLE 2.3 (Continued)

Multiple-Use Potential of Forest Land and Evaluation of Forestry Programs

Year	U.S. Department of Agriculture						State Agricultural Experiment Stations			Other Organizations			Total		
	Forest Service			Other Agencies											
	1	2	3	1	2	3	1	2	3	1	2	3	1	2	3
1987	17	15.5	2,327.2	0	0.0	0.0	70	13.4	2,377.2	26	3.2	371.4	113	32.1	5,075.8
1986	17	15.0	1,880.3	0	0.0	0.0	72	13.0	2,024.4	25	3.7	727.1	114	31.7	4,631.8
1985	15	32.2	4,160.6	0	0.0	0.0	76	14.8	1,721.4	17	4.7	541.2	108	51.7	6,423.2
1984	15	28.6	4,073.4	1	0.1	0.0	80	16.7	2,089.4	15	7.4	327.5	111	52.8	6,490.3
1983	16	33.7	4,788.7	1	0.0	0.0	77	24.5	2,145.9	19	2.8	259.9	113	61.0	7,194.5
1982	15	34.4	4,571.0	0	0.0	0.0	76	17.3	1,866.4	20	3.1	411.7	111	54.8	6,849.1
1981	17	37.4	5,350.8	0	0.0	0.0	69	15.5	1,650.4	25	4.0	503.6	111	56.9	7,504.8
1980	14	28.3	4,340.9	0	0.0	0.0	66	15.6	1,390.5	25	5.0	510.7	105	48.9	6,242.1
1979	13	27.4	3,188.4	0	0.0	0.0	63	14.3	1,185.2	23	4.6	430.2	99	46.3	4,803.8
1978	9	18.0	2,874.1	0	0.0	0.0	61	14.2	1,057.1	15	4.5	270.2	85	36.7	4,201.4
1977	9	18.5	1,631.0	0	0.0	0.0	62	11.5	814.4	14	1.7	117.7	85	31.7	2,563.1
1976	5	8.7	580.9	0	0.0	0.0	66	17.4	998.0	19	4.4	241.9	90	30.5	1,820.8
1975	3	5.2	430.9	0	0.0	0.0	56	16.0	874.4	15	1.7	100.0	74	22.9	1,405.3

[a]1 = Number of projects; 2 = Scientist years; 3 = Funding ($=000).

Note: Other U.S. Department of Agriculture agencies include Economic Research Service, Agricultural Research Service, Agricultural Cooperative Service, Cooperative State Research Service, Competitive Research Grant Office, and Small Business Innovative Research Grant Program. Other organizations include forestry schools, 1890/Tuskegee institutions, and cooperating organizations (e.g., industry). Data reflect conditions during federal fiscal years.

Source: Inventory of Agricultural Research. USDA–Cooperative State Research Service, Washington, D.C., 1988 (and earlier years).

Table 2.4   Forest Economics and Related Research Project Objectives, by Objective Category and Frequency, 1984-1985

Research Project Objective Category	Frequency	Percent
Timber supply/demand, pricing, and requirement assessment	33	12
Timber economy, economic importance, and economic impact documentation	26	10
International trade and market assessment	13	5
Industrial organization, structure, and capacity documentation	11	4
Timber utilization evaluation	9	3
Fuelwood and wood energy assessment	11	4
Investment decision-making technique development	27	10
Program and issue evaluation	11	4
Taxation and estate transfer evaluation	9	3
Administration and organizational evaluation	7	3
Production agent evaluation (labor, machines, transportation)	16	6
Silvacultural treatment evaluation	11	4
Forest protection and damage assessment	10	4
Timber yield, regime, and harvest-scheduling evaluation	8	3
Landowner investment rational assessment	8	3
Recreation and wildlife output evaluation	21	8
Timber/nontimber joint product assessment	7	3
Forest land/nonforest land use and allocation evaluation	4	1
Environmental (air, water) assessment	2	1
Other	23	9
Total number of project objective statements	267	100

Note: Based on review of 146 projects as identified from the Current Research Information System (CRIS) and related documents. A single research project may have multiple objectives that fall into more than one category.

Source: Based on Current Research Information System reports. DIALOG Information Services, Palo Alto, CA, 1985.

Of special interest is that of the 153 sector statements (or opportunities to make such statements), 70 percent did not refer to a particular sector. Researchers apparently wish greater latitude for their efforts than might occur if orientation to a particular sector were specified.

Strategic Opportunities for Future Research

Forest economics and related research has developed a considerable foundation over time. During its seven to eight decades of existence, it has matured in terms of definition and scope and has addressed a number of important problems facing the owners and administrators of public and private forests. But what of the future, especially the next five to ten years? Consider possible settings in which forest economics research may be undertaken in the next five to ten years, and possible strategic directions that such research might follow in response to these circumstances.

Table 2.5 Forest Economics and Related Research Project Orientations, by Geographic and Sector Category and Frequency, 1984-1985

Research Project Orientation	Frequency	Percent
Geographic Orientation		
Local	3	2
State		
Western state	16	10
Southern state	30	19
Northern state	17	11
Northeastern state	11	7
Total	74	47
Region		
West	14	9
South	11	7
North	0	0
Northeast	12	7
Total	37	23
National	1	a
Foreign	6	4
Timber type		
Eastern species	4	3
Western species	5	3
Total	9	6
Unspecified	28	18
Total number of project		
geographic statements	158	100
Sector Orientation		
Public programs and agencies		
Federal	16	10
State	6	4
Local	2	1
Public general	4	3
Total	28	18
Private		
Timber industry	25	16
Nontimber industry	11	7
Nonindustrial	19	13
Total	55	36
Unspecified	70	46
Total number of project		
sector statements	153	100

[a]Denotes less than one percent.

Note: Based on review of 146 projects as identified from the Current Research Information System (CRIS) and related documents. A single research project may be oriented toward more than one geographic or sector category.

Source: Based on Current Research Information System Reports. DIALOG Information Services, Palo Alto, CA, 1985.

The Setting

The setting for future research involving forest economics is uncertain to say the least—but such is the reality of the world of forestry. Some organizations have attempted to assess the future setting in which management and research activities will be undertaken, often defining such a setting in terms of issues. Consider three quite different organizations and their "view of the future," namely USDA–

Economic Research Service, USDA–Forest Service, and a leading U.S. wood-based company (Hammermill Corporation).

The Economic Research Service of the U.S. Department of Agriculture reviewed a number of conditions that could have a significant influence on future directions for agricultural economics research in the 1980s (Economic Research Service 1985). Although oriented to the agricultural sector, the agency's perceptions are relevant to natural resource use and management. Consider the following:

- *Growing linkages to the world economy*: Increasingly competitive trade between the United States and developing countries; growing importance of resource-trade policies of other countries; changing global monetary and macroeconomic policies developments that affect sector production (e.g., population, income distribution, inflation, exchange rates, interest rates, credit policies, technology, and oil and other commodity prices); and growing opportunity for export to low- and middle-income countries.
- *Increasing transformations within domestic economy*: Increasingly rapid substitution of capital, including new technology, for land and labor; declining number of producing units resulting in greater economic concentration; changing distribution of income among rural population; increasingly complex marketing links between producers and consumers; and growing economic concentration among secondary processors and marketing entities that affect market performance.
- *Expanding influence of technology on productivity and resource use*: Increasing role of technology in determining patterns of resource use and ability of the sector to compete (e.g., electronic information technology's effect on producer planning, management, and market assessment); and increasing concern over the ability of the resource base to sustain high levels of production (e.g., soil erosion, ground water depletion, toxic residuals).
- *Rising interest in rural economic development*: Increasing population growth in rural areas; growing prevalence of nontraditional rural industries and complex government systems; continuing rural poverty and rural unemployment; growing concern over financing and delivery of public rural services, physical facilities, and governing institutions.
- *Ongoing policy issues*: Continuing concern over the legitimacy of public programs and their beneficiaries; and misunderstanding of the appropriate public-private sharing of production risks.

The USDA–Forest Service identified in 1980 and 1985 a number of significant issues that were addressed by renewable resources programs (Forest Service 1980 and 1986). These issues present an additional perspective on the setting that will influence future directions for forest economics research. Consider the following:

- *Program issues identified in 1980*: Production of wood from nonindustrial private lands; production of softwood products from National Forest lands; wood fiber as an energy source; management and utilization of hardwoods; export and import of raw logs; pesticide use; consumer payment for recreational opportunities; financing capital development on public forests; eastern National Forests; multiresource planning and management of nonindustrial private forest lands and rangelands; forage for domestic livestock; minerals development on public forest lands; forestry assistance for nonfederal public lands; expanding wood supplies through improved technology and utilization; and management emphasis on fish and wildlife.

- *Program issues identified in 1985*: National Forest contribution to demand for minerals and energy; forage production from federal rangelands; supply of outdoor recreation opportunities from public lands; allocation of National Forest system lands to the National Wilderness Preservation System; departures from nondeclining yields on National Forest system lands; timber production at less than biological potential on commercial timberland; utilization of current wood resources to extend timber supply; fish and wildlife habitat management; nonmarket forest and range outputs in decisionmaking processes and economic analyses.

An industrial perception of the future setting in which forest economics research may well find itself was set forth in 1985 by Hammermill Corporation (Hammermill 1985). Among the conditions viewed as having more than a passing influence on company operations are

- *Fiscal and monetary policies*: Continued large-scale federal debts and restrictive monetary policies that will inhibit economic expansion.
- *Lower inflation*: Moderate inflation rates (4 percent or less) into the foreseeable future.
- *Strong U.S. dollar*: High demand for U.S. currency resulting in U.S. goods being expensive in world markets while imports continue to be inexpensive in U.S. markets.
- *Two-tiered economy*: Continuing decline in growth of heavy manufacturing industries while service industries (e.g., information processing, communications, retailing, financial systems, transportation, entertainment) continue significant growth.
- *Changing public attitudes*: Shift away from strong antibusiness sentiments to a realization that private enterprise is vital to the American way of life.
- *Environmental issues*: Basic environmental safeguards (e.g., laws, rules, investments) are in place. However, concern over the environment will continue indefinitely.
- *Population trends*: Population growth has slowed, more people living alone, families are smaller, higher proportion of population in work force, population better educated, migration to the South and West, baby boom generation entering peak spending years.
- *Government regulation*: Continuing deregulation in many economic sectors; less pressure for regulation in others.
- *Computer revolution*: Continued growth in computers' ability to store, retrieve, process, and communicate vast amounts of information. A "paperless" society, however, seems unlikely in the foreseeable future.

- *Manufacturing technologies*: High risk of pioneering radically new processes suggests few new breakthroughs in pulp and paper production. Labor-intensive operations (e.g., paper finishing) will be impacted by improved automation.
- *Forest management*: Fast-growing species in tropical and subtropical climates pose a possible long-term threat to industries using North American hardwoods. Rising forest yields—especially in the South—assure low-cost wood supplies for many years to come.
- *Globalization of wood-based industry*: Dramatic change from a domestic to a globally oriented business. Production, consumption, and prices will be further influenced by world situation. U.S. markets will open further to imports, while U.S. manufactures will look for more new foreign markets.
- *Strong economic base of industry*: Paper and allied product industry will continue in a favorable position. This is predicated on availability of an economical and renewable raw material; proven processing technology; high energy self-sufficiency; and location of major mills near international ports.
- *Profit opportunities*: Success will no longer require vast timberland holdings or a presence in all major consumer markets. Top industry performers will be more focused.
- *Supply-demand cycles*: Continuation of steady growth in demand for paper; capacity, however, will continue to expand in jumps. Industry will remain inherently cyclical because synchronization of demand and of capacity will be difficult.

Strategic Directions

Forest economics and related research can follow a number of avenues, all of which will be influenced by social and political conditions in the years ahead. Defining such avenues and subsequently proceeding to select those that have special merit can be a frightful exercise—but a necessary one. Listed below are twenty subject areas toward which forest economics and related research might possibly be directed. If an overall goal for research on these subjects were to be enunciated, it would certainly entail a desire to improve the effectiveness and efficiency with which the nation's public and private forests are used and managed. The subject areas have purposely been stated in a very general fashion—a laundry list of specific research projects being too extensive to develop here. The breadth of each area is also purposely very wide; too narrow a set of directions may well find the forest economics community defining itself out of future existence. Consider the following topical areas:

- Institutional arrangements directing use and management of forests
- Economic structure and performance of forest-based industries
- Development, dissemination and adoption of new technology
- Forecasting supply and demand of forest resources, products and services
- Social and economic growth of developing nations
- International trade in forest and related products
- Wood fiber production
- Timber harvesting

- Production and valuation of forest and wildland recreation
- Policy development and program administration
- Resource assessment, information management, and communications technology
- Forestry sector environmental effects
- Community and regional economic growth and development
- Taxation of forest products and resources
- Distribution and marketing of forest resource products
- Forest resource law and legal processes
- Management of fire in forested environments
- Management of insects and diseases in forested environments
- Structure and performance of nonindustrial private forests
- Production and valuation of water from forested watersheds

Summary and Conclusions

Forest economics research has played an ever-increasing role as a source of information sought by forest managers charged with the allocation of scarce resources. Its legitimacy as a focus for public and private research investments has been fostered by significant leadership and diligent reviews of definition, scope, and scientific orientation. Forest economics is in a continual state of redefinition as generated by changing definitions of "forestry" and of "economics." As a result, the scope of activities and the disciplines being applied have broadened considerably during the past six to seven decades. A more diverse set of organizations is involved in forest economics research than ever before, and the distinction between "application" and "research" has become blurred—with a significant tendency toward application.

Investments in forest economics research are significant—although one would hesitate to call them substantial. Research projects involving forest economics in the United States probably total more than 500, involve more than 500 researchers, and account for more than 200 scientist years of effort annually. Financial investments approach $30 million annually, half of which originates with state agricultural experiment stations. Agencies in the U.S. Department of Agriculture other than the Forest Service are accounting for a growing share of this investment. Overall, dollar investments and scientist years of effort expended on forest economics research have declined significantly in recent years.

Alternative uses of land as a problem area accounts for 33 percent of the financial investments made in forest economics research; economics of timber production and multiple land use problems are second and third, respectively. The USDA–Forest Service accounts for 78 percent of the scientist years devoted to research on problems concerning supply, demand, and price analysis of timber and nearly half of the scientist years devoted to multiple land use problems. Significant declines have occurred in USDA–Forest Service investments in marketing research—1978 investments exceeded 1987 investments by $2.9 million (current dollars) or nearly forty scientist years of research.

The objectives of forest economics research are geared most heavily toward assessment of demand, supply, and pricing of timber products, with objectives involving issue and administrative evaluations and development of techniques to assist decision-making being the second most common category. As to geographic focus, a specific state is the most common research orientation. Researchers are hesitant to specify a particular sector focus; when it is specified, the problems of the private sector are most commonly addressed.

Topics of interest to forest economics research will be determined by a variety of social, economic, and technical conditions. These will vary from greater linkages to the world economy to further economic and social adjustments in the domestic economy, and from an ever-growing influence of new technologies to a resurgence in rural economic development. Such conditions will lead to a forest economics research agenda that will invariably include topics such as economic development, industrial organization, adoption of new technologies, nonmarket forest outputs, resource assessment, supply-demand forecasting, and institutional and administrative arrangements for forest management. Specific problems demanding the attention of forest economics research will abound, increasing at more rapid rates as resources of all types become scarcer. Well-designed forest economics research programs— guided by imaginative researchers—should be in a position to address these challenges as they occur.

Literature Cited

Cooperative State Research Service. 1978. Manual of classification of agricultural and forestry research. Developed by Research Classification Subcommittee of the Agricultural Research Policy Advisory Committee. U.S. Department of Agriculture. Washington, DC.

Cooperative State Research Service. 1984. 1984-85 Directory of professional workers in state agricultural experiment stations and other cooperating state institutions. Ag. Handbook 305. U.S. Department of Agriculture. Washington, DC.

Cooperative State Research Service. 1988. Inventory of agricultural research: Fiscal Year 1987. U.S. Department of Agriculture. Washington, DC.

Clawson, M. C. 1977. Research in forest economics and forest policy. Resource for the Future. Washington, DC.

DIALOG Information Services. 1985. Current research information system reports. Palo Alto, CA.

Duerr, W. A. and H. J. Vaux. 1953. Research in the economics of forestry. Waverly Press. Baltimore, MD.

Economic Research Service. 1977. Looking forward: Research issues facing agriculture and rural America. U.S. Department of Agriculture. Washington, DC.

Economic Research Service. 1985. Economic Research Service in transition. U.S. Department of Agriculture. Washington, DC.

Ellefson, P. V. 1986. Decisions about public forestry programs: The role of policy analysts and analytical tools. Journal of Resource Management and Optimization 4(1):65-77.

Forest Service. 1980. A recommended renewable resources program: 1980 update. FS-346. U.S. Department of Agriculture. Washington, DC.

Forest Service. 1985. Manual of research work unit descriptions, funding and personnel. Forest Inventory and Economics Research Staff. U.S. Department of Agriculture. Washington, DC.

Forest Service. 1986. A recommended renewable resources program: 1985-2030. Final environmental impact statement. FS-403. U.S. Department of Agriculture. Washington, DC.

Hammermill Corporation. 1985. Annual Report: 1985. Erie, PA.

Lage, A. 1986. History of forest economics. Oral interviews with H.R. Josephson, M. Krueger and H.J. Vaux. Regional Oral History Office. The Bancroft Library. University of California-Berkeley. Berkeley, CA.

3

IMPACT EVALUATION
AND PLANNING OF
RESEARCH PROGRAMS

David N. Bengston

Economists have made significant progress in recent decades in understanding the contributions to social welfare made possible by new knowledge generated by many types of research. New knowledge that is embodied in new inputs and products creates technical change, or change in the techniques of production. Technical change has been shown to be a major contributor to economic growth and development. The role of public and private research in generating technical change has been examined extensively by economists, and the link between investment in research and growth of productivity has been repeatedly demonstrated in the agricultural sector and in various industrial sectors (Evenson et al. 1979, Griliches 1987). Recent studies have confirmed that forestry and forest products research also significantly affects productivity growth and generates rates of return comparable to investment in agricultural and industrial research (Jakes and Risbrudt 1987).

Progress in assessing the contributions of new social science knowledge has been disturbingly slow. This lack of progress is attributable largely to formidable problems in identifying and measuring the output of social science research and in determining the causality of change that occurs following such research. Knowledge produced by social science research is not embodied in tangible and readily measured items such as new inputs or products. Rather, if successfully adopted and implemented, it is embodied in improved decisionmaking or in more effective public and private policies. Conditions such as these pose serious problems to an economic evaluation: The tools of economic analysis are blunt instruments for assessing the value of an intangible commodity such as new social science knowledge.

Despite the conceptual and operational challenges, there is a need to evaluate the contributions of social science research, including forest resource economics and policy (FREP) research. This need arises from at least three sources. First is the demand for evidence of the value of research as required to justify continued budgetary support. As public agency budgets have tightened in recent years, competition for alternative uses of public funds has increased greatly. Public policy makers need clear evidence of the contribution of forest economics and policy research to society. The case for continued public support must increasingly be based on careful planning and hard evidence, not on anecdotal information, traditional budget allocations, or blind faith in the value of "science for science's sake." Evaluation can provide a solid analytical basis for justifying budgetary support.

Second, information about the value of different lines of research can contribute to more efficient allocation of research resources. To efficiently allocate available resources, research managers must ask questions such as: How should research resources be apportioned among different areas of inquiry, between basic and applied research, and between the development and diffusion of new knowledge? Evaluation can provide a basis for making more efficient use of limited resources available to forest economics and policy research.

Finally, research policy makers and administrators are faced with the task of setting future directions for research programs or organizations. Information about the potential economic impacts of research may be a valuable input into this decision process. Evaluations that address the following concerns can help provide guidance in determining future priorities for forest economics and policy research: What quantity of resources should society allocate to FREP research? How can we better anticipate emerging issues and problems so that research can respond in a timely fashion? How can we identify and redirect research projects or programs that may have outlived their usefulness? Answers to questions such as these could help make FREP research contribute more effectively and efficiently to social and economic development.

Several assessments of forest economics and policy research have addressed some evaluation needs, especially that of identifying future research priorities. Duerr and Vaux (1953) systematically explored the scope and methods of forest economics and identified information needs and promising areas for future research. The effort involved scores of contributors and covered 127 research topics. Researchers and research administrators involved in planning projects and programs were the primary intended audiences. Resources for the Future sponsored a 1977 symposium to identify research needs and priorities in forest economics and policy research (Clawson 1977). Emery Castle (1977) summarized and synthesized the symposium's discussions and identified research priorities based on three broad criteria: (1) importance of the problem, or relevance, (2) probability of research success, and (3) impact of the research. Forest economics and policy research was also assessed by Duerr and Duerr (1980) and Whaley and Bell (1982). These assessments were mainly concerned with the coverage of priority research topics and the quality or adequacy of research inputs and outputs.

Assessment of priority topics for forest economics research has undoubtedly contributed to research planning and priority setting, as well as to improving the quality and effectiveness of the research. Typically, however, such assessments have not explicitly dealt with the economic value of forest economics and policy research or, more broadly, the various contributions of such research to society. What follows addresses the problem of evaluating the contributions of forest economics and policy research. The approach is exploratory and by no means an exhaustive treatment of the subject. Most would agree that the potential social impacts of forest economics and policy research are great. Less clear, however, is how to rigorously evaluate actual or anticipated impacts. Because a universal recipe for evaluating a particular type of research is lacking, we must begin by providing a conceptual basis for understanding the contributions of forest economics and policy research rather than by specifying potential evaluation approaches.

Economic Evaluation of Research

A logical starting point for a discussion of the economic evaluation of research is the extensive agricultural literature relating to the topic. The origins of agricultural research evaluation can be traced to the work of Nobel laureate T. W. Schultz, who compared expenditures on agriculture research to the value of agricultural inputs saved due to more efficient production technology (Schultz 1953, pp. 114-122). Building on the efforts of Schultz, more than eighty evaluations of investment in agricultural research have been carried out over the past three decades (Bengston 1985). Several methods have been developed to evaluate agricultural research and extension, ranging from simple scoring models to sophisticated production function analyses (Schuh and Tollini 1979, Norton and Davis 1981). Such methods may evaluate past research (*ex post* techniques) or they may evaluate proposed research projects or programs (*ex ante* techniques). Ruttan (1982, pp. 242-243) summarized the results of many economic evaluations of agricultural research. Estimates of annual internal rates of return (IRRs) to research ranged from a negative 48 percent to a positive 110 percent, with most studies reporting returns of 20 percent to 50 percent.

Despite the impressive number of evaluations and evaluation approaches in agriculture, application has been limited in several respects (Bengston 1985). First, agricultural economists have overwhelmingly favored just two methods for evaluating research—the consumer and producer surplus approach and the production function approach account for about three-quarters of all evaluations. Several other approaches have been developed conceptually but never tested empirically (Cartwright 1971) or have been only partially implemented (Mahlstede 1971). Second, evaluations of agricultural research have been mostly *ex post* in nature, i.e., past research investments have been examined. This poses obvious problems in using evaluation information in decision making: "It is difficult, if not impossible, to take *ex-post* results and forecast the variables in the analyses to decide what to do in research and extension" (Hildreth 1980, p. 558). Third, agricultural evaluations have been extremely limited in terms of evaluation criteria. Only a handful of the agricultural studies have considered evaluation criteria other than contribution to economic efficiency. It is unlikely that single-measure evaluation approaches will be able to adequately assess the contributions of forest economics and policy research.

Finally, the coverage of agricultural research evaluation has been limited in terms of the type of research evaluated. The focus was almost exclusively on production-oriented research (primarily crop production). The outputs of production-oriented research are much easier to identify and value than the outputs of most other types of research. For example, the benefits of crop-breeding research can be identified in terms of higher yields per acre that are valued in the marketplace. New knowledge produced by social science research is much more elusive. Consequently, evaluation of social science research has received scant attention. Estimates of returns to economics and policy research—or other areas of social science research—are conspicuously absent from Ruttan's (1982) summary of empirical research evaluations. In view of these limitations, it seems unlikely that the bulk of the agricultural research evaluation literature can provide much useful guidance in evaluating the contributions of forest economics and policy research.

Agricultural research evaluation has strongly influenced the evaluation of research in other fields. Evaluators of industrial research have adopted evaluation approaches developed in agriculture, for example (Mansfield et al. 1982, pp. 118-119). Forest economists have also followed the lead provided by agricultural economists. In 1983, the USDA–Forest Service formally established a project to develop methods of evaluating forestry research (Lundgren 1986). Forestry research evaluations have also been undertaken at Duke University (Hyde 1985) and the University of Minnesota (Gregersen 1985). Jakes and Risbrudt (1988) summarized the results of recent economic evaluations of forestry research and found the reported IRRs to vary widely, as would be expected—different evaluation methods and assumptions were employed, and a variety of innovations and types of research in different time periods were evaluated. As was reported for agriculture, most IRRs from investments in forestry research fall in the 20 percent to 50 percent range. These relatively high rates of return confirm that some types of forestry research have significant economic impacts. Nevertheless, forestry research evaluations carried out to date are limited by many of the same shortcomings found in agriculture (e.g., narrowness in terms of evaluation methods, *ex post* orientation, single evaluation criteria). As in agriculture, the impacts of forest economics and policy research or other social science research have not been empirically estimated. A possible exception is an evaluation of management sciences in forestry. Valfer et al. (1981) estimated the returns on investments in five projects carried out by the management sciences staff of the USDA–Forest Service between 1962 and 1979. The IRRs ranged from 34 percent to 184 percent. However, the projects evaluated involved short-term efforts focused on specific management problems— not research projects per se.

Problems in Evaluating Social Science Research

Why is there a lack of economic evaluations focused on social science research, including forest economics and policy research? Part of the answer lies in the array of fundamental problems encountered when evaluating any type of research— problems that are compounded in the case of social sciences. Schuh and Tollini (1979) and Lundgren (1982) discuss such problems, including lack of a standardized unit of measure for new knowledge, lack of a well-defined market and market price for knowledge, how to deal with joint outputs of research, and how to treat the existing stock of knowledge. In addition to problems of such a nature, economic evaluation of social science research faces a number of very unique challenges. Norton (1987, p. 179), for example, points to the diversity and complexity of economics research with respect to the variety of goals that may be involved: "Some research projects are concerned with increasing or stabilizing aggregate income while others are directed toward facilitating a more equitable income distribution or improving health and safety. Some are aimed at multiple goals." Evaluating research that contributes to diverse social and economic goals is clearly more difficult than research involving a single objective or a narrower set of objectives, such as increasing technical efficiency in a particular production process. Because of the multitude of objectives often sought, comprehensive evaluation of the impacts of forest economics and policy research will require multiple evaluation criteria. Failure to account for the contribution of forest economics and policy research to a variety of economic and social goals may seriously understate the true value of the research.

The nature of the outputs of social science research is also a vexatious problem for research evaluation. Social science research—or, more precisely, the application of the outputs of such research—is less tangible than the output of research in the biological and physical sciences. As noted earlier, new knowledge produced by social science research may be embodied in improved decision making and new or improved policies. Improved decisions and policies often lack the concreteness that facilitates valuation when knowledge is embodied in physical inputs or products. As a result, most social science research does not have impacts that are readily observable in quantifiable terms.

Most intractable is the problem of establishing causality of change that occurs as a result of social science research. It may be impossible to determine if social science research that preceded and suggested institutional change actually brought about the change or whether the change would have come about anyway for political or other reasons. Estimation of research benefits is thus confounded by the often circuitous path by which social science research affects decisions and social welfare. Research on the utilization of social science research has shown that decision makers use social science research findings indirectly as a source of ideas and orientations to the world instead of applying specific findings to specific decisions (Weiss 1977, 1980; Sabatier 1987). This indirect use of social science research has been termed "knowledge creep" to indicate the incremental, subtle influence on decisions and policymaking.

Forest economics and policy research results can thus be thought of as one of many tributaries contributing to a stream of decisions and policy changes. Just as isolating the effects of one tributary on the course of a river may be a lost cause, evaluating the contribution of forest economics and policy research may exceed our ability to analyze empirically. A lesson from the causality problem is that evaluation of large aggregates of forest economics and policy research will be problematic. Occasionally an exemplary case can be found in which such research has unambiguously guided policy outcomes, but at an aggregate level, the evaluator's reach may exceed his or her grasp.

Economic Evaluation of Economics
and Policy Research

Despite the numerous difficulties faced when evaluating economic and policy research, scattered attempts have been made to do so. Three main approaches have been employed: (1) the consumer and producer surplus approach, (2) the decision theoretic approach, and (3) the descriptive case study approach.

Consumer-producer surplus—or net social benefit—analysis has a long history in the field of economics. In the context of research evaluation, the basic line of reasoning underlying the approach is that successful implementation of research findings reduces the marginal cost of producing some good or service. For example, forecasts of future market conditions may help producers reallocate resources in ways that are consistent with future conditions, thereby increasing allocative efficiency. A downward shift in the supply curve for the good in question is thus the graphical manifestation of a successful research and technology transfer effort. The shift in supply will result in changes in consumer and producer surpluses, which can be compared to research and technology transfer costs to obtain a measure of net social benefits attributable to the research effort.

Hayami and Peterson (1972) used the consumer-producer surplus approach to examine the marginal social returns of reducing the sampling error of U.S. Department of Agriculture crop and livestock statistics. They developed an inventory adjustment model and a production adjustment model for estimating the social benefits of improved information, and they calculated benefit-cost ratios for public investment in increasing the accuracy of agricultural production surveys. Hayami and Peterson found large marginal benefit-cost ratios, and they concluded that benefits from investment in increasing the accuracy of agricultural production statistics would greatly exceed costs. The consumer-producer surplus approach was also used by Volker and Deacon (1982) to determine the benefit-costs and cost-effectiveness of individual home economics research projects. The approach has not been applied to broad aggregates of economics or social science research, however.

The decision theoretic approach has been applied in value of information studies and in evaluation of social science research. As outlined by Eisengruber (1978), the approach assumes that several future states of nature are possible; that decision makers have some knowledge of the probability of future states occurring; that a variety of courses of action are available to the decision maker; and that the decision maker will choose the course of action that maximizes expected utility given a state of knowledge. If additional information becomes available through research, the probabilities of future states occurring will be modified (e.g., by Bayes' Rule). The value of the new information is calculated as the difference between the decision maker's maximum utility with and without the information, which can then be weighed against the cost of generating and disseminating the information. Eisengruber (1978) and Norton (1987) discuss several major difficulties with the decision theoretic approach, including estimating the probabilities of occurrence of the states of nature and determining the appropriate utility function for decisionmakers. For simplicity, a linear utility function implying risk neutrality has been assumed in empirical studies so that maximizing expected utility is identical to maximizing expected profits.

Norton and Schuh (1981) proposed an evaluation approach using Bayesian decision theory in combination with a consumer and producer surplus model to evaluate price outlook information—forecasts of commodity prices over the coming marketing year. The value of price outlook information for the case of soybean production in Minnesota was estimated to be more than $600,000 in 1978-1979. Norton and Schuh did not compare the value of this information to the associated research and dissemination costs. In an ambitious effort, Norton (1987) attempted to empirically measure the value of agriculture economics research (AER) in the areas of farm management, marketing efficiency, and price analysis. Norton's approach, adapted from Antonovitz and Roe (1986), defined an *ex post* measure of the value of information as the difference between optimal profits that would have been earned in a state of perfect information and the profits that were actually realized. Norton hypothesized that expenditures on agricultural economics research would explain variability between optimal and realized profits or reductions in allocative error on the part of producers. The results of empirically estimating the impact of agricultural economics research were disappointing, however. The coefficient on AER was insignificant, and the only variable that explained reductions in allocative error was a measure of extension expenditures.

A descriptive case study approach to research evaluation has virtue when there exists a direct link between a specific area of social science research and improved decision making and policy making. Several descriptive case study evaluations of economics and policy research have been carried out. They set out to build a solid, scholarly case detailing the contributions of a carefully delineated area of research, but do not try to quantify net social benefits. In some cases, it is concluded that research has had substantial positive impacts. For example, Ruttan (1984) discusses the major role played by agricultural economics research in the design and evolution of the direct payment approach to support commodity prices and farm incomes, and Adams (1983) documented the case of economics research that contributed to the creation of the cereal import facility at the International Monetary Fund. In order to be useful, Adams concluded that social science research must be timely and carefully translated and packaged for policy makers. Other case studies have been much less sanguine about the contributions of economics research. Hamermesh (1982) found that economics research in the area of unemployment insurance had little influence on public policy. Descriptive case study evaluations have found that labor economics research (Dunlop 1977, Ashenfelter and Solon 1982) and research on utility rates (Acton 1982) have contributed little to public policy.

A number of obvious problems arise from the use of descriptive case studies. For example, studies of this nature are generally limited to narrowly circumscribed subject areas and are usually void of explicit consideration of research costs. Furthermore, the potential for less than fully objective evaluation is always present, as it was in Heller's (1966, p. 1) extremely optimistic assessment of Keynesian macro policy of the early 1960s: "Two Presidents have recognized and drawn on modern economics as a source of national strength and Presidential power. Their willingness to use, for the first time, the full range of modern economic tools underlies the unbroken U.S. expansion since early 1961." And finally, even when tempered with a high degree of objectivity, descriptive case studies are suspect in that only highly successful cases may be selected for evaluation.

Research Supply and Demand:
A Framework for Evaluation

A major conclusion to be gained from the preceding discussions is that a broader evaluation framework is needed for forest economics and policy research. Most previous evaluation approaches have focused on contribution to economic efficiency to the exclusion of other evaluation criteria. Therefore they are unlikely to capture all but a small portion of the potential contributions of multiple-objective research such as forest economics and policy. Such shortcomings point to the need for a more comprehensive framework for assessing the impacts of social science research. What follows is a socioeconomic model of research supply and demand that is intended to provide such a framework. Drawing on several previous models (de Janvry 1973, Blase and Paulson 1972), the primary purpose of the model is to conceptualize the social value of new knowledge produced by forest economics and policy research. The model identifies the most salient features of the research and social systems at a macro level—especially the sources of demand for FREP research and factors influencing the supply of new FREP knowledge—and sheds some light on the interaction between the research system and the socioeconomic environment in which the system operates. The model (Figure 3.1) consists of two main components: the social system, from which the demand for research arises, and the research system, a subset of the social system in which the supply of new knowledge is generated. The research system in the model refers to public sector research (universities and government research organizations), although fully recognized is that private sector research, "informal" research, and learning by doing also make substantial contributions to supplies of new knowledge.

Change in Society and the Demand for Research

Change in society that originates outside of the research system is the hub of the model and the ultimate source of demand for forest economics and policy research. Without such change there would be no demand for new knowledge or research. Examples of change that is exogenous to the research system include

- Economic change (e.g., rising factor or product prices)
- Social change (e.g., changes in a population's age distribution)
- Political change (e.g., rising power of special-interest groups)
- Technological change (e.g., advances in computer technology)
- Cultural change (e.g., increasing environmental awareness)
- Ecological change (e.g., increasing incidence of a particular forest disease)
- Legal change (e.g., enactment of the Forest and Rangeland Renewable Resources Planning Act of 1974)

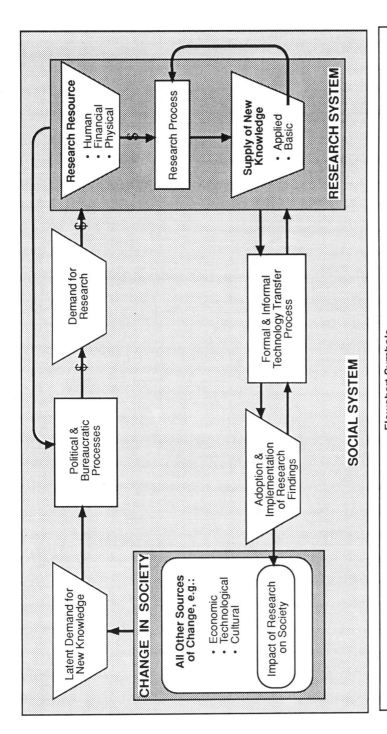

Figure 3.1 Socioeconomic model of research supply and demand

The research system is assumed to be responsive to the gamut of change in society. This is consistent with, but more inclusive than, models of induced innovation that focus on the role played by economic forces—factor prices, product prices, and other economic variables—in determining the rate and direction of technical change (Binswanger and Ruttan 1978). This is also consistent with Ruttan's (1984) hypothesis that the demand for new knowledge in the social sciences is derived primarily from the demand for institutional innovation and improvements in institutional performance. Ruttan defines institution broadly to include "both the behavioral rules that govern patterns of relationships and action as well as decision-making units such as government bureaus, firms, and families," and defines institutional innovation as changes "(a) in the behavior of a particular organization, (b) in the relationship between such organization and its environment, or (c) in the rules that govern behavior and relationships in an organization's environment" (Ruttan 1984, p. 549).

Change in society that induces a demand for new knowledge may be actual, anticipated, or desired. For example, actual change in stumpage prices would create a demand for current price information or price forecasts to facilitate the attainment of economic efficiency in the production of wood products (i.e., equating the value of the marginal product of each input to its unit cost). An anticipated increase in social conflict over the management of forest resources would create a demand for policy research to help manage or resolve disputes. A desire of public officials or industry to increase the quantity of stumpage marketed by nonindustrial private forest land owners would create a demand for new policies to achieve this goal, information about the price elasticity of stumpage supply, and so on.

Latent and Actual Demand for Research

The demand for new knowledge that arises from change in society is first a latent demand. Demand, of course, is the willingness and ability to pay a sum of money for some amount of a particular good or service at a particular time and place. Latent demand is defined as demand that is present or potential but not manifest; demand that is not clearly visible. As a result of actual, anticipated, or desired change, various groups within society develop latent demands for new knowledge to deal with or achieve the change. These latent demands eventually surface and are communicated to politicians and bureaucrats through pressure exerted by interest groups and the general public. As indicated by the feedback loop linking the research system to the political and bureaucratic processes (Figure 3.1), latent demands for new knowledge may also be communicated through research entrepreneurs. Researchers and research administrators sometimes engage in risk-taking entrepreneurial behavior by identifying and responding to latent demand for new knowledge and working to satisfy that demand.

Latent demands for new knowledge are transformed into a set of actual demands for public research through the political and bureaucratic processes. Nonentrepreneurial researchers respond to these actual demands through research funding opportunities. As observed by de Janvry (1973, p. 418), actual demand for research may be manifested in two forms: "(1) the budget allocated for research, both in its absolute size and in its allocation restrictions, and (2) a flow of information." Demand for research is not always objectively motivated. Some social science research (especially policy research or program evaluation) is sometimes used for political purposes by bureaucrats who commission the research. Research may be used to buy time and postpone decisions, substitute for risky political action, and gain political leverage (Rein and White 1977). In such cases, the latent demand for research arises within the political and bureaucratic processes.

Research Inputs and Outputs

The resource and information flows resulting from the political and bureaucratic processes represent key inputs into the research process, which can be conceptualized as a production process, transforming scarce inputs into outputs (Figure 3.1). Research inputs include stock resources such as human and physical capital, and flow inputs such as budgets and information. The primary output of the research process is new knowledge, which can be classified as either basic or applied. Applied research is carried out with a potential or actual application in mind and is often oriented toward problem solving. Basic or disciplinary research is intended to advance scientific knowledge. The demand for basic research is derived mainly from the need for theories and tools to conduct applied research. The results of basic research typically feed back into the research process and are used primarily by other researchers (see feedback loop in Figure 3.1).

Dissemination of Research Findings

The new knowledge produced by applied research is often a highly perishable commodity, and delays in transferring that knowledge to potential users may substantially reduce its utility and value. Applied research results are translated and packaged into forms that are usable to potential adopters in the technology transfer process. Formal technology transfer is usually carried out under the aegis of the research system, but technology transfer may be carried out informally outside of research through trade publications, newspapers, programs of continuing education, and other communication channels. As indicated by two-way information flows (see Figure 3.1), the technology transfer process should involve two-way interaction—both transferring of research results to users and communicating of user needs and latent demands to researchers.

The outcome of the technology transfer process is the adoption and implementation of research findings. Adoption and implementation are critical from the standpoint of valuing research—it is at this point that new knowledge produced by research first has an impact on society. As Boulding (1966, p. 9) has observed in reference to the impact of economics knowledge, "The only point at which knowledge can affect a social system is through its impact on decisions." Interestingly, thousands of studies concerning the adoption of innovations have been carried out over the past several decades (Rogers 1983), including a number of case studies in forestry. Many factors influence the adoption of innovations, including characteristics of the innovation, characteristics of the adopters, and characteristics of the environment in which diffusion and adoption take place.

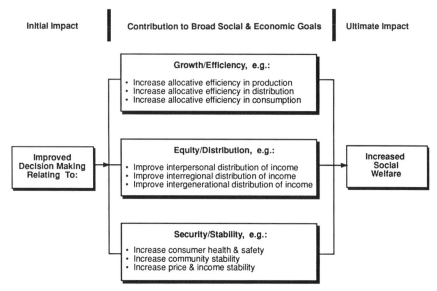

Figure 3.2 Potential source of value to society resulting from the application of new knowledge produced by forest economics and policy research

Impacts of Research

The impact of new forest economics and policy knowledge on the decisions of producers, consumers, and policy makers can take a variety of forms that influence the value of the research. In exceptional cases, new knowledge produced by forest economics and policy research may have a direct impact on decisions and policies (e.g., by improving the allocation of resources or by creating or modifying policies). More commonly, however, research has an indirect impact on decisions through "knowledge creep" and the "enlightenment function" of policy and social science research (Weiss 1977, 1980). Another type of indirect impact is helping to identify important problems and issues, "for whether it solves problems or not, [policy] research helps those involved in the making of policy to decide what the problems are" (Rein and White 1977, p. 136). Finally, much research has a latent or potential impact, such as basic or disciplinary research that feeds back into the research process or applied research that remains "on the shelf."

A comprehensive, if highly simplified, way of looking at the impacts of forest economics and policy research is presented in Figure 3.2. Improved decision making as a result of research may relate to three fundamental and broad social goals: growth/efficiency, equity/distribution, and security/stability. These are the triumvirate of goals frequently mentioned in discussions of the justification for public policy (Musgrave 1959). Each broad goal entails many dimensions or subgoals and a large number of specific objectives. The growth/efficiency goal is concerned with increasing the national capacity to satisfy people's wants. Krutilla and Haigh (1978) have argued that this should be the principal goal of public involvement in forestry, and it has undoubtedly received the most emphasis in research planning and evaluation. The equity/distribution goal refers to distribution of the benefits resulting from productive activities and to equality of rights of individuals within society. Equity considerations have significant political and social importance for all public decisions but are usually ignored in the evaluation of research (Bengston and Gregersen 1988). Finally, the security/stability goal is concerned with the reduction of threats or potential threats to society, ranging from issues of human health and safety to conservation of natural resources. Increased economic stability—such as reduction of fluctuations in aggregate economic activity—is a widely recognized component of this goal. Research contributions toward the attainment of each of these three broad social goals—and the endless list of specific objectives subsumed under them—constitutes an increase in social welfare. A comprehensive framework such as this is needed to adequately assess the contributions of forest economics and policy research as well as related social science research. Evaluation criteria should be developed that relate to each category of social goals.

A final characteristic of the model (Figure 3.1) is that the impacts of research often induce further latent demands for new knowledge. For example, Iverson and Alston (1986) analyze the historical development of forest management planning models in the USDA–Forest Service, tracing the movement from single-resource scheduling models to integrated, multiresource planning models. They discuss changes in society that have given rise to the latent demand for forest planning models, including increasing demands for all forest outputs, environmental activism, and land management legislation. As each generation of models was introduced, interest groups reacted to the actual or anticipated impacts of implementation in forest management, creating latent demands and political pressure for new and more integrated planning approaches. Public outcry at the results and limitations of existing models induced latent demand on the part of resource managers for more realistic models. Thus, the impacts of research become a source of change in society from which latent demands for further research arise.

Implications for Planning and Evaluating
Forest Resource Economics Research

A number of issues relating to the evaluation of forest economics and policy research have been explored here. The complexity and difficulty of such an evaluation has not been discounted—the general neglect of evaluation of social science research suggests the difficulty of the task. Because of the many conceptual issues and operational problems involved, evaluating any area of social science research is likely to stretch evaluation methods and the abilities of evaluators to their limits. In the spirit of improving the research planning and evaluation environment, consider the following points.

Limitations of Single Criterion Evaluations

Past research evaluation approaches are quite limited in what they can contribute to the evaluation of forest economics and policy research. Perhaps most restrictive is the focus on economic efficiency as a sole evaluation criterion. Single criterion evaluations of social science research—which may contribute to multiple social and economic goals—cannot adequately address the concerns of policy makers. Commenting on single criterion evaluations, Franz (1981, p. 9) states that

> the objectives or decision criteria OMB [Office of Management and Budget] uses deal with the contribution of agricultural research in meeting such national objectives as creating a positive balance-of-trade position, improving the quality of the environment, adding to the general health and safety of the population, and enhancing the general quality of life of the population. . . . When one restricts himself or herself to such a narrow evaluation base, that person runs the risk of saying nothing or very little about the fundamental questions of need and the appropriate federal role.

Promising Evaluation Approaches

Several evaluation approaches that may prove helpful in evaluating forest economics and policy research have been touched upon here. For example, researchers investigating the utilization of social science knowledge and research have learned much in recent years about the microprocesses by which policy makers learn about and incorporate social science knowledge. For example, the channels through which social science research is communicated to policy makers have been found to be exceptionally diverse: they include conferences, consultant briefings, conversations with colleagues, field reports, and the media. In the legislative sphere, interest groups have been found to be key communicators of social science research results as they make the case for or against specific legislative proposals. Research of this nature could be useful in designing realistic and pragmatic evaluation approaches for forest economics and policy research.

Evaluation approaches employing multiple criteria (e.g., scoring models, simulation models, mathematical programming) may also provide useful guidance. Several research evaluations utilizing such methods have been proposed—they have seldom been tested empirically. Ruttan (1982) addresses multiple criteria and related approaches to evaluation, including several promising experimental research resource allocation models that may be useful for planning and managing ongoing research programs.

Problems with aggregate evaluation of forest economics research are troublesome, especially the problem of establishing causality of change. A useful approach to the problem has been suggested by Gregersen et al. (1983). They evaluated the economic impacts of seven major timber utilization technologies using an *ex ante* benefit-cost approach. The projected benefits from the adoption of the seven technologies were compared to total expenditures for all forest products research. The analysis revealed that the entire research budget could be economically justified by the benefits associated with the seven technologies, which resulted from only a part of the total research program. Such an approach could be used in an aggregate economic evaluation of any area of research in which program-wide benefits are difficult to estimate but benefits due to individual research projects or thrusts may be obtained.

Research Supply and Demand Model

Finally, the socioeconomic model of research supply and demand presented earlier has several implications for valuation of forest economics and policy research and for research planning. First, the elasticity of supply of forest economics and policy research is likely to be a key determinant of the value of such research. Ruttan (1984, p. 552) has argued that the supply of economics knowledge for institutional innovation is relatively elastic: "Economists respond rapidly to changes in the economic and political environment. Advances in economic thought are becoming an increasingly effective substitute for trial and error in the design and reform of economic institutions and economic policy." Forest economists, as producers of new forest economics and policy knowledge, should strive for a highly elastic supply of this knowledge. Research administrators have a vital role in creating responsive research systems through their influence on the institutional environment in which research takes place, this includes nurturing research entrepreneurship and creating appropriate incentives for scientists (e.g., rewarding scientists for interaction with users or potential users, technology transfer activities).

Timing as it affects the value of forest economics and policy research is another implication of the model. Research findings must reach appropriate decision makers in a useful form and in a timely manner if they are to have an impact on decisions. Attention to the forces that create the demand for forest economics and policy research will increase the likelihood of timely information being produced and will increase the value of research efforts. Forest economics and policy researchers should invest more time and effort communicating with users and policy makers to identify research needs and emerging issues. If incorporated into research planning, such information could increase the timeliness—and hence the value— of forest economics and policy research.

Finally, the model also has implications for an important research policy issue: distribution of the benefits of research. The effectiveness with which latent demands for new knowledge are translated into actual demand for research in the political and bureaucratic processes is a key to the eventual distributional impacts of research. If relatively large or wealthy user groups (e.g., owners of large timberland areas, major corporations) are more successful in communicating their latent demands for new knowledge, the actual demand for research will reflect the utility function of such groups (Grabowski 1979). Having exerted disproportional influence in setting the research agenda, these groups will capture the lion's share of research benefits. Groups that are not effective in communicating their latent demands will gain little or may experience a loss of welfare due to technical or institutional change resulting from research. The lesson for research administrators and scientists is the importance of soliciting input from all user groups when formulating research programs and setting research priorities. In addition to the potentially regressive distributional implications, relying on the most vocal interest groups or on subject-matter experts to set future research directions may also reduce the value of a research program. Innovation that is "directed" by a small group rather than induced by broad social and economic forces is less likely to be consistent with economic and cultural endowments and is less likely to be an effective contributor to economic and social development.

Summary and Conclusions

What can we conclude about evaluating the contributions of forest economics and policy research? Does forest economics and policy research make a difference? Attempts to empirically assess the effects of forest economics and policy research will be problematic, especially at the aggregate level. Rigorous evaluation of forest economics and policy research as a whole may require more than our current ability to conceptualize and empirically analyze. But it is important to bear in mind that unless society is seriously underinvesting in forest economics and policy research, we would expect the economic impacts of such research to be relatively small because the field itself is small. If investments in forest economics and policy research are reasonably efficient, the marginal cost of such research should approximate the marginal benefits resulting from it. The cost of forest economics and policy research is diminutive compared to the value of economic activity in the forest-based sector. Thus, forest economists could earn their keep by having minor and scarcely detectable positive impacts on decisions and policy formulation concerning a very large and important sector.

Perhaps the key indication of the potential value of forest economics and policy research is its responsiveness to demands for new knowledge or what Thurow (1977, p. 80) has called the "client relationship" that economics as a profession has with society: "Economists have not been generally instrumental in shaping society's agenda, but they have been willing to work on that agenda—whatever it is." Forest economists and policy researchers have been willing to work on society's agenda and direct their talents toward important issues of public policy. A focus on continued contributions to this agenda should be the touchstone in any efforts to guide future directions for forest economics and policy research.

Literature Cited

Acton, J. Paul. 1982. An evaluation of economists' influence on electric utility rate reforms. American Economic Review 72(2):114-119.

Adams, R. H. Jr. 1983. The role of research in policy development: The creation of the IMF cereal import facility. World Development 11(7):549-563.

Antonovitz, F. and T. Roe. 1986. A theoretical and empirical approach to the value of information in risky markets. The Review of Economics and Statistics 68(1):105-114.

Ashenfelter, O. and G. Solon. 1982. Employment statistics: The interaction of economics and policy. American Economic Review 72(2):233-236.

Bengston, D. N. 1985. Economic evaluation of agricultural research: An assessment. Evaluation Review 9(3):243-262.

Bengston, D. N. and H. M. Gregersen. 1988. Income redistribution impacts of technical change: A framework for analysis and empirical evidence. In: Management of Technology I by T.M. Khalil, B.B. Bayraktar, and J.S. Edowomwan (eds.). Interscience Enterprises, Ltd. Geneva, Switzerland.

Binswanger, H. P. and V. W. Ruttan. 1978. Induced innovation: Technology, institutions and development. Johns Hopkins University Press. Baltimore, MD.

Blase, M. G. and A. Paulson. 1972. The agricultural experiment station: An institutional development perspective. Agricultural Science Review 10(second quarter):11-16.

Boulding, K. E. 1966. The economics of knowledge and the knowledge of economics. American Economic Review 56(2):1-13.

Cartwright, R. W. 1971. Research management in a department of agricultural economics. Ph.D. thesis. Purdue University. Lafayette, IN.

Castle, E. N. 1977. Research needs in forest economics and policy: An interpretative and evaluative summary. In: Research in forest economics and forest policy by M. C. Clawson (ed.). RFF Research Paper R-3. Resources for the Future. Washington, DC.

Clawson, M. (ed.). 1977. Research in forest economics and forest policy. RFF Research Paper R-3. Resources for the Future. Washington, DC.

de Janvry, A. 1973. A socioeconomic model of induced innovations for Argentine agricultural development. Quarterly Journal of Economics 87(3):410-423.

Duerr, W. A. and J. B. Duerr. 1980. Social sciences in forestry as a field of research: What is the Forest Service doing and how well? Unpublished report to the USDA–Forest Service, Forest Resource Economics Research Staff. Washington, DC.

Duerr, W. A. and H. J. Vaux (eds.). 1953. Research in the economics of forestry. Charles Lathrop Pack Forestry Foundation. Washington, DC.

Dunlop, J. 1977. Policy decisions and research in economics and industrial relations. Industrial Labor Relations Review 30(3):275-282.

Eisengruber, L. M. 1978. Developments in the economic theory of information. American Journal of Agricultural Economics 60(5):901-905.

Evenson, R., P. Waggoner, and V. Ruttan. 1979. Economic benefits from research: An example from agriculture. Science 205(Sept. 14):1101-1107.

Franz, M. 1981. Evaluation information on agricultural research. In: Evaluation of Agricultural Research, G. W. Norton et al. (eds.). Misc. Publication 8-1981. Agricultural Experiment Station. University of Minnesota. St. Paul, MN.

Grabowski, R. 1979. The implications of an induced innovation model. Economic Development and Cultural Change 27(4):723-734.

Gregersen, H., J. Haygreen, I. Holland and D. Erkkila. 1983. Impacts of forest utilization research—an economic assessment. Final Report to Forest Products Laboratory (USDA-FPL-81-0395). USDA Forest Service. Madison, WI.

Gregersen, H. M. 1985. The University of Minnesota forestry research evaluation program. In: Forestry research evaluation: Current progress, future directions by C. D. Risbrudt and P. J. Jakes (eds.). General Technical Report NC-104. USDA Forest Service. North Central Forest Experiment Station. St. Paul, MN.

Griliches, Z. 1987. R&D and productivity: Measurement issues and econometric results. Science 237(July 3):31-35.

Hamermesh, D. S. 1982. The interaction between research and policy: The case of unemployment insurance. American Economic Review 72(2):237-241.

Hayami, Y. and W. Peterson. 1972. Social returns to public information services: Statistical reporting of U.S. farm commodities. American Economic Review 62(1):119-130.

Heller, W. W. 1966. New dimensions of political economy. Harvard University Press. Cambridge, MA.

Hildreth, R. J. 1980. Concluding remarks. In: Research and extension productivity in agriculture by A. A. Araji (ed.). Department of Agricultural Economics and Applied Statistics. University of Idaho. Moscow, ID.

Hyde, W. F. 1985. A proposed forestry research evaluation project. pp. 34-48. In: Forestry research evaluation: Current progress, future directions by C. D. Risbrudt and P. J. Jakes (eds.). General Technical Report NC-104. USDA Forest Service, North Central Forest Experiment Station. St. Paul, MN.

Iverson, D. C. and R. M. Alston. 1986. The genesis of FORPLAN: A historical and analytical review of Forest Service planning models. General Technical Report INT-214. USDA Forest Service, Intermountain Research Station. Ogden, UT.

Jakes, P. J. and C. D. Risbrudt. 1988. Evaluating the impacts of research. Journal of Forestry 86(3):36-39.

Krutilla, J. V. and J. A. Haigh. 1978. An integrated approach to National Forest management. Environmental Law 8:373-383.

Lundgren, A. L. 1982. Research productivity from an economic viewpoint. In: Proceedings of the 1981 convention of the Society of American Foresters. SAF Pub. 82-01. Society of American Foresters. Bethesda, MD.

Lundgren, A. L. 1986. A brief history of forestry research evaluation in the United States. In: Evaluation and Planning of Forestry Research by D. P. Burns (compiler). General Technical Report NE-GTR-111. USDA Forest Service. Northeastern Forest Experiment Station. Broomall, PA.

Mahlstede, J. P. 1971. Long-range planning at the Iowa Agricultural and Home Economics Experiment Station. In: Resource allocation in agricultural research by Walter L. Fishel (ed.). University of Minnesota Press. Minneapolis, MN.

Mansfield, E., A. Romeo, M. Schwartz, D. Teece, S. Wagner and P. Brach. 1982. Technology transfer, productivity and economic policy. W. W. Norton and Company. New York, NY.

Musgrave, R. A. 1959. The theory of public finance: A study in public economy. McGraw-Hill Publishers. New York, NY.

Norton, G. W. 1987. Evaluating social science research in agriculture. Paper presented at the National Symposium on Evaluating Agricultural Research and Productivity. January 29-30, 1987. Atlanta, GA.

Norton, G. W. and J. S. Davis. 1981. Evaluating returns to agricultural research: A review. American Journal of Agricultural Economics 63:685-699.

Norton, G. W. and G. E. Schuh. 1981. Evaluating returns to social science research: Issues and possible methods. In: Evaluation of agricultural research by G. W. Norton et al. (eds.). Misc. Publication 8-1981. Agricultural Experiment Station. University of Minnesota. St. Paul, MN.

Rein, M. and S. H. White. 1977. Can policy research help policy? The Public Interest 49(Fall):119-136.

Rogers, E. M. 1983. Diffusion of innovations (Third Edition). The Free Press. New York, NY.

Ruttan, V. W. 1982. Agricultural research policy. University of Minnesota Press. Minneapolis, MN.

Ruttan, V. W. 1984. Social science knowledge and institutional change. American Journal of Agricultural Economics 66(5):549-559.

Sabatier, P. A. 1987. Knowledge, policy-oriented learning, and policy change. Knowledge: Creation, Diffusion, Utilization 8(4):649-692.

Schuh, G. E. and H. Tollini. 1979. Costs and benefits of agricultural research: State of the arts. Staff Working Paper 360. World Bank. Washington, DC.

Schultz, T. W. 1953. The Economic Organization of Agriculture. McGraw-Hill Publishers. New York, NY.

Thurow, L. C. 1977. Economics, 1977. Daedalus 106(Fall):79-94.

Valfer, E. S., M. W. Kirby, and G. Schwarzbart. 1981. Returns to investments in management sciences: Six case studies. General Technical Report PSW-52. USDA–Forest Service. Pacific Southwest Forest and Range Experiment Station. Berkeley, CA.

Volker, C. B. and R. E. Deacon. 1982. Evaluation of benefits from research in home economics. Home Economics Research Journal 10(4):321-331.

Weiss, C. H. 1977. Research for policy's sake: The enlightenment function of social research. Policy Analysis 3(Fall):531-545.

Weiss, C. H. 1980. Knowledge creep and decision accretion. Knowledge: Creation, Diffusion, Utilization 1(3):381-404.

Whaley, R. S. and E. F. Bell. 1982. Health of Forest Service economics research. Journal of Forestry 80(6):347-349, 364.

4

ORGANIZATIONAL INVOLVEMENT AND MANAGEMENT OF RESEARCH PROGRAMS

Larry W. Tombaugh

Forest economics and policy research is an especially important activity within the nation's forestry community. The number of organizations supporting or conducting such research is significant as are the number of researchers applying assorted principles of economics and policy. Of importance to any research planning effort is a clear understanding of the institutional landscape in which research is undertaken and the manner in which research programs are designed and coordinated.

Forest economics and policy research surfaces a number of important definitional problems, including the distinction between forest economics research and forest policy analysis. In what follows, a separation between the two will not be made; doing so would be like trying to draw clear boundaries between basic and applied research. Also to be recognized is that forest economics has roots in two quite different disciplines—economics and forestry. The long-run viability of the field requires that the "economics" in forest economics draw upon sound, accepted principles of economic theory and that the "forestry" in forest economics guide economics research toward topics of concern to the forestry interests of society. Balancing such functions is typically the responsibility of research administrators. Several thoughtful scholars have expressed concern that short-run policy issues of immediate concern to forestry agencies and corporations are steering the forest economics research system—much to the detriment of the system's long-term intellectual development. At risk may be linkages between advances in fundamental economics and the increasingly complex management of renewable forest resources. The forest economics community should be cautious of the consequences of depleting the intellectual capital upon which the field of forest economics is based.

In addition to being rooted in the fundamental discipline of economics, forest economics research is a component of a larger—much larger—set of activities known as forestry research. Describing organizational issues of forest economics research without putting them into the broader context of forestry research would not be useful. They are closely intertwined. Just as forest economics is part of forestry research, forestry research is an integral part of forestry. Although a continuing stream of questions arise in forestry research that necessarily lead scientists into research on fundamental processes, forestry research ultimately is directed toward applied problems arising from the management and utilization of forest resources.

Forest economics research and policy analysis are strongly influenced by the biological characteristics of forest resources, by the attitudes and ethos that have evolved in the United States concerning the appropriate uses of these resources, and by the institutions that have been created to express such attitudes and ethos. As an applied social science, forest economics is—or certainly should be—at the very center of debate and controversy that so often characterizes contemporary forest resource management.

Forests occupy 737 million acres, or 33 percent of the land area of the United States (Forest Service 1982a). The resource is amazingly diverse in terms of species and species mixes, proximity to population centers, productive potential, and uses to which it can be put. Such biological, climatic, topographic, and locational diversity necessitates widely different management strategies (National Association of Professional Forestry Schools and Colleges and Cooperative State Research Service 1985). For example, problems faced in the management of forests and wildlife in the old-growth Douglas-fir region of Oregon and Washington differ radically from the management problems represented by newer, second-growth forests of the Lake States. These management challenges are different still from those involved in managing wilderness areas in fragile mountainous areas of Colorado or loblolly pine plantations in Mississippi.

Ownership patterns of the nation's forest resource are also highly diverse. A single federal agency—USDA–Forest Service—is the custodian of 19 percent of the nation's forested land. Over half of the nation's forested land is commercial timberland, owned by hundreds of thousands of private, nonindustrial landowners— including farmers. Forest industries own 14 percent. The remaining forest land is owned by other federal agencies and state, county, and municipal governments (Forest Service 1982a). Forest land ownership patterns vary substantially from region to region. Most public forest land is in the West, whereas most private forest land is in the South and North. Each ownership group faces unique forest management issues and problems.

Finally, diversity characterizes the array of goods and services produced by the nation's forests. A recent report points out that, in one fashion or another, forests contribute significantly to the social and economic fabric of the United States (National Association of Professional Forestry Schools and Colleges and Cooperative State Research Service 1985). They attract millions of visitors annually for hunting, fishing, hiking, camping, or sight-seeing. Forest-based tourism and recreation are often the primary local industries for many communities. Outdoor recreation equipment and sporting goods represent a multi-billion dollar enterprise. In addition, water, wildlife, and livestock are also important products of forest and associated range lands. All major rivers in the West and most in the East originate in forested mountains. These rivers furnish drinking water, hydroelectric power, and irrigation for homes, industries, and farms. Forests and range lands also provide forage and shelter for livestock and wildlife and essential habitat for endangered plants and animals. One of the most important forest benefits in terms of employment and direct economic contributions is wood products. An estimated 500 products are derived from wood. The industry that manufactures such products is a significant contributor to the U.S. economy and is likely to become even more so in the future. The U.S. Congress Office of Technology Assessment has stated that "the United States is well positioned to satisfy both domestic and a major share of future global forest products requirements" (Office of Technology Assessment 1983).

The enormous diversity of forest resources, and their ownership and competing uses, has had a profound influence on the nature and diversity of institutions that provide new knowledge for managing and utilizing forest resources. Being a part of forestry research, forest economics and policy research is likewise supported and conducted by a wide variety of organizations.

Forest Economics Research Organizations

The Forest Service within the U.S. Department of Agriculture and research programs associated with state agricultural experiment stations in land-grant universities conduct most forest economics and policy research in the United States (Ellefson 1987). Several public universities outside the land-grant system also maintain strong research programs in forest economics and policy, as do at least three private universities—Harvard, Yale, and Duke. Other federal agencies that either conduct or support such research include the USDA–Economic Research Service, several agencies within the U.S. Department of the Interior, the U.S. Department of Treasury, the Congressional Research Service, and the U.S. Congress Office of Technology Assessment. The USDA–Cooperative State Research Service is a major actor in forest economics and policy research. It discharges legislatively mandated responsibilities by funding research through eligible universities.

The USDA–Forest Service has conducted economics research almost since its establishment in 1905. The agency's research mission was significantly expanded and formalized by the enactment of the McSweeney-McNary Act of 1928. Among other charges, the law instructed the secretary of agriculture to "conduct such investigations, experiments, and tests as he may deem necessary . . . to determine and promulgate the economic considerations for the management of forest land and the utilization of forest products" (Dana 1956). The Act also authorized periodic surveys of the nation's timber resource. Such surveys have created a data base that has proven to be invaluable to forest economists and policy researchers.

The McSweeney-McNary Act was superceded by the Forest and Rangeland Renewable Resources Research Act of 1978, which, in turn, specified a broadened USDA–Forest Service research mission. Three provisions of the act are especially relevant to forest economics research. First, the act made clear that USDA–Forest Service research activities should not be restricted to issues associated with the national forest system. The charge to the secretary of agriculture was "to conduct, support, and cooperate in investigations, experiments, tests, and other activities the Secretary deems necessary to obtain, analyze, develop, demonstrate and disseminate scientific information about protecting, managing, and utilizing forest and rangeland renewable resources in rural, suburban, and urban areas" (Forest Service 1983). The law did not mention the national forest system. Second, the law gave the USDA–Forest Service a very broad subject-matter mandate. It specified that USDA–Forest Service research is to include, but not be limited to, such broad areas as renewable resource management, environmental research related to renewable resources, renewable resource protection, utilization and renewable resource assessment research. Although forest economics and policy research is not specifically identified in the law, it is implicit in several of these broad categories. Third, the law authorized the secretary of agriculture to make competitive grants to federal and state agencies, universities, businesses, individuals, and other institutions to conduct renewable resources research. Although competitive grants for forestry have not had a significant influence on forest economics and policy research, they have certainly affected research in forest biology and utilization. Such influence is expressed by creation of an environment in which forest scientists must compete for resources on the basis of scientific excellence. As a mechanism for encouraging excellence, the forestry competitive grants program initiated in Fiscal Year 1985 has been a valuable supplement to the more project-oriented programs that have developed in forestry research.

USDA–Forest Service appropriations for research were $113.8 million in 1985. Of this total, the agency invested $7.85 million on activities defined as economics or policy research—68.4 scientist years working on thirty separate projects (Ellefson 1987). Most of these funds were disbursed to USDA–Forest Service researchers administratively affiliated with one of eight USDA–Forest Service Experiment Stations or to the USDA–Forest Service's Forest Products Laboratory in Wisconsin. A staff office in Washington, D.C. is responsible for overall coordination and policy direction of USDA–Forest Service economics research. Station directors and project leaders, however, are given considerable autonomy to select research areas and to manage specific research programs.

Collectively, most forest economics and policy research in the United States is conducted by universities. Forestry research is conducted in three categories of universities, namely, land-grant universities, which are public institutions associated with state agricultural experiment stations; public universities not affiliated with a state agricultural experiment station or the land-grant system; and private universities.

Important to forest economics research is the unique federal-state partnership in agriculture that was created by the historic Hatch Act of 1887. The act was based on the premise that a strong, efficient food-producing system is in the national interest and that the nation's various states are in the best position to determine research needed to solve site-specific problems. From such a perspective emerged the state agricultural experiment stations (SAES), which exist at land-grant universities. The Hatch Act also authorized the first major federal commitment to science in the form of research funds allocated (on the basis of a formula) to each state agricultural experiment station. An important feature of the act was that federal contributions are to be at least matched by nonfederal expenditures for agricultural research in each state; thus establishing a partnership between the federal government and the states. The federal Hatch funds coupled with state appropriations have provided a stable base of support for agricultural research for over 100 years. They are credited with playing a major role in developing the technological base that has enabled U.S. agriculture to become the most productive in the world.

The importance of forestry in the United States and the long time scales often involved with forestry research gradually made clear the need to develop a stable program of support for forestry research. In 1962, the Cooperative Forestry Research Act (McIntire-Stennis Act) established a program of forestry research support based on the allocation of federal funds according to a formula. It, too, required that federal expenditures be at least matched by nonfederal contributions. The key legislative sponsors of the Act were Senator John C. Stennis of Mississippi and Representative Clifford G. McIntire of Maine. The Cooperative Forestry Research Act has had an enormous influence on forestry research at universities. Because research support for forest economics had often been difficult to obtain and sustain, the act has had a major positive influence on these research programs.

The USDA–Cooperative State Research Service is the federal agency that is responsible for maintaining the federal share of the federal-state partnership in forestry research and for administering the McIntire-Stennis Cooperative Forestry Research program. The agency also serves an important coordinating function for forestry research. Both through the Current Research Information System (CRIS), a computerized data base on forestry and agricultural research, and through its professional staff, the agency is constantly in touch with forestry researchers throughout the nation and is able to make useful suggestions for research coordination.

Although federal funds have been essential to the development of strong forestry research programs, states have contributed the larger share to the partnership, and the relative magnitude of the state funding continues to increase. In Fiscal Year 1985, for example, an estimated $88.3 million was expended on forestry research by the nation's universities. Of this total, only $12.4 million (14 percent) was from the federal portion of the McIntire-Stennis program as allocated by the USDA–Cooperative State Research Service. Another $2.2 million (2.5 percent) was made available to universities through Hatch Act appropriations. Other federal agencies (e.g., USDA–Forest Service and the U.S. Department of the Interior agencies) contributed 14.9 percent, or $13.2 million. Nonfederal agencies, most of which are state agencies, were responsible for $60.1 million (68 percent) of the nation's university-based forestry research program (Meadows 1988).

Universities receiving funds through programs administered by the USDA–Cooperative Research Service conduct a significant proportion of the forest economics and policy research in the United States (Ellefson 1987). In 1985, 39.4 percent of the scientist years dedicated to forest economics and policy research was housed in universities. They accounted for 42.3 percent of the $30.5 million invested and 78 percent of the projects. The USDA–Forest Service accounted for 26.2 percent of the scientist years and 25.8 percent of the expenditures. Other agencies of the U.S. Department of Agriculture accounted for 29.7 percent of the scientist years and 27 percent of the expenditures. Most, but by no means all, university-based forest economics and policy research is conducted at land-grant universities. Sixty institutions participate in the McIntire-Stennis program; only 83 percent are associated with the land-grant university system. Furthermore, several private universities—most notably Harvard, Yale, and Duke—do not receive federal formula funds but historically have excelled in economics and policy research.

The U.S. Department of the Interior is another executive branch agency of the federal government that supports or conducts forest economics and policy research. The USDI–Bureau of Land Management, the USDI-National Park Service, the USDI-Fish and Wildlife Service, and the Office of the Secretary support economic and policy analyses on a variety of topics relating to their respective management responsibilities.

Forest economics and policy research is also conducted by three U.S. congressional agencies. All three direct their efforts to issues that are of particular interest to legislators or to legislative committees. The Office of Technology Assessment "provides an independent and objective source of information about the impacts, both beneficial and adverse, of technological applications and identifies policy alternatives for technology-related issues" (Office of Federal Register 1985). The office has conducted studies on such diverse topics as the role of technology in the forest products industry and technologies that might be useful in sustaining tropical forests (Office of Technology Assessment 1983, 1984). Both the General Accounting Office and the Congressional Research Service, a branch of the Library of Congress, conduct investigations about specific issues of legislative concern. The Congressional Research Service works exclusively for Congress, conducting research, analyzing legislation, and providing information at the request of committees, members, and their staffs. It has produced over seventy studies and analyses of forestry issues since 1983. Though not readily available to the general public, the reports can be obtained from legislators or authors.

Private nonprofit research organizations have played major national roles in furthering forest resource economics and policy research. Attesting to its interest in forest economics and policy research, Resources for the Future, Inc. (RFF) sponsored a 1977 symposium devoted to the subject and, subsequently, published a book entitled *Research in Forest Economics and Forest Policy* (Clawson 1977). Since then, RFF has continued to maintain an active forest economics and policy program that has produced several scholarly works that have contributed to the theory and practice of forest economics. Another Washington, D.C., nonprofit agency with an interest in forest economics and policy research is the Conservation Foundation. Among the Foundation's recently published reports are *Policies for Lake States Forests* (Shands and Dawson 1984) and *The Lake States Forests: A Resources Renaissance* (Shands 1988). Both documents have provided a recognizable stimulus to strengthening forestry coalitions in the Lake States.

A variety of private special-interest organizations also conduct forest economics and policy research, almost all of which is directed at specific issues. In the mid-1980s, for example, controversy raged over the relationship between the price received for national forest timber on specific sales and the cost of making the timber available for harvest. This so-called "below-cost" timber sale issue generated a spate of economic analyses by such disparate groups as the National Forest

Products Association, the Wilderness Society, and the Natural Resources Defense Council. Finally, most large forest products companies and several private consulting organizations regularly conduct sophisticated economic and policy analyses in support of their commercial objectives. Such analyses are usually not made public.

Research Planning and Coordination

The conduct of forest economics research in the United States is a highly decentralized enterprise. This operating style puts a premium on research planning and coordination. Forestry research in general and forest economics research in particular is planned and coordinated by a variety of formal and informal mechanisms. Most of the coordination that does take place is among programs and investigators in publicly supported agencies—particularly the USDA–Forest Service, the USDA–Cooperative State Research Service, and the nation's forestry schools.

Probably the most ambitious forest research planning effort extended over a two-year period and culminated in 1978 with a report entitled "National Program of Research for Forests and Associated Rangelands" (U.S. Department of Agriculture and National Association of State Universities and Land Grant Colleges 1978). The effort was initiated in July 1976 at the request of the Agricultural Research Policy Advisory Committee, a joint committee of the U.S. Department of Agriculture and the National Association of State Universities and Land-Grant Colleges (NASULGC). At that time, the research programs of U.S. forestry schools were collectively represented by the Association of State College and University Forestry Research Organizations (ASCUFRO), later to become the National Association of Professional Forestry Schools and Colleges (NAPFSC). The deputy chief for research of the USDA–Forest Service and the president of ASCUFRO were designated as cochairpersons of the steering committee appointed to conduct the effort. The steering committee, in turn, asked the Renewable Natural Resources Foundation (an organization of eleven natural resources professional and scientific societies) to bring together a select group of distinguished scientists, research administrators, and educators to examine the conduct of forestry research. The recommendations from these deliberations were published in a 1977 volume entitled "A Review of Forest and Rangeland Research Policies in the United States." The steering committee worked with four regional planning groups—also cochaired by university and USDA–Forest Service representatives—to conduct research-needs conferences in the North Central, Northeastern, Western, and Southern regions. Representatives of governmental, professional, environmental, conservation, and other organizations attended the conferences and, in total, identified over 2,000 research issues. A major capstone conference was held in Washington, D.C. in 1978 to identify researchable topics of national concern.

Evaluating all of the above inputs, seven national task forces identified 51 major areas in need of increased research attention. Several dealt specifically with forest economics and policy research and included the following (U.S. Department of Agriculture and National Association of State Universities and Land Grant Colleges 1978):

- Improve techniques for determining future demand for renewable natural resources and services.
- Identify and analyze land use preferences and factors affecting changes in land use patterns.
- Develop quantitative and other methods to determine production interactions and tradeoffs among major forest and associated rangeland uses.
- Improve capability to determine costs and benefits resulting from alternative timber management practices for a wide variety of forest types and sites.
- Determine effects of public policies on the production of timber from public and private lands.
- Produce more information on the economics of forest products utilization techniques.
- Provide more accurate marketing information to meet the needs of timber owners, wood processors, and product distributors.
- Increase capability to evaluate trends in forest products consumption and trade.
- Produce methods to quantify social and economic benefits from recreation on forests and rangelands.
- Investigate factors that determine trends in demand for various kinds of forest-based recreation activities.
- Determine factors underlying wilderness attractiveness, wilderness concepts and preferences held by the general public, and costs involved with wilderness.
- Produce economic data of use to potential commercial recreation producers.

The 1978 "National Program of Research for Forests and Rangelands" was updated in 1982, partly as an effort to fulfill USDA–Forest Service planning activities required by the Forest and Rangeland Renewable Resources Planning Act of 1974 (RPA). The report was entitled "1980-1990 National Program of Research for Forests and Associated Rangelands" (Forest Service 1982b). Using the same categories of research used in the 1978 report, the update projected scientific manpower needs by region for both the USDA–Forest Service and the nation's forestry schools. The total number of scientist years devoted to forest and associated rangeland research for the USDA–Forest Service, State Agricultural Experiment Stations, and forestry schools in 1980 was 1,640. By 1990 the number of scientist years required to accommodate regional research goals was expected to total 3,277 (Forest Service 1982b). The report showed graphically how this proposed doubling of scientific capability would be shifted within the national research program and among regions.

The only other coordinated national planning effort for forestry research since 1982 dealt with basic research. A National Task Force on Basic Research in Forestry and Renewable Natural Resources was created in 1979. Members were selected from the USDA–Forest Service, schools of forestry, and the forest products industry. At that time, the forestry research community was concerned that the scientific foundations for future applied research had not kept pace with the need for such research. The Task Force report (National Task Force on Basic Research in Forestry and Renewable Natural Resources 1982) stated,

Forestry is truly living not only on "borrowed time," but on borrowed resources. We are spending the principal at a faster pace than the interest is accruing. The scientific foundations for future applied research and management decisions are not being adequately replenished. Should this situation continue, we will inevitably confront decreased forest productivity, decreased wood utilization efficiency, deteriorating timber, water, and wildlife conditions, and drastically limited opportunities for recreation.

Economics and related managerial sciences were among those identified as needing increased attention from basic research.

As previously stated, the USDA–Forest Service is required to plan and coordinate its forestry research activities. Such is required by the Forest and Rangeland Renewable Resources Planning Act of 1974 as amended by the 1976 National Forest Management Act. Plans for USDA–Forest Service research must be developed and periodically updated in conjunction with the overall assessment and planning for the management and use of the nation's forest and rangelands as required by law.

Also mandated by law are formal mechanisms for planning and coordinating research conducted with funds appropriated through the McIntire-Stennis Cooperative Forestry Research Program. The original McIntire-Stennis bill created two bodies. The Cooperative Forestry Research Advisory Committee was composed of representatives of forest products industries and public agencies, including the USDA–Forest Service. It was designed to assure that a wide spectrum of public and private interests and concerns were being considered in the conduct of research supported by the program. The second advisory group was the Cooperative Forest Research Advisory Board, which was composed of forestry school representatives. It advised the secretary of agriculture on the appropriateness of the formula used to allocate the research funds among states. These two bodies were combined (Farm Bill of 1981 (P.L. 97-98)) into the Cooperative Forestry Research Advisory Council. This council advises the secretary of agriculture on plans and accomplishments of the McIntire-Stennis program. It assures a regular source of input at the national level from representatives of forest products industries, voluntary natural resources and environmental groups, state agricultural experiment stations, government resource agencies, forestry schools, and private individuals.

National-level research planning and coordination activities are necessary to help assure that research is being directed toward important problems and that the appropriate balance among disciplines is struck. Substantive coordination within specific disciplines, however, comes about via interactions among researchers. Science progresses through the slow accumulation of knowledge acquired by the work of individual investigators or teams of investigators building systematically upon the discoveries of others. Essential to a healthy scientific endeavor, then, are mechanisms to assure effective communication among research workers.

Different disciplines have evolved different modes of communication. Scientific journals represent one important mode. No journals have been specifically developed as outlets for the results of forest economics and policy research. Forest economists do, however, publish regularly in the *Journal of Forestry* and *Forest Science*. Articles in such periodicals serve a number of purposes, including transmitting information to noneconomists who may be able to put the findings to practice, enabling authors to calibrate the quality or importance of their research via review by peers prior to publication, and providing fresh ideas and insights to other investigators. The importance of having access to high-quality, rigorous technical journals cannot be minimized for forest economics or any other branch of research.

Technical meetings represent another mode of interaction among forest economics and policy researchers. Meetings have the advantage of allowing researchers to present preliminary ideas and to obtain "real-time" feedback. These types of interactions help assure that researchers are attuned to the latest findings or techniques and that duplication of effort is either reduced or is planned for purposes of verification. Several arenas exist in which forest economists can present ideas to their peers. Perhaps the most formal are the working group technical sessions at the annual national convention of the Society of American Foresters and the divisional and subject group meetings of the International Union of Forestry Research Organizations. Both organizations provide outstanding opportunities for active researchers to compare ideas on research directions and approaches.

In the 1950s, forest economists recognized the need for, and the importance of, additional informal interactions among colleagues. Technical journals play an important role in disseminating research findings, but forest economists have a strong desire to share ideas on research approaches while research is being conducted. To satisfy this desire, informal groupings of forest economists began to develop in four different parts of the nation. The history of the regional forest economics meetings is somewhat murky. Records indicate, however, that the Midwest Forest Economists first started meeting in about 1954; the tradition of an annual meeting has been carried on to date. The Northeast Forest Economists traditionally alternate their annual meeting between the Harvard Forest and forestry schools throughout the Northeast. The Southern Forest Economics Workers (SOFEW) has frequently sponsored a session in conjunction with the annual meeting of the Southern Economics Association in addition to having a spring meeting. Probably the largest of the regional groupings is the Western Forest Economists. The group is popularly referred to as "Wemme" because it meets annually in Wemme, Oregon, on the slopes of Mt. Hood. Although each of these groups evolved somewhat differently, they all began as informal sessions of a few active researchers in which the free expression of preliminary ideas and vigorous debate about research approaches were distinguishing characteristics. They have grown into fairly large meetings with formal agendas and technical papers describing research results. Although formality may have substituted for richness to some extent, the regional meetings of forest economists still serve as a most effective means of fostering communications among researchers.

Summary and Conclusions

Forest economics and policy research is conducted by forest economists in a wide variety of agencies. This decentralized approach has the advantage of permitting research to be focused on problems and issues of immediate concern to the institution funding or managing research. The disadvantage is that pressure to solve immediate and pressing problems tends to derail investment in more fundamental inquiries designed to link forestry research with theoretical research being conducted in the field of economics. Over the years, mechanisms have been developed to provide funding for very basic research in some areas. For example, the U.S. Department of Agriculture's competitive grants program has been established as a means of improving the quality of basic research in forest biology and utilization. No such program exists for forest economics research. If forest economics is to continue as an area that provides significant information for policy development in an increasingly sophisticated and complex natural resource milieu, more attention should be given to competitive funding arrangements. Such could be an effective mechanism for complementing base programs that now exist and for permitting the best forest economics researchers to continue their efforts to push back the conceptual frontiers of the field of forest economics.

Literature Cited

Clawson, M. C. (ed.). 1977. Research in forest economics and policy. Resources for the Future. Washington, DC.

Dana, S. T. 1956. Forest and range policy. McGraw-Hill Publishers. New York, NY.

Ellefson, P. V. 1987. Forest economics and policy research: Investments of the past and opportunities for the future. In Proceedings of Joint Conference of Midwest Forest Economists and Southern Forest Economics Workers. Asheville, NC.

Forest Service. 1982a. An analysis of the timber situation in the United States: 1952-2030. Resource Report No. 23. U.S. Department of Agriculture. Washington, DC.

Forest Service. 1982b. 1980-1990 national program of research for forests and associated rangelands. General Technical Report WO-31. U.S. Department of Agriculture. Washington, DC.

Forest Service. 1983. The principal laws relating to Forest Service activities. Ag. Handbook No. 453. U.S. Department of Agriculture. Washington, DC.

Meadows, J. C. 1988. Personal correspondence.

National Association of Professional Forestry Schools and Colleges and Cooperative State Research Service. 1985. University-based forestry research: Unlocking the future. Cooperative State Research Service. U.S. Department of Agriculture. Washington, DC.

National Task Force on Basic Research in Forestry and Renewable Natural Resources. 1982. Our natural resources: Basic research needs in forestry and renewable natural resources. Forest, Wildlife and Range Experiment Station. University of Idaho. Moscow, ID.

Office of the Federal Register. 1985. The United States government manual 1985-1986. National Archives and Records Administration. Washington, DC.

Office of Technology Assessment. 1983. Wood use: U.S. competitiveness and technology. U.S. Congress. Washington, DC.

Office of Technology Assessment. 1984. Technologies to sustain tropical forest resources. U.S. Congress. Washington, DC.

Shands, W. E. and D. H. Dawson. 1984. Policies for the Lake States forests. The Conservation Foundation. Washington, DC.

Shands, W. E. (ed.). 1988. The Lake States forests: A resource renaissance. The Conservation Foundation. Washington, DC.

U.S. Department of Agriculture and the National Association of State Universities and Land Grant Colleges. 1978. National Program of Research for Forests and Associated Rangelands. Washington, DC.

PART TWO

Strategic Research Directions

5

INSTITUTIONAL ARRANGEMENTS DIRECTING USE AND MANAGEMENT OF FORESTS

Perry R. Hagenstein

The forestry community's ability to provide a sustained flow of goods and services from the nation's forests is dependent on the availability of well developed and effectively operating institutions. From a broad social perspective, such institutions are of two fundamental types. There are social rules (expressed both in law and social mores) employed by society to guide the conduct of its members. Included are rules as specific as antitrust laws or zoning regulations (even though interpretations may differ) and as indefinite as common understanding of responsibilities to future generations. In addition to social rules, institutions can also take the form of organizations and processes which society creates to perform a variety of important functions. Institutional forms of this sort include government organizations and processes (e.g., courts, public hearings) as well as private organizations (e.g., corporations, cooperatives, financial institutions).

Institutional frameworks important to forestry affect the allocation of resources and the distribution of benefits and costs. Since land is vital to public and private forestry enterprises, property rights and related issues assume special importance. Also of vital importance to forestry is capital. As such, the institutional framework used to meet the investment needs of forestry must have an important position on the research agenda of any organization. The allocation of labor to forestry has little that distinguishes it from the allocation of labor in other areas. Here it will not be of major concern.

Current Status of Research on Institutions

Research on institutions has been mainly descriptive in nature. The great institutional economists (e.g., Commons, Ely, Veblen) stressed the importance of institutions as a means of affecting economic behavior, but their science was largely observational and descriptive. So it is with most current institutional research. Current interest in research on institutions is often undertaken in response to change—or the possibility of change. There is, however, little incentive to invest heavily in research on institutions that seem to change very slowly. Property rights, for example, are often assumed to be immutable, even though they also have evolved. As a result, much of the economics and policy research on institutions and their role in forestry relates to developing countries, where rapid institutional change is possible and highly relevant. In the United States, recent attention to forestry-related institutions has focused on regulatory practices. The increase in environmental quality regulations has provided fertile ground in which to define the economic effects of new and proposed regulations. Most of such research has emphasized allocative effects. Distributive effects have largely been ignored.

Research on institutions important to forestry serves a number of purposes, two of which are especially noteworthy. Such research keeps the forestry community apprised of the continually evolving institutional framework that affects forestry and of the allocative and distributive effects implied by such changes. Furthermore, such research provides information on the effects of institutional changes that are proposed to serve specific purposes. A solid and up-to-date understanding of the existing institutional environment and its economic and social effects is important to a variety of forestry interests.

Future research concerning institutions will in all likelihood continue to be descriptive; data-based testing of hypotheses about behavior will be of secondary interest. However, organizations and persons guided by institutions, and those desirous of changing such institutions, require a sound understanding of the institutional mechanisms by which they can affect behavior. Such an understanding is founded on information resulting from an attentive and perceptive research program involving descriptive analyses. The choice of topics for such a program can be guided in part by immediate problems and issues that result from institutions and institutional change. For example, problems arising from the flurry of environmental laws and regulations established over the past two decades suggest numerous areas of needed research. To a degree, these problems have been reflected in recent research efforts. But research attention should also be directed at keeping the forestry community abreast of long-term changes in property rights and the response of various government and private institutions to such changes. Such research should be undertaken, even though no immediate issue or problem has arisen.

Research focused on institutional arrangements for directing the use and management of forests could pursue a number of strategic directions. Those that would seem most appropriate for the future are:

- Develop improved understanding of concepts of public interest in private land use and management.
- Acquire greater knowledge of institutions directing the public's interest in private forest lands.
- Obtain improved information about institutions used to allocate interests in public forests.
- Develop improved understanding of institutions used to allocate capital to forestry.

• Generate improved knowledge about intergovernmental institutions directing use and management of forests.

Develop Improved Understanding of Concepts of Public Interest in Private Land Use and Management

The concept of private property rights in forest land is narrowing as land becomes an increasingly scarce resource and as the public's interest in private forest land assumes broader proportions. Increasing restrictions on the actions of private forest land owners are evidence that this is so. Most such restrictions are aimed at externalities associated with timber production, especially water and air quality. But there are also increasing restrictions aimed at the production of public goods such as aesthetics and wildlife. The extent to which the public exercises interest in private property is primarily a political determination, an interest whose limits are continually being redefined in law. The scope and intensity of such limits is an important consideration to the allocation and distribution of land and related resources. Concepts of property rights do change. In so doing, they may vary according to the region of the nation in question and may also vary according to the perceived importance of particular kinds of externalities and public goods. Landowners, public officials, and others interested in the use and management of forest lands should be aware of changing property rights concepts and the implications they pose for the use and management of forest lands.

Concepts of public interest in forest land should be the subject of significant research of a descriptive nature. Such research should define the character of property rights, identify important regional and other differences, and note trends in the way in which property rights are viewed relative to the public's interest in forest land. Research should also provide economic information that can be used in analyzing land use allocations, forecasting changes in land use, identifying optimum allocations, and weighing the distributive effects of such changes. The following directions for research are relevant.

Define and provide economic measures of externalities and public goods pertinent to the use and management of private forest lands. Changes in private property-rights concepts in forest land are based on the forestry community's understanding of the importance of externalities, and on the nature of public goods resulting from the use and management of forest land. Knowledge about such changes is necessary for informed judgments about proposed modifications in institutions— modifications often suggested as means of achieving more efficient and equitable use and management of private forest lands. Straightforward descriptions of the implications of changes in statutory and case law affecting property rights in forest land are needed. Especially helpful would be comparisons among states or other jurisdictions. Such descriptions are a necessary starting point for research on the implications of institutional change. Also needed are economic measures of the effects of changes in perceived property rights. Such measures would aid in assessing the importance of institutional changes and in guiding subsequent allocation and distribution decisions. Areas involving regulatory matters should receive special attention; they often have the greatest potential for economic and social impact. Other areas should be monitored and research efforts appropriately directed.

Describe issues involved in on-site public use of private lands. As growth occurs in the importance of public uses of private forest lands (e.g., recreation, hunting, and fishing), concerns arise over issues such as access and landowner liability. The legal status of such issues is evolving; it also varies from state to state. Generous descriptions and carefully thought out comparisons of such concerns are needed. Descriptions and comparisons would be greatly facilitated by more effective measures of the use and value of recreation on private forest lands. Ultimately such measures would facilitate landowner allocation decisions and would enhance policy analyses. They could also strengthen examinations of issues involving landowner liability and access to private lands. Also emerging as an issue is the relationship between public access to private lands for recreation purposes versus landowner access to preferential property taxes. The issue links preferential tax treatment (e.g., current use assessments, timber yield taxes) to access to private forest land for public recreation purposes. The arguments for such linkages and the economic implications implied are in need of careful description and analysis.

Acquire Greater Knowledge of Institutions Directing the Public's Interest in Private Forest Lands

A range of institutional mechanisms is available for pursuing the public's interest in private forest lands. In the past, education and technical assistance to landowners were the main choices. In recent years, the scope of available mechanisms has increased dramatically. Economic incentives, for example, have grown in importance—especially incentives offered via lowered property and income taxes. Public regulation of forestry practices on private lands has become especially significant since the late 1960s, particularly in relation to forest practices affecting water quality. The public has also used fee and partial acquisition of land and rights in land to accomplish public purposes. Each of these mechanisms has unique characteristics and numerous advantages and disadvantages. Although the choice of institutional mechanism used to pursue a particular public interest goal is often political in nature, there are differences among such mechanisms in terms of costs and efficiency. Because of growing public interest in private forests, these differences need to be described and compared.

Compare costs and effectiveness of alternative institutional mechanisms for directing public interest in private forests. Comparative descriptions of various institutional mechanisms currently used to secure the public's interest in private forests are needed. Such descriptions should identify situations in which each mechanism is used, the level of government that uses them, the extent to which each is used, and trends that are significant to the use and management of forests. Descriptions of this sort would prove useful as a basis for further analyses. Since the effectiveness of institutional mechanisms used to pursue public interests in forest land is judged according to ability to achieve an intended purpose, evaluation of relevant experience is needed to determine accomplishments and reliability in typical situations. An example of an economic institution (incentive) deserving additional examination in terms of effectiveness is treatment of timber income as a capital gain for federal income tax purposes. Although preferential tax treatment of this nature was ended in 1986, it is likely to rise again as a legislative proposal. In addition to effectiveness, both public and private costs of various institutional mechanisms need to be assessed in terms of efficiency. Some research has addressed financial and technical assistance programs used to encourage timber production. Additional efficiency-type research should be focused on the use of regulatory mechanisms as means of securing the public's interest in private forest land.

Define and assess equity implications of various institutional arrangements for directing public interest in forest land. The incidence (income distribution effects) of benefits and costs associated with various mechanisms for securing public interest in forest land needs careful definition and evaluation. At the very least, there is a need to identify the distribution of benefits and costs associated with specific programs, of benefit and cost distributions between generations, and of benefits and costs of programs distributed among geographic regions. Who bears the direct and indirect costs of alternative institutions under relevant circumstances? How and to what extent does the public benefit from such institutions? Because most approaches (short of public acquisition of fee interest in forest land) encourage, rather than require, particular behavior, the predictability of results are often questionable. This adds weight to the importance of research on equity considerations.

Obtain Improved Information About Institutions Used to Allocate Interests in Public Forests

Interests in various uses of public forest lands are allocated in various ways— from almost purely market-type allocations to strictly administrative allocations. Especially relevant are allocative mechanisms used to distribute resources and uses of public forest lands among private claimants, and mechanisms used to allocate public forest lands among alternative uses, thereby determining how much of each resource will be produced and made available for use.

Evaluate alternative mechanisms for allocating resources among private claimants. Mechanisms for allocating resources among private claimants can vary from price allocations determined via competitive bidding procedures (e.g., federal timber sales) to free use of forest land outputs allocated on the basis of queuing (e.g., dispersed recreation). Variations also exist within such extremes, as can be see in allocations by administrative means (e.g., grazing rights) and limitations on the use of competitive bidding (e.g., small business set-asides, sustained yield units for federal timber). The trend, if any, is toward increased use of competitive bidding as a means of setting prices for allocation purposes.

A comparative review of different means of allocating various resources from federal and state forest lands should be carried out. This would set the stage for undertaking research on the efficiency and equity concerns involving alternative allocation mechanisms. Research has been performed on competitive bidding for national forest system timber and on grazing fees on federal lands. Since these allocative mechanisms are subject to considerable change over time, monitoring is needed. Especially important is the need for research directed at alternative ways of allocating recreational use of federal lands. Research of such nature will gain in importance as future levels of recreation distributed by current allocative methods outrun the ability of public lands to supply recreational opportunities. Considerably more research attention should also be directed at the distributive effects of current allocative mechanisms—and alternatives to them. This is especially so for administrative allocation mechanisms involving grazing rights and set-aside timber sales. An important policy question is: What does the public, as owner of such resources, gain at the expense of failing to collect full economic rents for the resources?

Evaluate alternative mechanisms for allocating uses of public forests. Levels of resource use on public lands are determined by a combination of factors. In addition to basic productivity, inventories, and interactions among resources, institutional rules and practices are also important. They affect both the allocation of investments to public lands and the allocation of resources and uses from them. The ubiquitous mechanism used to allocate uses of public forests is planning—at both national level and local (e.g., forest or district level in the national forest system) levels. While the function of planning is generally understood to be that of resource allocation, the role of market forces in planning is not always clear. Also, the relation of national to local plans and of prescriptive to contingency plans is often hazy. Similarly, the place of budgeting and appropriations in allocating resources is often unclear. Research to improve the role of planning and administrative decisions in the allocation of forest resources in a changing market economy should be undertaken. The products of such research could be useful for improving efficiency of public forest land management.

Develop Improved Understanding of Institutions Used to Allocate Capital to Forestry

Forestry is a very capital-intensive activity and, as such, is very sensitive to changes in interest rates and changes in the manner in which investment capital is made available. Important to forestry investments have been a number of significant changes in the financial structure of the nation. For example, forest land ownership has been changing in relation to the availability of capital for forestry investments. Industrial forestry firms are moving away from the previously common practice of integrating timberland ownership with processing facilities. Several major firms have effectively separated processing facilities from timberland ownership and have assumed a posture of relying on the open market for wood supplies. Consequently, funds generated by processing and marketing forest products are no longer available for forestry investments. This is not to say that new sources of capital for forestry have not surfaced. Pension funds and insurance companies have accumulated capital at a dramatic rate; both sources are increasingly interested in long-term forest land and timber investments. The formation of limited partnerships—sometimes used by forest products firms as an anti-takeover mechanism—has led to changes in owners' objectives and access to additional capital. The rise of interstate banking, and the attendant increase in the size of banks, has also affected the availability of funds for forestry—perhaps in a positive way. The extraordinarily high real interest rates of the early and mid-1980s (rates two and three times historic levels) often made long-term investments (e.g., forestry) questionable; thus curtailing capital available for forestry purposes. Very large federal government deficits have also been a factor contributing to declines in available capital. Surely they have constrained federal government investments in public forest lands and in supporting investments in private forest lands.

Changes in the nation's financial structure affecting forestry investments should be described and evaluated. To keep the descriptions relevant, focus should be on the forestry investment aspects of financial institutions. Once adequate descriptions of financial institutions relevant to forestry become available, attention can then be turned to evaluations of efficiency and of equity effects.

Describe the role of financial institutions germane to investments in forestry on private lands. Research should be undertaken to describe the practices of banks, pension funds, landowning corporations, and other private institutions that provide funds for forestry investments. Of special importance should be financial institutions that focus on forest practice investments—not just acquisition of land and timber. Included in such descriptions should be the practices of public sources of funds for forestry on private lands, e.g., public loan guarantee programs, insurance programs, and federal and state land banks. Also needing description are mechanisms by which forest landowners raise capital, e.g., equity and debt issues, and internal cash flows. The characteristics of each source can affect its availability to forestry, e.g., cost of capital (interest or payout to equity owners), capital availability, and terms under which capital is acquired. All such characteristics change with great frequency. Rules or general practices that are relevant to sources of capital include debt-equity ratios, type of collateral, insurance requirements, and form of organization (i.e., corporations, partnerships, cooperatives, proprietorships).

Descriptions of how financial institutions—broadly defined—function in providing investment capital for forestry should be carried out. Attention should be assigned to identifying factors that constrain the flow of capital to forestry, the reasons for such limits, and their relative importance. Effects on the efficiency of forestry investments should be identified and their importance assessed. Distribution effects—identifying who has access to capital and limitations on such access—should be of secondary concern.

Describe the role of institutions as they affect investments in forestry on public lands. The primary source of capital for public land forestry is legislative appropriations. Such, however, is not the only source. For example, regular federal appropriations for forestry are augmented by earmarked funds such as Knutson-Vanderberg funds and by purchaser constructed roads. Similar funding sources exist at the state level. In addition, proposals for various forms of revolving funds are regularly suggested for the National Forest System.

Mechanisms used to fund investments in public land forestry should be described and evaluated. Comparisons among government entities, including states, could be especially useful. Particular attention should be given to the criteria for judging investment opportunities, how the burden of costs is borne, the extent to which investments can be allocated efficiently, and the assurance with which investment funds can be obtained. In addition, variations in primary mechanisms for funding investments on public forest lands should also be researched. Focus on the latter should include public or quasi-public corporations and joint public-private ventures. Descriptions are needed of the manner in which such arrangements might operate for forestry activities on public lands and how such approaches compare with current methods of financing. And last, the politics of financing forestry investments on public lands should be addressed by descriptive research. The legislative power of appropriation, the regional distribution of legislators and expenditures, the degree of control over nonappropriated funds, and the relationship of funding mechanisms to the various uses of public lands must be recognized. Their implications for efficiency and equity in spending should be determined.

Generate Improved Knowledge of Intergovernmental Institutions Directing Use and Management of Forests

Various levels of government involvement in forestry have become increasingly complex. Long-standing are federal-state relationships involving forest fire management, insect and disease control, and provision of assistance to private landowners. More recently, the federal government has assumed a posture of assisting states in planning their forestry activities. Also significant are federal efforts to have National Forest System plans recognize the forestry roles of state and local governments. Intergovernmental relationships of this sort can substantially affect the use of forest land as well as the planning and delivery of forest management funds and services. Carefully developed descriptions of these relationships—with an emphasis on trends and change—is needed. Implications for the use and management of forest lands as a result of such change should be identified and described.

Describe intergovernmental relationships affecting forest land in general. Current status and trends in intergovernmental arrangements concerning the use and management of forest lands, in general, should be described. Primary focus should be on federal-state and interstate arrangements, especially interstate compacts, formal cooperative agreements, and informal agreements. Specific examples are the continuing federal-state forest fire control agreements and the Lake States Forestry Alliance. The extent and impact of these arrangements (including funding levels and stability) should be determined.

Assess interactions between state and local planning and zoning programs and the use of federal forest lands. The relationship between state and local government planning and zoning programs and planning and management activities on forests of the national forest system has grown in importance. Legal and administrative requirements that bind actions on federal lands to the interests of state and local government should be described and evaluated. Attention should be assigned to the manner in which resource use and management are affected, the way in which costs and revenues are shared, and trends in these relationships.

Summary and Conclusions

Institutional arrangements guiding the use and management of public and private forests are fundamental to society's interest in achieving high levels of sustained benefits from forests. Such arrangements must be creatively developed and must operate in an efficient and effective manner. The forest economics research community has given only modest attention to evaluation of existing institutions, and has given virtually no attention to development and testing of new and imaginative institutional alternatives. Research investments in the latter are certainly warranted. Some of the more significant avenues to be followed by such research include development of improved understanding of institutions directing public interest in private forest land; institutions used to allocate interests in public forests; institutions used to allocate capital to forestry; intergovernmental institutions directing use and management of forests; and concepts of public interest in private forest land use and management.

Bibliography

Ackerman, Bruce A. 1975. Economic foundations of property law. Little, Brown and Company. Boston, MA.

Bosselman, Fred, David Callies and John Banta. 1973. The taking issue. U.S. Council on Environmental Quality. Washington, DC.

Brubaker, Sterling (ed.). 1984. Rethinking the federal lands. Johns Hopkins University Press. Baltimore, MD.

Clawson, Marion. 1975. Forests for whom and for what? Johns Hopkins University Press. Baltimore, MD.

Clawson, Marion. 1983. The federal lands revisited. Johns Hopkins University Press. Baltimore, MD.

Council on Environmental Quality. 1971. The quiet revolution in land use control. Washington, DC.

Cubbage, Fredrick W., Thomas M. Skinner, and Christopher D. Risbrudt. 1985. An economic evaluation of the Georgia rural forestry assistance program. Res. Bulletin 322. College of Agriculture. University of Georgia. Athens, GA.

Culhane, Paul J. 1981. Public lands politics. Johns Hopkins University Press. Baltimore, MD.

Derthick, Martha. 1974. Between state and nation. The Brookings Institution. Washington, DC.

Ellefson, Paul V. and Frederick W. Cubbage. 1980. State forest practice laws. Environmental Policy and Law. Sept:125-133.

Ellefson, Paul V. and Christopher D. Risbrudt. 1987. The forestry incentives program: Assessment of federal fiscal incentives for forestry. Evaluation Review: A Journal of Applied Social Research 11(5):660-669.

Ellefson, Paul V. 1988. Private forest, public interest. Habitat: Journal of Maine Audubon Society 5(4):28-30.

Freeman, A Myrick III, Robert A. Haveman and Allen V. Kneese. 1973. The economics of environmental policy. John Wiley and Sons. New York, NY.

Gray, Gerald J., and Paul V. Ellefson. 1987. Statewide forest resource planning programs: An evaluation of program administration and effectiveness. Agr. Exper. Stat. Bulletin 582. Agricultural Experiment Station Bulletin. University of Minnesota. St. Paul, MN.

Haefele, E. T. 1973. Representative government and environmental management. Johns Hopkins University Press. Baltimore, MD.

Haefele, E. T. 1974. The governance of common property resources. Johns Hopkins Press. Baltimore, MD.

Healy, R.G. and J.S. Rosenberg. 1979. Land use and the states. Johns Hopkins Press. Baltimore, MD.

Henly, Russell K. and Paul V. Ellefson. 1987. State administered forestry programs: Current status and prospects for expansion. Renewable Resources Journal 5(4):19-23.

Kent A. Price (ed.). 1982. Regional conflict and national policy. Johns Hopkins Press. Baltimore, MD.

Kneese, Allen V. and Charles L. Schultze. 1975. Pollution, prices, and public policy. The Brookings Institution. Washington, DC.

Mills, Edwin S. 1978. The Economics of Environmental Quality. W. W. Norton and Company. New York, NY.

Nothdurft, William E. 1984. Renewing America: Natural resource assets and state economic development. Council of State Planning Agencies. Washington, DC.

Peterson, George L. and Allen Randall. 1984. Valuation of wildland resource benefits. Westview Press. Boulder, CO.

Rawls, John. 1971. A theory of justice. Harvard University Press. Cambridge, MA.

Royer, Jack P. and Christopher D. Risbrudt (ed.). 1983. Nonindustrial private forests: A review of economic and policy studies. School of Forestry and Environmental Studies. Duke University. Durham, NC.

Schultze, Charles L. 1977. The public use of private interest. The Brookings Institution. Washington, DC.

Sedjo, Roger A. (ed.). 1983. Governmental interventions, social needs, and the management of U.S. forests. Resources for the Future. Washington, DC.

Sedjo, Roger A. (ed.). 1985. Investments in forestry: Resources, land use, and public policy. Westview Press. Boulder, CO.

Sinden, J.A. and A.C. Worrell. 1979. Unpriced values: Decisions without market prices. John Wiley and Sons.

U.S. Public Land Law Review Commission. 1970. One-third of the nation's land. Washington, DC.

U.S. Water Resources Council. 1983. Economic and environmental principles and guidelines for water and related land resource implementation studies. U.S. Water Resources Council. Washington, DC.

6

ECONOMIC STRUCTURE AND PERFORMANCE OF FOREST-BASED INDUSTRIES

Jay O'Laughlin

Forest-Based Industrial Organization Research

The performance of the nation's forest-based industry is determined by the conduct of the firms operating within the industry, which in turn is a function of how the industry is structured. In the context of industrial organization economics, structure generally refers to the number and size distribution of firms operating within an industry. The structure-conduct-performance paradigm is the distinguishing characteristic of industrial organization economics. Industrial organization economists include both theoreticians (see works of Baumol 1982, Bresnahan and Schmalensee 1987, Jacquemin 1987, and Ricketts 1987) and researchers addressing specific industries and the public policies focused on them (see works of Acs 1984, Filipello 1985, and Hall 1987a). The *Journal of Economic Literature* includes the economics of technological change within the subject matter of industrial organization.

Industrial organization economics can be applied to many areas of interest to forest-based industrial performance. As set forth by the *Journal of Economic Literature*, the following outline depicts the range of topics that can be included under the major heading "Industrial Organization; Technological Change; Industry Studies."

Industrial Organization and Public Policy
  Market structure: industrial organization and corporate strategy
  Public policy toward monopoly and competition
  Public utilities and costs of government regulation of other industries in the
    private sector
  Public enterprises
  Economics of transportation
  Industrial policy

Economics of Technological Change
Technological change; innovation; research and development
Industry Studies
    Industry studies: manufacturing
    Industry studies: extractive industries
    Industry studies: distributive trades
    Industry studies: construction
    Industry studies: services and related industries
    Nonprofit industries: theory and studies
Economic Capacity

Headings concerning market structure, technological change, and manufacturing are especially significant. They are topical areas within which most economic structure studies of the wood-based industry have been undertaken.

Knowledge of an industry's structure is complicated by the dynamic nature of economic organizations operating at various levels within an industry (i.e., firm, industry, nation, worldwide). "Restructuring" is the currently popular term used to describe such dynamics, a term which basically infers a reorganization to cope with changes in the environment in which economic organizations operate. As Lindheim (1985) says, the decade of the 1980s may well be viewed by historians as the era of restructuring. Research addressing organizational economics of forestbased industries is an especially promising area, partly because dynamic change in the industrial landscape necessitates periodic monitoring, and partly because industrial organization of the forest-based industry is greatly affected by regional and local factors. The major challenge to researchers in the industrial organization of forest-based industries is to form a linkage between descriptions of structure, conduct, and industrial performance—linkages that have policy making implications.

Wood-Based Industrial Research

Productive resources—labor, land, and capita—are assembled by firms that compose industries. The efficiency and effectiveness with which they do so have far-reaching performance implications, often extending to the survival of nations. Together, two major wood-based industries—lumber and wood products and paper and allied products— account for seven percent of the national income generated by U.S. manufacturing industries. Nationwide, only six other manufacturing industries have a larger economic impact (Ellefson and Stone 1984). According to the Office of Technology Assessment (1983) of the U.S. Congress, no other manufacturing industry appears to have a brighter outlook for the future than the wood-based industry.

The significance of economic structure as an important determinant of forest products markets has recently been recognized in forest economics textbooks. Furthermore, Ellefson and Stone (1984) have written the definitive work on the industrial organization of forest-based industries of the early 1980s. Their work reinforces the need for research in the performance dimension. Only 86 pages (20 percent) of the book is devoted to industrial performance. Descriptive analyses of industrial structure and conduct compose 35 percent and 41 percent, respectively. Introductory and concluding comments account for 3 percent; no discussion is given to policy approaches regarding the industry. Other recent studies of structural issues affecting the wood-based industry include Adams and Haynes 1985, Baudin and Westlund 1985, Bilek and Ellefson 1986, 1987, Boyd 1983, Buongiorno et al. 1981, Callahan 1985, Dickerhoof 1986, Eckstein 1983, Ellefson 1985, Irland 1987, Moslemi 1986, Nautiyal and Singh 1985, Nautiyal et al. 1985, Rideout 1984, and Slatin 1985.

A focus on performance dimensions is the major direction for future research concerning wood-based industrial organization. It will, however, be necessary to continue to monitor changes in structure and conduct dimensions at the national level and to conduct baseline studies of industry structure-conduct-performance at regional, state, and local levels. Fortunately, at the national level there exists a statistical base for doing the necessary descriptive research (Phelps 1980, Ellefson and Stone 1984). Once familiar with information sources, especially the U.S. Department of Commerce Census of Manufactures, information bases for smaller scale research can be assembled. An example of a recent descriptive study of industry structure at the regional level is "The South's Fourth Forest" report (Forest Service 1988). An analytical state-level industry structure study which hints at policy analysis was recently completed in Texas (O'Laughlin and Williams 1988). At the state level, the path toward policy analysis for economic development has been paved in Minnesota (Ellefson et al. 1985).

The need for performance-oriented industrial organization research can be expressed at many levels, including international trade, regional economic development, and corporate shareholders' wealth. All such dimensions are affected by current restructuring efforts. The redesign of a corporation or organization either through refinancing or redirection of business efforts is a restructuring. The merger or divestiture of entire corporate entities or parts thereof are also restructuring efforts. The governance of such conduct with public policies is restructuring. Ways of dealing with interregional and international competition are restructuring. All such restructuring topics have affected and will continue to affect the wood-based industry. All of them have researchable dimensions.

Strategy-Structure Connections

Chandler (1962), a business historian, was the first to recognize that organizational structure is directly related to an organization's strategy. The linkage is important since the strategy of individual firms ultimately determines industrial performance. Only recently have industrial organization economists incorporated this important concept into their subject matter (Caves 1980). The marriage of business management (i.e., strategy and organizational structure) and economics (i.e., industrial organization structure-conduct-performance) means, in part, that firms with similar strategies are likely to conduct themselves in a similar manner. Industrial performance is somewhat easier to ascertain if groups of similar firms can be described and aggregated based on strategy as well as structural characteristics (O'Laughlin and Ellefson 1982a, Cleaves and O'Laughlin 1986a, 1986b).

Strategies of wood-based firms have been described and summarized (O'Laughlin and Ellefson 1982b). The key characteristic for future research is, once again, the dynamics of change. Recent research by Rich (1986) provides evidence of firms shifting strategically toward marketing and away from traditional strengths in timber resources and manufacturing of commodity products. During the deep recession of the early 1980s (the beginning of the restructuring era), the best financial performers in the industry were paper-makers without extensive corporate-owned timber resources (Ross 1985). It is doubtful, however, that the leading forest products firms of tomorrow will be those without control over timberlands today (O'Laughlin 1986).

Regional Patterns of Structural Change

Regional economics is a specialized subject concerned with economic development; thus it has (or should have) strong ties to industrial organization economics. Much of the regional economics research involving the wood-based industry has focused on developing input-output multipliers to describe intra-industry linkages (e.g., Kaiser 1972, Schallau and Maki 1983). But regional economics has more to contribute to an understanding of industrial organization, and vice versa. Schallau (1985) posed a simple question that challenged the conventional wisdom of many analysts: "What's this about forest industry moving to the South?" The answer to the question is "It isn't" (Schallau and Maki 1986). The wood-based industry of the Pacific Northwest is alive and well (Schallau 1986). So, too, is the southern wood-based industry (Schallau et al. 1987). In the latter series of studies, which reported analyses of structural changes and trends in 13 southern states between 1970 and 1980, the concept of "economic base" is used to describe the role of a state's wood-based industry.

Schallau's (1985, 1986, 1987) research provides examples of regional economic methods beyond economic input-output multipliers. At this point, however, the potential contribution of such methods needs to be better defined. Application of the "economic base" concept is subject to different interpretations. For example, as asserted by Schallau et al. (1987) is a state not "self-sufficient" in forest products because the forest products industry is not part of the economic base? The answer has important implications for forest resource management and economic development. It is not fully addressed in the cited state-level studies. The measures used to determine economic base industries can also be subject to question. Schallau et al. (1987) used the percentage of forest industry's total employment in a state and compared it to that occurring at the national level. Is that all there is to determining the economic base? In research describing the wood-based industry in Pacific Northwest states (Keegan and Polzin 1987), the criterion used to determine economic base industries was not even defined. Nonetheless, the objectives and methods of regional economic studies have something to offer industrial organization economics studies; further research in this area should be encouraged. The payoff could be a better understanding of industrial plant location decisions. The latter are fundamentally strategic decisions, yet are made in response to attributes or characteristics of the structure of an industry in a particular region (Cleaves 1983).

The preceding brief review of forest-based industrial organization research has attempted to point out major gaps in knowledge. Research involving economic structure of forest-based industries would prove most fruitful if focused on the following strategic directions:

- Develop an improved understanding of economic restructuring within the wood-based industry.
- Develop improved means of defining the wood-based industry and various segments of the industry.
- Develop means of linking wood-based industrial structure with regional economic development.
- Develop an improved understanding of timberland as a managerial and economic factor.
- Develop an improved understanding of the structural and performance consequences of technological innovations.
- Develop an improved understanding of corporate strategies as related to economic performance.
- Develop an improved understanding of wood-based structural dimensions in an international setting.
- Develop an improved understanding of the economic structure, conduct, and performance of nonwood-based forest industries.

What follows is an expansion of the above areas, including identification of specific research topics. Citations from recent industrial organization research are presented. Almost all have been published since Ellefson and Stone's (1984) bench mark study. Such citations were gleaned from *Social Sciences in Forestry* (Albrecht et al. 1987) and the *Journal of Economic Literature*.

Develop an Improved Understanding of Economic Restructuring Within the Wood-Based Industry

"Restructuring" is often suggested to be the economic by-word of the 1980s. The concepts and implications therein can have substantial significance for the wood-based industry. Numerous researchable topics exist at various levels, i.e., international, macro, industry, and firm.

Assess worldwide economic waves and their implications for wood-based industrial organization. Is the world economy in the grip of a 50-year "long-wave" trough that must inevitably occur and can only be broken after the occurrence of fundamental changes in a variety of institutions? Is such a perspective realistic, and, if so, what are the implications for the wood-based industry? Rostow (1985) researched the subject (without specific reference to wood-based industries), with results that are less than revealing. Robinson (1988) has scratched the surface of long-wave theory implications for U.S. forestry. In a similar vein, Walker and Vatter (1986) compared the recession of the early 1980s to the Great Depression. Their analysis might be a good starting point for further research in this area.

Evaluate the U.S. position in worldwide forest products trade. The U.S. has expressed significant interest in worldwide trade in wood-based products. What are the implications of such an interest? What (if any) comparative advantages might the U.S. wood-based industry enjoy in world markets? Evaluation of international trade has critical implications for timber resources strategies, which relate directly to raw material costs, labor wage rates, and productivity.

Evaluate industrial policy options to relieve periodic adverse effects of economic recessions. Of major concern is specification and evaluation of industrial policy options that could reduce the adverse impact of major economic fluctuations on wood-based industries. Much debate has occurred over this subject in general (see Backhaus 1985, Curzon Price 1986, Hayes and Wheelwright 1984, Lavoie 1985, Miller et al. 1984, Moriarty 1986, National Research Council 1986, Norton 1986, Redburn et al. 1986, Rothwell and Zegveld 1985 and Solo 1984). Definitive research focused on the wood-based industry is needed.

Assess institutional arrangements for improving the flow of capital for plant and equipment and research and development purposes. Of concern are mechanisms that would encourage capital investments in various activities, including research and development. Adams and Haynes (1986) identified such research as important to improving U.S. timber market modeling efforts. During a time when research spending may be crucial for survival in international markets, the forest products industry is among the lowest investors of all U.S. industries (Marbach and Smith 1988).

Evaluate structure and conduct implications of various antitrust policies and programs. One of the industry's largest companies (Louisiana-Pacific Corporation) was created by an antitrust action of the federal government (O'Laughlin 1981). In addition, wood-based companies of the 1970s became known as price-fixers, paying fines totaling more than $750 million. Companies have apparently become more sensitive to price-fixing activities; antimerger policy has been relaxed under the Reagan administration. However, increased international competitiveness looms larger on the horizon than does the nebulous criterion of freely competitive domestic markets. The role of antitrust as an inhibitor to wood-based firms facing international competition is largely unknown.

Evaluate industry-level performance effects of mergers, takeovers, white knights, greenmail, and related activities. A variety of conduct actions have restructured and tended to consolidate the wood-based industry. The full implications of such conduct is in need of systematic evaluation in terms of effect on industrial performance. Recent research sheds some light on the structural effects of mergers and related public policies (see Alexander 1986, Backhaus 1985, Curry and George 1983, Gruchy 1985, Mukhopadhyay 1985, and Stigler 1982). Additional research specific to the wood-based industry is needed.

Assess the industry-level joint venture as an organizational mechanism for the wood-based industry. Several large joint ventures have been disbanded by firms in the wood-based industry. Why did these restructuring experiments fail? How might they be more effectively used in the future? Research by Reynolds and Snapp (1986) provides some insight to this phenomenon.

Evaluate potentially optimal organizational arrangements for wood-based firms. Is there an optimal organizational arrangement for wood-based firms? Which structural arrangements foster better financial performance? Hall (1987b) offers some advice. How should timber-related activities be handled in the corporate organizational structure (Eckstein 1983)?

Develop Improved Means of Defining the Wood-Based Industry and Various Segments of the Industry

Boyer (1984, p. 768) makes an important observation: "The modern movement to make industrial organization theory more useful by making it richer in variables and more realistic in its framework depends critically on the existence of a logical criterion to define a firm's industry." The observation was made in the context of searches to determine a logically consistent criterion for defining industries in antitrust and regulatory proceedings. Such also has relevance to data gathering and descriptive research and to analytical research on industry structure. Reinforced is the need to resolve a number of obvious definitional inconsistencies in existing structural research focused on wood-based industries.

Define and evaluate potential application of strategic group concepts. What are the "strategic groups" of firms operating within the wood-based industry? Given the size and diversity of scope within manufacturing firms, the structural approach to industry definition sought by Boyer (1984) may not be attainable. For regulatory purposes, Boyer criticizes the conduct-based "strategic group" approach promoted by Porter (1979) and Caves (1980). For other purposes, however, the concept of similar corporate strategies defining subgroups within an industry of firms producing similar outputs is the most theoretically and pragmatically consistent classification scheme presently available for defining the industry within which a firm operates.

Evaluate alternative combinations of Standard Industrial Classifications (SICs) appropriate to the wood-based industry. Which industry SICs should and should not be included in wood-based industries? For example, convenience leads to inclusion of Mobile Homes (SIC 2451) in Lumber and Wood Products (SIC 24); the U.S. Department of Commerce's Bureau of the Census statistics have done so since 1972. However, does the mobile home industry (SIC 2451) really belong in SIC 24 along with logging camps, sawmills, and veneer mills? Although many industrial structure studies include mobile homes (Phelps 1980, Forest Service 1988, Foster [undated]), several recent studies exclude mobile homes as a wood-based industry (e.g., Schallau and Maki 1986, Schallau et al. 1987, O'Laughlin and Williams 1988). Exactly which SIC industries should be considered part of the wood-based industry—and why?

Evaluate alternative means of segmenting the wood-based industry for purposes of description and analysis. There is a need to segment the wood-based industry and to base structural analyses on these segments. Suggested segmentation of the industry is often based on Phelps's (1980) classification scheme for SICs. The industry is divided into three major segments: primary "forest-based" manufacturing; secondary "wood-based" manufacturing; and tertiary "wood-dependent" industries. Obviously, a distinction needs to be made between "forest-based" and "wood-based" industries. "Forest-based" is a suggested term for industries that process raw timber (e.g., logging, sawmilling, veneer mills, and pulpmills). "Wood-based" is a suggested term for industries that depend on substantial inputs from primary forest-based industries; it has been used to describe the aggregate of primary and secondary manufacturing industries (Ellefson and Stone 1984).

A valid criticism of most forest industry structural studies is that secondary manufacturing industries are lumped together with primary manufacturing industries. While this inflates the apparent significance of the industry in a region, the nexus between forest resources and manufactured wood products is lost. The significance of the difference between primary and secondary is that primary forest-based industries are dependent upon the forest resources of a particular region; secondary industries are not. The distinction may not be important for economic development, but is most important for forest resource development. Inputs to secondary wood-based industries can easily be imported from other states; timber generally cannot be economically transported any great distance.

Phelps (1980) has a suggested list of secondary "wood-based" manufacturing industries. As an example of the need to separate primary and secondary industries, consider that in 1984 the four leading states in value added by wood-based manufacturing were California, Wisconsin, Oregon, and Pennsylvania (O'Laughlin and Williams 1988). Yet California trails Oregon in lumber production (they rank second and first, respectively). California does not make the list of the top 15 states in structural wood panel production or pulpwood consumption (O'Laughlin and Williams 1988). How can California be the leading value added state? How can Pennsylvania rank fourth, and not be among the top 15 states in any of the aforementioned primary industries? The answer lies in the importance of secondary wood products manufacturing in these populous states.

Tertiary "wood-dependent" industries can be used to make an assessment of total economic activity that depends on timber production. Phelps (1980) describes industries that should be included in this grouping. Some (chemicals and textiles) are manufacturing industries, others (transportation, construction, and wholesale and retail trade) are not. Printing and publishing, a manufacturing industry that is dependent on paper, is not included, but likely should be.

Develop Means of Linking Wood-Based Industrial
Structure with Regional Economic Development

One of the most important performance considerations is the contribution of an industry to regional economic development. Research focused on such subjects should link descriptive structural studies with analytic assessments of the industry's contribution to regional employment and income. While certainly not a new area of research, it is an important one (Gray et al. 1987, Kaltenberg and Buongiorno 1986, Lichty et al. 1985, Long and Hines 1984, McCoy and Chang 1983, and Merrifield and Haynes 1985).

Determine logically consistent definitions of regions which can be used for analytical purposes. Fundamental to linking industrial structure to regional development are workable definitions of the regions in question. For example, is

Kentucky in the South, or isn't it? Kentucky is included as a southern state in virtually all USDA–Forest Service reports requiring regional definitions (Foster [undated], Schallau et al. 1987, Forest Service 1987) except forest research geographic organization (where Kentucky is in the Northeastern region) and assessment of timber supply and demand (where Kentucky is in the North Central region). While this likely causes little consternation, there are difficulties in attempting to resolve forest land use figures between data in RPA assessments (Forest Service 1980, 1984) and timberland statistics (Forest Service 1982, 1988) because Kentucky has to be passed back and forth.

Evaluate regional economic impacts of alternative means of organizing the wood-based industry. The generation and identification of "correct" economic impact multipliers for a particular region and the techniques for applying them are areas of research that should be expanded to include industrial organization concerns. In this vein, two fundamentally important questions arise: What is the economic impact of increased timber output in regions where timber supply is a constraint on expansion of the industry (O'Laughlin and Williams 1988)? And how can such regions be identified?

Refine methods of identifying wood-based industries as part of a region's economic base. Attention to the following points would alleviate problems with economic base research: specification and refinement of methods to determine whether wood-based industries are part of a region's "economic base"; and the implications of such methods, especially in terms of timber self-sufficiency within a region.

Develop methods of more effectively linking industrial structure with regional economics. Analytical and related methods for linking the structure of an industry to a regional economy are not well developed nor have existing approaches been tested in a wood-based industrial setting. Needed is identification of methods used by specialists in regional economics that could be applied to studies of wood-based industrial structure.

Develop an Improved Understanding of Timberland as a Managerial and Economic Factor

Timberland as a strategic investment variable is vital to the wood-based products industry (Bingham 1985, O'Laughlin and Ellefson 1982b, Yoho 1985). Restructuring of timberland ownership has managerial and economic implications that can affect future timber supplies and regional economies. Such restructuring is deserving of significant research attention.

Assess corporate timberland ownership in terms of cost and profit centers. Arguments are frequently made in favor of or against timberland ownership as a center of costs or a center of profits (Yoho 1983, p. 13). Such debates beg the real questions: What are the economics of the cost of industrial wood (Yoho 1983, p. 9)? And what are the quantitative implications of the risk reduction tactic of fee-owned timberland?

Needed is research to:

- *Evaluate long-term versus short-term raw material procurement strategies.* Market analysis research of this type is vitally important at the mill timbershed level. It may be best carried out by corporate planners rather than government or academic researchers. However, new methods developed by noncorporate researchers could lead to more effective ways of undertaking mill timbershed research.
- *Evaluate the applicability of the "investment center" concept to industrial timberland ownership.* The "investment center" concept may present an effective alternative approach to measuring the contribution of timberlands to an organization (Anthony and Dearden 1980). The investment center is a subset under the profit center concept, whereby the profit contribution is compared to the assets employed in earning it. The concept and its accounting methods offer some promise in answering Berkwitt's (1979) question: Does the profit center concept work? Although seemingly illogical (investors cannot buy individual pieces of a corporation), the concept is useful for maintaining managerial control and assessing departmental performance. The recognized danger (Berkwitt 1979) is that overall corporate objectives (e.g., survival in an industry dependent on timber raw materials) may be sacrificed to the pursuit of short-range profitability.

Evaluate alternative institutional arrangements conducive to timberland investments and management. Yoho (1983) identified a number of means of encouraging industrial timberland ownership and management, labeled generally as "creative financing" or "monetization" of timberland investments. What specific arrangements are likely to occur, and what effects will they have on future timber supplies and manufacturing capacity in a region? Needed is research to:

- *Evaluate timberland leasing arrangements.* Although additional supplies of industrial wood are unlikely to become available through long-term leases of nonindustrial private timberlands (Meyer et al. 1986), leasing arrangements should be continually assessed as one means of assuring industrial supplies of wood.
- *Evaluate limited partnership arrangements.* Although there was a flurry of interest and some substantial restructuring of timberland ownership with limited partnerships in the early and mid-1980s (Harris and Reddick 1986), the Tax Reform Act of 1986 has seriously limited this restructuring vehicle (Howard and Lacy 1986). The implications of partnership arrangements for timberland ownership should, however, be assessed.
- *Evaluate opportunities for tax-exempt institutional investors.* The vast majority of pension fund wealth is invested in stocks and bonds (Condrell and Yoho 1986, Bilek and Ellefson 1986, Goetzl and Royer 1982). These institutions have begun to invest in timberland; in the future they may be expected to acquire more timberland in their investment portfolios. Dyer (1984, p. 60) estimates that if private pension funds diversified by putting 10 percent of their portfolios into investments in timberlands, they would own all private timberland in the U.S. There are a multitude of researchable questions in this area.

- *Evaluate foreign ownership of U.S. timberland.* Condrell and Yoho (1986) have focused some attention on foreign ownership of U.S. timberlands. A number of important researchable questions can be posed. What does Sir James Goldsmith (an Anglo-French industrialist) know about U.S. timberlands that many segments of our financial community apparently do not know? Over the past decade, Goldsmith has acquired and dismantled two major forest products companies (Diamond International and Crown Zellerbach) by casting off the milling and marketing facilities and keeping the timberlands. Futhermore, why did Bowater (Great Britain) and Jefferson Smurfit (Ireland), companies new to the U.S. forest products industry in the 1980s, seek involvement in the ownership of U.S. timberland? And is it possible that a major U.S. paper and forest products company could be renamed "Nippon Forest Products" by some new owners? What implications might there be for timberland ownership and management strategies?

Assess regional and local implications of forest land use changes as they affect timber production. As suggested by Yoho (1983) and pursued by Waggener (1985) and Alig (1986), more knowledge about the economics of land use changes and land values as they relate to timber production is needed at regional and local levels.

Develop an Improved Understanding of the Structural and Performance Consequences of Technological Innovations

Technical change, innovation, and investments in research and development can have a significant bearing on the future structure of the wood-based industry. Such implications are often difficult to foresee but are undeniably important. Technological change in the wood-based industry has been the subject of considerable recent research, including Anderson 1987, Bengston and Strees 1986, Georgescu 1985, Gorte and Fletcher 1984, Greber and White 1982, Gregersen et al. 1984, Hoover 1984, Margl and Ellefson 1985, 1987, Martinello 1985, Ringe and Hoover 1987, Skog and Haynes 1987, Stier 1980a, 1980b, 1985, Strees 1984, and Tillman 1985. Recent research by industrial organization economists (Gort and Wall 1984, 1986) is especially interesting.

Assess the structural implications of new processing technologies. Research on new processing technologies is an area especially rich for future research, and one that is necessary for improving timber market models because of demand-side effects. New and substitute products (e.g., "spaceboard," oriented strand boards, laminated veneer lumber) may make current definitions of timber supplies obsolete. This may already be happening in certain segments of the paper industry where recycled waste paper is the primary source of fiber. Fort Howard Paper is a firm that relies completely on recycled wastepaper. The firm is a perennially profitable leader in the industry.

Evaluate structural implications of integrated wood utilization. Labeling the logistics of integration, Yoho (1983) identified integrated wood utilization as an area of needed research. Greber and Wisdom (1985) and Newman (1987) have addressed the subject; there is, however, room for additional research on process integration, especially since Lin (1986) contends that lumber and pulpwood production are independent of each other. Process integration is different than horizontal, vertical (Globerman and Schwindt 1986), or forward integration (Barnes and Sinclair 1985). Process integration was identified by Cleaves (1983) as a key variable influencing the location of forest industry manufacturing facilities. The implications of integrated utilization for restructuring within the industry are substantial. Due to competitive factors, existing firms are much more likely to make capital investments closer to their existing facilities than elsewhere (Cleaves and O'Laughlin 1985).

Develop an Improved Understanding of Corporate Strategies as Related to Economic Performance

Corporate strategies have obvious implications for economic performance. They cannot, however, be treated strictly from a neoclassical macroeconomic theory approach (i.e., that all firms behave in a similar fashion). Instead, they must be researched at the firm level where decision-making is a function of the individual firm's personality. Distinctions between firms ("personality of individual firms") have been developed via research involving corporate strategy and business management. Especially noteworthy in this respect are Caves 1980, Dugger 1987, Earl 1984, Gort and Singamsetti 1987, Marris and Mueller 1980, Ricketts 1987, and Williamson 1981.

Evaluate the performance implications of strategic group concepts. Although each firm is an individual entity, some are more alike than others. Although the concept of a strategic group is chiefly useful as a scheme for classifying firms, the concept has important performance implications because strategy is a conduct variable. Member firms may be expected to conduct themselves like other firms in the same group.

Assess the strategic implications of timberland ownership. Timberland ownership is an especially important strategic variable because of industry's dependence on timber and the long production period required to produce the raw material. Ownership of timber resources declined as a perceived corporate strength during the 1980s (Rich 1986). Viewed as a strength during the high inflation period of the late 1970s, timberland became a liability in the eyes of Wall Street during the recession of the 1980s. What are the long-term implications of such a viewpoint for an industry that depends on timber resources for its existence?

Evaluate strategic directions that emphasize marketing of wood-based products. The shift from commodity production to seeking out distinctive market niches has been documented by Rich (1986). This development follows what Yoho (1983) identified as a key research need for the companies in the industry—an emphasis on market development.

Evaluate strategic options involving capital investments. What affects investment levels in new plants and equipment and in research and development? If answers to such questions were known, changes in industrial structure could be more readily determined and more effective timber market models developed. Companies that make such investments are likely to be survivors in the long run; historic research in this area would no doubt be revealing. A research starting point would be to determine a nonearnings-based measure that could be used to identify companies likely to make investments in new technology. One possibility is the ratio of long-term investment or debt to total assets. Such research would have implications for regional economic development.

Develop an Improved Understanding of Wood-Based Structural Dimensions in an International Setting

Changes in the structure of the wood-based industry is a global phenomenon that extends well beyond U.S. borders. Many other nations are restructuring their forestry sectors in order to deal with global competition. See, for example, Croon 1986, Dykstra and Kallio 1986, Ellefson 1989, Halket 1988, Johansson 1986, Mickwitz 1986, Pace 1986, Ryti 1986, Steele and Williamson 1987, Vaatja 1987, and White 1987. Global restructuring is not limited to the wood-based sector; much can be learned from other sectors (Caves 1985, Curzon Price 1986, Drucker 1986, Jacquemin 1984, Rogerson 1987, Silk 1987, and Van Hulle 1986). Industrial structure in the global dimension deserves research along the following lines.

Define and assess wood-product trade flows in world markets. Fundamental to evaluation of world markets for wood-products is a clear definition and understanding of existing trade flows. Among the organizations that compile information about trade flows is the Food and Agriculture Organization of the United Nations. Activities such as Wisdom's (1988) are deserving of continued emphasis.

Evaluate comparative advantages in raw materials and wood-based products. Comparative advantage is a simple but powerful concept that is often overlooked in international and structural wood-based industry studies. Recent research dealing specifically with comparative advantage includes research by Cassing and Hillman, Culem 1984, Jung 1981, Mason 1984, and Oregon Economic Development Department 1986. Sedjo's (1983) comparative analysis of plantation forestry is revealing and has been extended by Sedjo and Lyons (1985). Analyses of this sort should be extended to include other factors that determine where future industrial surplus and deficit regions will occur. Presumably, global modelers are doing this, but within a different, less comprehensible framework.

Evaluate opportunities for and implications of the U.S. becoming a net exporter of wood-based products. Will the U.S. become a net exporter of forest products? If so, when and under what circumstances? The U.S. Congress Office of Technology Assessment (1983) indicated the U.S. would become a net exporter of forest products by 1990. And former President Carter's Presidential Statement of Forest Policy set forth the same goal. If such aspirations are to be realized, research is needed to identify barriers to the export of wood products and to determine policy options capable of achieving national interests in the exporting of wood-based products.

Evaluate the U.S. position in world markets, especially with regard to comparative advantages (raw material and technology) and product development and promotion. What is (and what will be) the niche for the U.S. as a producer and consumer of wood-based products in a world economy? The answers depend on many things, including monetary exchange rates, trade agreements, and structure of international markets. The structure of industrial organizations on both national and international scales will have an important influence. In this respect, two prominent industrial structure and conduct variables deserving of additional research are: comparative advantages in raw material supplies and production technology; and product development and promotion. The latter is a firm conduct variable that is market development oriented. It can be an especially important factor in determining the global role of the U.S. wood-based industries.

Develop an Improved Understanding of the Economic Structure, Conduct, and Performance of Nonwood-Based Forest Industries

Forests are capable of producing a wide range of outputs including timber, water, wildlife, recreation, and forage. Most research involving economic structure within forestry has focused on enterprises and industries that manufacture and distribute products made from timber. In many cases, however, private nonwood-based forest industries have developed (e.g., tourism, recreational camping, private hunting reserves, Christmas tree producers, maple syrup producers, downhill and cross-country skiing). Such industries are an important part of the nation's forestry sector. Unfortunately, they have received limited, if any, attention from economics research involving industrial structure, conduct, and performance. At a minimum, the economic structure of nonwood, forest-based industries should be identified. Subsequent efforts should focus on the manner in which such industries conduct their business activities. The overall intent is to design economic structures that will permit the performance of such industries to improve.

Summary and Conclusions

The economic structure, conduct, and performance of industries dependent on the nation's forests are topics of special importance for research. A major challenge to research involving industrial organization of forest-based industries is to link descriptions of structure, conduct, and industrial performance in ways that facilitate public and private policy making. Not to be overlooked is the need to continue efforts to monitor changes in structure and conduct dimensions at the national level and to conduct baseline studies of industry structure-conduct-performance at regional, state, and local levels. Specifically needed is research that will improve understanding of restructuring within the wood-based industry; definitions of the wood-based industry and segments thereof; linkages between structure and regional economic development; timberland as a managerial and economic variable; structural consequences of technological innovations; corporate strategies as related to performance; structural dimensions in an international setting; and structure and performance of nonwood-based forest industries. Economics research focused in such directions will go far toward improving the manner in which the nation's many forest industries organize and conduct their activities.

Literature Cited

Acs, Z. J. 1984. The changing structure of the U.S. economy: Lessons from the steel industry. Praeger. New York, NY.

Adams, D. M. and R. W. Haynes. 1985. Changing perspectives on the outlook for timber in the United States. Journal of Forestry 83(1): 32-35.

Adams, D. M. and R. W. Haynes. 1986. Development of the Timber Assessment Market Model [TAMM] system for long-range planning by the United States Forest Service. In: Proceedings of Division 4. XVIII World Congress. International Union of Forest Research Organizations. Ljubljana, Yugoslavia.

Albrecht, J., et al. 1987. Social sciences in forestry: An annotated bibliography. College of Natural Resources. University of Minnesota. St. Paul, MN.

Alexander, R. J. 1986. Is the United States substituting a speculative economy for a productive one? Journal of Economic Issues 20(2): 365-374.

Alig, R.J. 1986. Econometric analysis of the factors influencing forest acreage trends in the Southeast. Forest Sciences 32:119-134.

Anderson, W. C. 1987. Technical changes that solved the southern pine lumber industry's small-log problem. Forest Products Journal 37(6):41-45.

Backhaus, J. 1985. Public policy toward corporate structures: Two Chicago approaches. Journal of Economic Issues 19(2): 365-373.

Barnes, C. S. and S. A. Sinclair. 1985. Forward integration in the building products market and its impact on the financial performances of selected forest products firms. Forest Products Journal 35(11/12):68-74.

Baudin, A. and A. Westlund. 1985. Structural instability analysis; the case of newsprint consumption in the United States. Forest Science 31:990-994.

Baumol, W. J. 1982. Contestable markets: An uprising in the theory of industry structure. American Economic Review 72(1):1-15.

Bengston, D. N. and A. Strees. 1986. Intermediate inputs and the estimation of technical change: The lumber and wood products industry. Forest Science 32:1078-1085.

Berkwitt, G. J. 1979. Do profit centers really work? In: Rice, G. H., Jr., ed., Industrial organizations: Selected readings. Lone Star Press. Austin, TX.

Bilek, E. M. and P. V. Ellefson. 1986. Stock ownership of selected public U.S. wood-based corporations. Forest Products Journal 36(3):13-16.

Bilek, E. M. and P. V. Ellefson. 1987. Organizational arrangements used by U.S. wood-based companies involved in direct foreign investment: An evaluation. Bulletin 576-1987. Agricultural Experiment Station. University of Minnesota. St. Paul, MN.

Bingham, C. W. 1985. Rationale for intensive forestry investment: A 1980s view. In: Investments in forestry: Resources, land use, public policy by R. A. Sedjo. Westview Press. Boulder, CO.

Boyd, R. 1983. Lumber transport and the Jones Act: A multicommodity spatial equilibrium analysis. Bell Journal of Economics 14(1):202-212.

Boyer, K. D. 1984. Is there a principle for defining industries? Southern Economic Journal 51(4):761-770; and Reply to Comments, 52(2):542-546.

Bresnahan, T. F. and R. Schmalensee. 1987. The empirical renaissance in industrial economics: An overview. Journal of Industrial Economics 35(4):371-378.

Buongiorno, J., J. C. Stier and J. K. Gilless. 1981. Economies of plant and firm size in the United States pulp and paper industries. Wood and Fiber 13(2):102-114.

Callahan, J. C. 1985. Structural changes in the hardwood face-veneer industry. Forest Products Journal 35(6):11-16.

Cassing, J. H. and A. L. Hillman 1986. Shifting comparative advantage and senescent industry collapse. American Economic Review 76(3):516-523.

Caves, R. E. 1980. Corporate strategy and structure. Journal of Economic Literature 18(1):64-92.

Caves, R. E. 1985. International trade and industrial organization: Problems, solved and unsolved. European Economic Review 28(3):377-395.

Chandler, A. D. 1962. Strategy and structure: Chapters in the history of American enterprise. MIT Press. Cambridge, MA.

Cleaves, D. A. 1983. An analysis of plant location decisions in wood-based companies. Ph.D. dissertation. Texas A&M University. College Station, TX.

Cleaves, D. A. and J. O'Laughlin. 1985. Forest inventory, plant locations, and firm strategies in the southern plywood industry. In: Proceedings of 1985 Southern Forest Economics Workshop and 1984 Southern Economic Association by F. W. Cubbage (ed.). Raleigh, NC.

Cleaves, D. A. and J. O'Laughlin. 1986a. Analyzing structure in wood-based industry: Part I. Identifying competitive strategy. Forest Products Journal 36(4):9-14.

Cleaves, D. A. and J. O'Laughlin. 1986b. Analyzing structure in wood-based industry: Part II. Categorizing strategic diversity. Forest Products Journal 36(5):11-17.

Condrell, W. K. and J. G. Yoho (comp.). 1986. Timberland marketplace for buyers, sellers and investors and their advisors. Center for Forestry Investment. School of Forestry and Environmental Studies. Duke University. Durham, NC.

Croon, I. 1986. Structural changes in the Swedish forest industry. Svensk Papperstidning 89(11):6-8.

Culem, C. 1984. Comparative advantage and industrial restructuring: The Belgian case, 1970-1980. Cahiers Economiques de Bruxelles 103(3):457-484.

Curry, B. and K. D. George. 1983. Industrial concentration: A survey. Journal of Industrial Economics 31(3):203-255.

Curzon Price, V. 1986. Industrial and trade policy in a period of rapid structural change. Aussenwirtschaft 41(2/3):201-223.

Dickerhoof, H. E. 1986. Changing U.S. softwood timber markets and international trade trends, with implications for domestic forest products production. Forest Products Journal 36(10):35-40.

Drucker, P. F. 1986. The changed world economy. Foreign Affairs 64(4):768-791.

Dugger, W. M. 1987. Corporate hegemony and market mythology. Challenge 29(6):55-58.

Dyer, C. 1984. Role of pension funds in timberland investment—why Eastern Airlines invested. In: Timberland marketplace for buyers, sellers and investors and their advisors by W. K. Condrell and J. G. Yoho. Center for Forestry Investment. School of Forestry and Environmental Studies. Duke University. Durham, NC.

Dykstra, D. P. and M. Kallio. 1986. Introduction to the IIASA forest sector model. In: Proceedings of Division 4. XVIII World Congress. International Union of Forest Research Organizations. Ljubljana, Yugoslavia.

Earl, P. F. 1984. The corporate imagination: How big companies make mistakes. Sharpe Publishers. Armonk, NY.

Eckstein, L.W., Jr. 1983. Consideration of organizational structure and design for improved efficiency in southern wood departments. Southern Journal of Applied Forestry 7(1):3-6.

Ellefson, P. V. 1985. Economic structure of the eastern hardwood industry: Past and future trends. In: Proceedings of Conference on Eastern Hardwoods: The Industry and the Markets. Forest Products Research Society. Madison, WI.

Ellefson, P. V. 1989. U.S. wood-based industry: Economic structure in a worldwide context. In: Concise Encyclopedia of Wood and Wood-based Materials by Schniewind (ed.). Pergamon Publishers. Oxford, England.

Ellefson, P. V., B. J. Lewis and R. A. Skok (eds.). 1985. Forest industry as a force in economic development: Options for Minnesota's future. Miscellaneous Publication No. 33. Agricultural Experiment Station. University of Minnesota. St. Paul, MN.

Ellefson, P. V. and R. N. Stone. 1984. U.S. wood-based industry: Industrial organization and performance. Praeger Publishers. New York, NY.

Filippello, A. N. 1985. The U.S. chemical industry—restructuring for the future. Business Economics 20(4):44-49.

Forest Service. 1980. An assessment of the forest and rangeland situation in the United States. FS-345. U.S. Department of Agriculture. Washington, DC.

Forest Service. 1982. An analysis of the timber situation in the United States, 1952-2030. Forest Resource Report 23. U.S. Department of Agriculture. Washington, DC.

Forest Service. 1984. America's renewable resources: A supplement to the 1979 assessment of the forest and range land situation in the United States. FS-386. U.S. Department of Agriculture. Washington, DC.

Forest Service. 1987. Report of the Forest Service, fiscal year 1986. U.S. Department of Agriculture. Washington, DC.

Forest Service. 1988. The South's Fourth Forest: Alternatives for the Future. Forest Resource Report 24. U.S. Department of Agriculture. Washington, DC.

Foster, B. B. (undated). Value added in the wood-based industries of the 13 southern states, 1982. Cooperative Forestry Technology Update (unnumbered). Southern Region. USDA Forest Service. Atlanta, GA.

Georgescu, D. M. 1985. Cybernetization and robotization of the operating processes in the wood and building materials industry. Economic Computation and Economic Cybernetics Studies and Research 21(1):55-68.

Globerman, S. and R. Schwindt. 1986. The organization of vertically related transactions in the Canadian forest products industries. Journal of Economic Behavior and Organization 7(2):199-212.

Goetzl, A. and J. P. Royer. 1982. Financing forestry investments: Attracting private capital to timber and timberland. Workshop proceedings. National Forest Products Association. Washington, DC.

Gort, M. and R. Singamsetti. 1987. Innovation and the personality profiles of firms. International Journal of Industrial Organization 5(1):115-126.

Gort, M. and R. A. Wall. 1984. The effect of technical change on market structure. Economic Inquiry 22(4):668-675.

Gort, M. and R. A. Wall. 1986. The evolution of technologies and investment in innovation. Economic Journal 96(383):741-757.

Gorte, J. F. and W. W. Fletcher. 1984. Technology, timber and the future. Renewable Resources Journal 2(2/3):16-19.

Gray, G. J., P. V. Ellefson and D. C. Lothner. 1987. Lake States wood product production and consumption: Trends and perspectives. Northern Journal of Applied Forestry 4:193-197.

Greber, B. J. and D. E. White. 1982. Technical change and productivity growth in the lumber and wood products industry. Forest Science 28(1):135-147.

Greber, B. J. and H. W. Wisdom. 1985. A timber market model for analyzing roundwood product interdependencies. Forest Science 31:164-179.

Gregersen, H. M., J. Haygreen, A. Strees, and I. Holland. 1984. Factors affecting productivity change in the forest products sector. In: Proceedings of 1983 convention. Society of American Foresters. Bethesda, MD.

Gregory, G. R. 1987. Resource economics for foresters. Wiley and Sons Publishers. New York, NY.

Gruchy, A. G. 1985. Corporate concentration and the restructuring of the American economy. Journal of Economic Issues 19(2):429-439.

Halket, J. C. 1988. Comment: Volatility still in public sector restructuring of forest industry]; and Recent events: Forestry Corporation reports. New Zealand Forestry 32(4):2-4.

Hall, J. V. 1987a. Financial fragility and restructuring of the U.S. petroleum industry. Economic Forum 16(1):69-80.

Hall, R. H. 1987b. Organizations: Structures, processes, and outcomes. 4th ed. Prentice-Hall. Englewood Cliffs, NJ.

Harris, T. G., Jr. and M. S. Reddick. 1986. The role of Master Limited Partnerships in forest industry restructuring. In: Proceedings of 1986 southern forest economics workshop by E. E. Mathews (ed.). SOFEW. Raleigh, NC.

Hayes, R. H. and W. C. Wheelwright. 1984. Restoring our competitive edge: Competing through manufacturing. Wiley and Sons Publisher. New York, NY.

Hoover, W. L. 1984. Resource driven technological change in the southern hardwood lumber industry. In: Proceedings of 1984 southern forest economics workshop by R. W. Guldin (ed.). SOFEW. Raleigh, NC.

Howard, T. E. and S. E. Lacy. 1986. Forestry limited partnerships: Will tax reform eliminate benefits for investors? Journal of Forestry 84(12):39-43.

Irland, L. C. 1987. Canada-U.S. forest products trade: Tensions in a maturing market. Forest Products Journal 37(2):21-29.

Jacquemin, A., ed. 1984. European industry: Public policy and corporate strategy. Oxford University Press. New York, NY.

Jacquemin, A. 1987. The new industrial organization: Market forces and strategic behavior. MIT Press. Cambridge, MA.

Johansson, B. 1986. A structural change model for regional allocation of investments. In: Systems analysis in forestry and forest industries by M. Kallio et al. (eds.). Elsevier Publishers. New York, NY.

Jung, Y. 1981. Comparative advantage and productive efficiency of Korea in the textiles, clothing, and footwear industries. Journal of Economic Development 6(2):133-164.

Kaiser, H.F. 1972. Multiregional input-output model for forest resource analysis. Forest Science 18(1):46-53.

Kaltenberg, M. C., and J. B. Buongiorno. 1986. Growth and decline of the paper industry: An econometric analysis of U.S. regions. Applied Economics 18:379-397.

Keegan, C. E., III, and P. E. Polzin. 1987. Trends in the wood and paper products industry: Their impact on the Pacific Northwest economy. Journal of Forestry 89(11):31-36.

Lavoie, D. 1985. Rebuilding America: A blueprint for the new economy: Review article. Comparative Economic Studies 27(3):99-113.

Lichty, R. W., D. N. Steinnes and D. A. Vose. 1985. Strategic planning of economic development based on an analysis of the extent and pattern of importation. Regional Science Perspectives 15(1):46-62.

Lin, W.T. 1986. Analysis of lumber and pulpwood production in a partial adjustment model with dynamic and variable speeds of adjustment. Journal of Business and Economic Statistics 4(3):305-316.

Lindheim, J. B. 1985. Currents & contexts for the forest products industry. In: The forest industry in perspective: Public opinion and the forest products industry—the outlook in 1985. American Forest Institute. Washington, DC.

Long, R. B. and G. A. Hines. 1984. Analyzing a forest resource based region facing economic change. Forest Products Journal 34(10):51-58.

Marbach, W.D. and E. T. Smith. 1988. R&D scoreboard: A perilous cutback in research spending. Business Week 3057 (June 20):139-162.

Margl, R. A. and P. V. Ellefson. 1985. Trends in U.S. wood-based industrial technology: An evaluation of assigned patents. Staff Paper Series No. 50. Department of Forest Resources. University of Minnesota. St. Paul, MN.

Margl, R. A. and P. V. Ellefson. 1987. Assigned patents: Technology trends in 15 U.S. wood-based companies. Forest Products Journal. 37(1):47-50.

Marris, R. and D. C. Mueller. 1980. The corporation, competition, and the invisible hand. Journal of Economic Literature 18(1):32-63.

Martinello, F. 1985. Factor substitution, technical change, and returns to scale in Canadian forest industries. Canadian Journal of Forest Research 15:1116-1124.

Mason, D. H. 1984. Foreign capital in U.S. timberland investment: The world-wide comparative advantage of North America, particularly the U.S. South and Pacific Northwest. In: Timberland marketplace for buyers, sellers and investors and their advisors by W. K. Condrell and J. G. Yoho (eds.). School of Forestry and Environmental Studies. Duke University. Durham, NC.

McCoy, D. R. and S. J. Chang. 1983. The secondary wood-using industries in Kentucky: An economic analysis. Bulletin 719. Agricultural Experiment Station. University of Kentucky. Lexington, KY.

Merrifield, D. E. and R. W. Haynes. 1985. A cost analysis of the lumber and plywood industries in two Pacific Northwest sub-regions. Annals of Regional Science 19(3):16-33.

Meyer, R. D., W. D. Klemperer and W. C. Siegel. 1986. Cutting contracts and timberland leasing. Journal of Forestry 84(12):35-37.

Mickwitz, G. 1986. The forest industry—sunset or restructure? Ekonomiska Samfundets Tidskrift 39(2):65-68.

Miller, J. C., III, et al. 1984. Industrial policy: Reindustrialization through competition or coordinated action? Yale Journal on Regulation 2(1):1-37.

Moriarty, B. M. 1986. Regional industrial change, industrial restructuring, and U.S. industrial policy. Review of Regional Studies 16(3):1-10.

Moslemi, A. A. 1986. Strong medicine for the forest industry: Has it healed the patient? Forest Products Journal 36(5):9-10.

Mueller, D. C. 1986. Profits in the long run. Cambridge University Press. New York, NY.

Mukhopadhyay, A. K. 1985. Technological progress and change in market concentration in the U.S., 1963-77. Southern Economic Journal 52(1):141-149.

National Research Council. 1986. Toward a new era in U.S. manufacturing: The need for a national vision. Commission on Engineering and Technical Systems, Manufacturing Studies Board. National Academy Press. Washington, DC.

Nautiyal, J. C. and B. K. Singh. 1985. Production structure and derived demand for factor inputs in the Canadian lumber industry. Forest Science 31:871-881.

Nautiyal, J. C., B. K. Singh and O. Menezes. 1985. Market structure and economic performance of forest products industry in Ontario and Canada. Canadian Journal of Forest Research 15:115-125.

Newman, D.H. 1987. An econometric analysis of the Southern softwood stumpage market:1950-1980. Forest Science 33(4):932-945.

Norton, R. D. 1986. Industrial policy and American renewal. Journal of Economic Literature 24(1):1-40.

Office of Technology Assessment. 1983. Wood use: U.S. competitiveness and technology. OTA-ITE-210. U.S. Congress. Washington, DC.

O'Laughlin, J. 1981. Antitrust in the era of diversification: The wood-based industry as a case example. Unpublished paper presented at 51st Annual Conference of the Southern Economic Association. New Orleans, LA.

O'Laughlin, J. 1986. The changing raw material base for forest products in the southern U.S.—Consequences and adjustments; including material describing the economic structure of the U.S. paper and forest products industry, 1985. Unpublished paper presented to a seminar sponsored by the Swedish Pulp and Paper Association. Stockholm, Sweden.

O'Laughlin, J. and P. V. Ellefson. 1982a. New diversified entrants among U.S.wood-based companies: A study of economic structure and corporate strategy. Bulletin No. 541-1982. Agricultural Experiment Station. University of Minnesota. St. Paul, MN.

O'Laughlin, J. and P. V. Ellefson. 1982b. Strategies for corporate timberland ownership and management. Journal of Forestry 80(12):784-788, 791.

O'Laughlin, J. and R. A. Williams. 1988. Forests and the Texas economy. Bulletin B-1596. Agricultural Experiment Station. Texas A & M University. College Station, TX.

Oregon Economic Development Department. 1986. The Oregon advantage for the wood products industry. Oregon Department of Economic Development. Salem, OR.

Pace, N. 1986. Changing international pressures affecting U.S. forest industry. In: Proceedings of the 1986 southern forest economics workshop by E. E. Mathews (ed.). SOFEW. Raleigh, NC.

Phelps, R. B. 1980. Timber in the United States economy, 1963, 1967, and 1972. General Technical Report WO-21. USDA Forest Service. Washington, DC.

Porter, M. 1979. The structure within industries and companies' performance. Review of Economics and Statistics 61(2):214-227.

Redburn, F. S., T. F. Buss, and L. C. Ledebur (eds.). 1986. Revitalizing the U.S. economy. Greenwood Press and Praeger Publishers. New York, NY.

Reynolds, R. J. and B. R. Snapp. 1986. The competitive effects of partial equity interests and joint ventures. International Journal of Industrial Organization 4(2):141-153.

Rich, S. U. 1986. Recent shifts in competitive strategies in the U.S. forest products industry and the increased importance of key marketing functions. Forest Products Journal 36(7/8):34-44.

Ricketts, M. 1987. The new industrial economics: An introduction to modern theories of the firm. St. Martin's Press. New York, NY.

Rideout, D. 1984. Structural changes in the timber economy of southeast Alaska. Canadian Journal of Forest Research 14:498-505.

Ringe, J. M. and W. L. Hoover. 1987. Value added analysis: A method of technological assessment in the U.S. forest products industry. Forest Products Journal 37(11/12):51-54.

Robinson, V. 1988. The relevance of forest economics research: A new perspective on a basic tenet. In: Proceedings of 1988 southern forest economics workshop by R. C. Abt (ed.). SOFEW. Raleigh, NC.

Rogerson, R. 1987. The economics of worldwide stagflation: A review essay. Journal of Monetary Economics 19(1):129-136.

Ross, L. 1985. Are timber and timberland evaluations important to the financial community? In: Recognition of timber and timberland asset values for financial reporting by W. Sizemore and R. Hoskins (eds.). School of Forestry and Environmental Sciences. Duke University. Durham, NC.

Rostow, W. W. 1985. The fifth Kondratieff upswing and the fourth industrial revolution: Their meaning for forestry. In: Investments in forestry: Resources, land use, and public policy by R. A. Sedjo (ed.). Westview Press. Boulder, CO.

Rothwell, R. and Zegveld, W. 1985. Reindustrialization and technology. Sharpe Publishers. Armonk, NY.

Ryti, N. 1986. Trends and likely structural changes in the forest industry worldwide. In: Systems analysis in forestry and forest industries by M. Kallio et al. (eds.). Elsevier Publishers. New York, NY.

Schallau, C. 1985. What's this about the forest industry shifting to the South? In: Proceedings of the 1985 Southern Forest Economics Workshop by F. W. Cubbage (ed.). SOFEW. Raleigh, NC.

Schallau, C. 1986. Forest products in the Pacific Northwest—a declining industry? Forest Products Journal 36(10):18-20.

Schallau, C. H. and W. R. Maki. 1983. Interindustry model for analyzing the regional impacts of forest resource and related constraints. Forest Science 29(2):384-394.

Schallau, C. H. and W. R. Maki. 1986. Economic impacts of interregional competition in the forest products industry during the 1970's: The South and the Pacific Northwest. Research Paper PNW-350. Pacific Northwest Forest and Range Experiment Station. USDA Forest Service. Portland, OR.

Schallau, C. H., W. R. Maki, B. B. Foster and C. H. Redmond. 1987. Texas' forest products industry: Performance and contribution to the state's economy, 1970 to 1980. Research Paper PNW-389. Pacific Northwest Forest and Range Experiment Station. USDA Forest Service. Portland, OR.

Sedjo, R. A. 1983. The comparative economics of plantation forestry: A global assessment. Research Paper 0-8018-3107-5. Resources for the Future. Washington, DC.

Sedjo, R. A., ed. 1985. Investments in forestry: Resources, land use, and public policy. Westview Press. Boulder, CO.

Sedjo, R. A. and K. S. Lyons. 1983. Long term forest resource trade, global timber supply and intertemporal comparative advantage. American Journal of Agricultural Economics 64(5):1010-1016.

Shepherd, G. 1986. Industrial restructuring: The European textile industry. Revue d'E-conomie Industrielle 31(1):68-78.

Silk, L. 1987. The United States and the world economy. Foreign Affairs 65(3):458-476.

Skog, K. and R. Haynes. 1987. The stumpage market impact of timber utilization research. Forest Products Journal 37(6):54-60.

Slatin, B. 1985. Changes in the economic structure of the paper industry: 1974-1984. Tappi Journal 68(2):30-38.

Solo, R. A. 1984. Industrial Policy. Journal of Economic Issues 18(3):697-714.

Steele, T. and T. Williamson (comp). 1987. The continuing challenge: Competition and new products in the world forest products markets. Northern Forestry Center. Alberta Department of Forestry, Lands and Wildlife. Canadian Forestry Service, Northern Forestry Center; Edmonton, Alberta. Canada.

Stier, J. C. 1980a. Estimating the production technology in the U.S. forest products industries. Forest Science 26(3):471-482.

Stier, J. C. 1980b. Technological adaptation to resource scarcity in the U.S. lumber industry. Western Journal of Agricultural Economics 5(2):165-175.

Stier, J. C. 1985. Implications of factor substitution, economies of scale, and technological change for the cost of production in the United States pulp and paper industry. Forest Science 31:803-812.

Stigler, G. J. 1982. The economists and the problem of monopoly. American Economic Review 72(2):1-11.

Strees, A. 1984. Productivity change in the U.S. forest industries. Ph.D. thesis. University of Minnesota. St. Paul, MN.

Tillman, D. A. 1985. Forest products: Advanced technologies and economic analyses. Academic Press. New York, NY.

Van Hulle, C. 1986. Corporate restructuring. Tijdschrift Voor Economie en Management 31(4):417-433.

Vaatja, M. 1987. Finland's forest industry stronger than ever before after rapid structural transformation. Svensk Papperstidning 90(12):9-10, 13.

Waggener, T. R. 1985. The economics of shifting land use margins. In: Investments in forestry: Resources, land use and public policy. Westview Press. Boulder, CO.

Walker, J. F. and H. G. Vatter. 1986. Stagnation—performance and policy: A comparison of the depression decade with 1973-1984. Journal of Post-Keynesian Economics 8(4):515-530.

White, D. E. 1987. New Zealand forestry: Privatization gets a test. Journal of Forestry 85(3):41-43.

Williamson, O. E. 1981. The modern corporation: Origins, evolution, attributes. Journal of Economic Literature 19(4):1537-1568.

Wisdom, H.W. 1988. Impact of international trade on the Southern economy. In: Proceedings of the 1988 Southern Forest Economics Workshop by R. C. Abt (ed.). SOFEW. Raleigh, NC.

Yoho, J. G. 1983. Economics research—critical needs for industry in the South. In: Proceedings of 13th annual southern forest economics workshop. SOFEW. Raleigh, NC.

Yoho, J. G. 1985. Continuing investments in forestry: Private investment strategies. In: Investments in forestry: Resources, land and public policy by R. A. Sedjo (ed.). Westview Press. Boulder, CO.

7

DEVELOPMENT, DISSEMINATION, AND ADOPTION OF NEW TECHNOLOGY

Allen L. Lundgren

Economic activities involving forestry entail the use of knowledge, skills, techniques, tools, equipment, and related inputs to transform raw materials into usable goods and services. Collectively, these inputs and their management constitute a technology. In an economic sense, resources in forestry are defined by available technologies. What is not considered a forest resource at a particular point in time could very well become an extremely important resource given the advent of a new technology. For example, once viewed as a weed tree in the Lake States, aspen is now a valued raw material for the pulp and paper and the structural particleboard industries. New pulping and wood processing technologies were the primary cause of this transition.

Technology plays a critical and often overlooked role in economic analyses of forestry activities. All such analyses assume that a set of technologies will be used in the production, distribution, and utilization of goods and services. A particular set of technologies determines the input-output relationships of these processes. At any one time, several different technologies may be available for use, each requiring a different array of inputs that produce a different array of outputs. The forest economist evaluates alternative technologies to determine effectiveness and efficiency. A given set of technologies will determine, to a large extent, the potential effectiveness and efficiency of a given production-distribution-utilization system as a means of transforming resources and raw materials into useful goods and services.

Technology is dynamic in modern societies. Technological innovation (the process by which new technologies are developed, disseminated, adopted, and utilized) changes the way in which forest and related resources are managed and used. New technologies include not only new machines, equipment, and materials but also include new methods, procedures, techniques, and approaches to accomplishing desired objectives. Technological innovations in forestry present new options and new alternatives to be evaluated by forest economists.

Because technological innovation plays an especially important role in the forestry sector, research concerning technological innovation is needed to enhance understanding of how technology influences economic activity and how costs and benefits of forestry activities are affected by technological innovations. Research is also needed to improve the forestry community's understanding of forestry research activities and the processes by which new technologies are developed, disseminated, adopted, and utilized. Such an understanding will enable improvements in the effectiveness and efficiency of such processes.

Prior to the early 1980s, very little systematic research on technological innovation in forestry had been undertaken. Neither Duerr and Vaux (1953) in *Research in the Economics of Forestry* nor Clawson (1977) in *Research in Forest Economics and Forest Policy* identify technology as a major topic for forest economics research. Although recent research has investigated specific aspects of technological innovation in forestry (e.g., forestry research), yet to be adequately addressed is how new technologies are developed, disseminated, adopted, and used in forestry. Lacking is knowledge and information needed to improve technological innovation systems. Also lacking are evaluations of technological innovations—evaluations which can be used to improve decision making involving the funding and management of forestry research and technological innovation.

Process of Technological Innovation in Forestry

Technological innovation in forestry is the adoption and use of new technologies to improve the management and utilization of forest and related resources and to improve the production, distribution, and utilization of the goods and services derived from or related to such resources. The technological innovation system in forestry includes several essential components (Figure 7.1):

- Research and the knowledge research contributes to the technological innovation process.
- Technological innovation process, including the development, dissemination, adoption, and utilization of technologies in the forestry sector.
- Changes brought about within the forestry sector by the adoption and use of new technologies in producing, distributing, and consuming forestry-related goods and services.
- Social, economic, and environmental impacts resulting from the changes induced by technological innovations in the forestry sector.

Forestry research is an especially important part of the technological innovation process. Research activities produce the scientific knowledge that is used to develop new forestry technologies. Not all knowledge produced by forestry research is applied immediately; some (i.e., basic research) is used as an input to subsequent forestry research. The distinction between basic and applied research, however fuzzy, is important to research evaluation. In general, basic research can best be evaluated by the subjective judgement of peer groups within a particular scientific discipline. In contrast, evaluation of applied research must explicitly consider changes in the costs and benefits brought about by the adoption and utilization of tech-

Research:

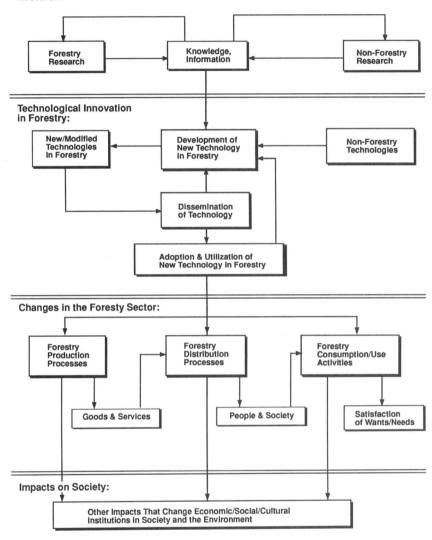

Figure 7.1 The technological innovation process in forestry

nological innovations and must recognize ultimate effects on the economy, the environment, and society.

Forestry research can be viewed as a production process (Lundgren 1982). Research uses scarce financial resources, skilled people, limited facilities, and knowledge gained from previous research as inputs to an often lengthy and complex process that transforms such resources into new knowledge. Economists can study the effectiveness and efficiency of alternative ways to plan, organize, and conduct research in forestry. The results are of interest to administrators, managers, re-searchers, and users of research products, including consumers ultimately affected by the products of research.

Marketplaces for establishing the exchange value of knowledge produced by research do not exist. As such, forest economists studying research activities face formidable obstacles when attempting to estimate research benefits. Such is especially so since many of the goods and services produced by forests are not exchanged through market systems (e.g., wildlife, scenic beauty, recreation). Establishing research benefits for new knowledge and technologies related to such goods and services poses special problems.

Evaluations of technological innovations in forestry must recognize that non-forestry research (e.g., research in soils, water, wildlife, ecology, botany, anthropology, and rural sociology) frequently has application to forestry. Furthermore, technological development in fields outside of forestry can also influence forestry operations and activities. For example, electronics research has provided new communication and data processing technologies useful to the development of new technologies used in forest resource management.

Forestry research administrators and managers, faced with mounting financial constraints, have become increasingly interested in methods for improving the effectiveness and efficiency of forestry research programs. Forestry researchers, faced with the need to justify research in terms of potential impacts on society, have sought better ways to estimate the costs and benefits of research efforts. And research users have shown an increasing interest in influencing the future direction of numerous forestry research programs. All such groups have a common interest in economic evaluations of forestry research. Such evaluations can provide important information that helps guide research investments.

Technological innovation (development, dissemination, adoption, utilization) is also an important segment of the technology development and adoption system. Technologies used in forestry may be developed primarily within forestry (e.g., new silvicultural techniques, new genetic strains of tree species, new wood pulping processes) or may incorporate slightly modified technologies developed in nonforestry sectors (e.g., scientific laboratory equipment, chemicals used in pulping processes). However, a great many technologies used in forestry are developed outside the forestry sector and are used for forestry purposes with only slight modification (e.g., radios, computers, vehicles). Some technologies developed outside of forestry are especially significant technological innovations to forestry since they cause notable changes in forest resource use and management (e.g., all-terrain vehicles).

New technologies are usually adopted and implemented within existing orga-nizations that already use a wide range of technologies and that operate under a wide array of formal and informal rules and procedures. The introduction of new technologies into existing organizations may require significant adjustments in order to mesh new technologies with existing operations. Required may be reassignment of jobs, retraining of people, construction of new facilities, readjustment of work patterns and additional development work to modify the new technology—all of which can add significantly to the expense of acquiring a new technology. Such difficulties are compounded by the fact that potential benefits from new technologies are often not well known; their outcome is highly uncertain. Factors such as these present real barriers to technological innovation. In order for innovation processes to proceed smoothly, strategies to overcome such barriers must be established and implemented.

The development of new technologies may entail the use of scientific knowledge and practical experience gained from ongoing operations and activities. Such practical experience is seldom formally documented and may reside in the technological skills gained from years of cumulative experience. New technologies that are developed usually undergo some type of testing or pilot trials. Those that survive are disseminated to potential users.

Potential users of new technologies must be aware of such technologies if they are to use them. Dissemination of new technologies is an important step in the technology innovation system. Such may take place through technical advertising, articles in trade publications, catalogs, personal contacts by product sales representatives, or personal contacts with users of the technology. Many new technologies may be directly adopted as means of changing the way goods and services are produced. Others may be an input to subsequent scientific research, thus causing change in the manner in which research is carried out (e.g., scientific instruments, laboratory equipment, and computers). Some may be used in developing still more complex or advanced technologies.

Changes in the forestry sector and impacts on society are also major considerations in the technological innovation system. Adoption and use of new technologies by the forestry sector change the process by which forestry-related goods and services are produced, distributed, and used. Changed may be the kinds and amounts of inputs used, the kinds of goods and services produced, the way in which goods and services are distributed, and the kinds of consumption and use activities available to people. Such changes are likely to alter the effectiveness and efficiency of forest production processes as well as the manner in which economic evaluations of such processes are carried out.

Forest sector changes brought about by new technologies can have direct economic, social, and environmental consequences for society. Impacts may include changes in employment, income distribution, population distribution, and demands for various goods and services; changes in resource use patterns; adverse environmental impacts; and changes in laws, regulations, customs, patterns of behavior, and new social institutions. For example, improved technologies for monitoring and detecting low levels of environmental contaminants (e.g., organic chemicals) have led to increasingly stringent laws and regulations governing pollution. Such changes in society's institutions can in turn lead to indirect effects or second-order consequences (Bauer 1969).

Changes brought about by new technologies can have substantial impacts on the effectiveness and efficiency of activities occurring in the forestry sector. As such, the potential impacts of technological innovations should be an essential part of evaluations of forestry activities. Because benefits and costs of technological innovation are often distributed unequally among different groups of society and different areas of a state or nation, distributional effects of technological innovation should be an important part of economic analyses (Bengston and Gregersen 1988). Furthermore, evaluation of technological innovation processes should be comprehensive. Research focused on a single segment of the system (e.g., research) can produce misleading results. Innovation is a dynamic, multifaceted process that requires comprehensive analysis. Furthermore, since the forestry sector is closely linked to other sectors of the economy, economic analyses of technological innovation within forestry should recognize and evaluate the potential effects of technological innovations brought from outside the forestry sector.

Knowledge available for identifying and describing technological problems and for improving the development, dissemination, adoption, and utilization of new technology in forestry is far from adequate. There exists relatively little documented information about how research is linked to technological innovation, what factors have an influence on the innovation process, what effects technological innovation has on adopting individuals and organizations and what economic, social, and environmental impacts result from technology-induced change. Without an adequate understanding of the technology innovation system, research undertaken to improve the process may well be ineffective, and research on evaluations of forestry operations and activities may neglect important aspects of technological innovation.

Research on technological innovation in forestry remains in infancy; much work needs to be done. Forest economics research should focus attention on the following strategic directions:

- Develop an improved understanding of the role forestry research plays in technological innovation.
- Develop a better understanding of technology development, dissemination, adoption, and utilization.
- Develop a better understanding of the impacts of technological innovation.

Develop an Improved Understanding of the Role Forestry Research Plays in Technological Innovation

Forestry research plays an important role in technological innovation. However, our understanding of the costs and benefits of forestry research is incomplete and fragmentary. A more comprehensive program to study and evaluate research in forestry is needed. Over the years a number of studies have focused on various topics concerning forestry research, including:

- General surveys of forestry research programs and needs (Kaufert and Cummings 1955; Buckman, Skok, and Sullivan 1982; Krugman and Cowling 1982).
- Investigative reports (Comptroller General 1972, 1981).
- Research funding, priorities, and resource allocation (Bethune and Clutter 1969; Claxton and Rensi 1972; Babcock 1974; Bullard and Straka 1986; Fox 1986).

- Research policy (Renewable Natural Resources Foundation 1977; Whaley and Bell 1982).
- Research organizations and research processes (Schreyer 1974; Lingwood and Morris 1976; Jakes and Van Dyne 1987).
- Evaluations of research projects and programs (Callaham 1981; Herrick 1982; Risbrudt and Kaiser 1982; Rose 1983; Bengston 1984; Callaham 1985; Seldon 1985; Haygreen et al. 1986; Westgate 1986; Fege 1987).
- Research in developing countries (Bengston, Gregersen, and Lundgren 1987; Bengston and Gregersen 1988; Bengston et al. 1988).

Most early studies of forestry research were relatively isolated efforts undertaken by relatively few individuals. Recently, however, a growing interest has occurred in systematically evaluating forestry research, an interest that stems in large measure from an increasing need to justify research programs. Within the USDA–Forest Service, a new research effort was initiated in 1980 to develop improved methods for evaluating forestry research (Lundgren 1983). The effort initiated a number of research studies aimed at better understanding the process and products of forestry research.

A recent review of the status of forestry research evaluation in the United States (Lundgren 1986) examined research studies covering a wide range of topics, including research organizations, administration, and management; the research process, planning and conducting research; evaluations of individual research projects and programs; and dissemination of research results and technology transfer. The review concluded that although considerable progress has been made in studying forestry research, knowledge of the forestry research process remains sketchy and incomplete. Especially needed is the development of evaluation methods applicable to research evaluation.

Conferences and workshops relating to the evaluation of forestry research were held in 1982 (Hyde 1983), in 1984 (Risbrudt and Jakes 1985), in 1985 (Burns 1986), and in 1986 (Risbrudt 1987). They are reflective of a growing interest in economic evaluations of forestry research. Such conferences pointed to major gaps in knowledge about forestry research, especially the lack of methods for and experience in evaluating research on nonmarket goods and services in forestry (e.g., recreation, watersheds, and wildlife) and the lack of a comprehensive understanding of forestry research systems.

Several approaches can be used to evaluate forestry research (Jakes and Leatherberry 1986). Research evaluation is not burdened by a shortage of generalized evaluation methods, but rather by the lack of evaluation framework appropriate for a particular research activity and an interested clientele. Gregersen and Lundgren (1986) suggest a framework that stresses the need to specify the clients for the evaluation and the evaluation objectives based on client needs; identify the evaluation measures and criteria; select an appropriate analytical model; conduct the evaluation; and interpret and present the results in a manner useful to the client.

Although a number of evaluations of forestry research has been carried out in recent years, most have focused on wood products and other market-oriented goods and services. For example, Bengston (1984) evaluated the economic impacts of structural particleboard research; Haygreen et al. (1986) evaluated the economic impacts of timber utilization research; Seldon (1985) evaluated softwood plywood research; and Westgate (1986) evaluated containerized forest tree seedling research. Bengston and Jakes (1987) summarized the returns to investment in forestry research reported by several recent studies. Internal rates of return ranged from 14 percent to 40 percent for wood products and engineering research, and from 9 percent to 111 percent for timber management research. However, many forestry outputs are nonmarket goods and services, which are difficult to quantify and value. Lacking are methods for evaluating research on nonmarket goods and services such as aesthetics, recreation, watershed management, and wildlife habitat. Unfortunately, research evaluations focused on agricultural products are of limited usefulness since most such evaluations have also concentrated on market-oriented products (Bengston 1985b).

Several issues regarding evaluation of forestry research are of particular importance to forest economics research:

- Contributions of forestry research to the forestry sector. Lacking is knowledge about the contributions of forestry research to the forestry sector, particularly nonmarket goods and services.
- Funding for forestry research. Lacking are effective methods for developing and justifying forestry research programs in terms of economic, social, and environmental costs and benefits.
- Managing forestry research. Lacking are effective methods for managing forestry research, for determining research priorities, for allocating resources among research programs, and for improving organizational effectiveness and efficiency.
- Relationships among research organizations. Lacking is an understanding of how forestry research affects and is affected by nonforestry research and foreign research, and how such relationships can be enhanced.
- Forestry research and international development. Lacking is an understanding of the forestry research needs in developing countries and the role U.S. forestry research programs and institutions should play in international development.

Recognizing the above issues, a number of specific research topics should be undertaken to enhance understanding of forestry research activities. The following should be seriously considered.

Document the contributions of forestry research to the forestry sector, especially research concerning nonmarket goods and services. Evaluations of forestry research contributions to productivity and economic growth should go beyond traditional areas of wood products and capital and labor productivity. Studies are needed to identify and evaluate the costs and benefits of various forestry research fields, especially outdoor recreation, watershed management, and silviculture. More intense investigation of research directed to such fields would provide useful information about research costs and benefits.

Develop improved methods for justifying forestry research programs in terms of economic, social, and environmental costs and benefits. Special attention should be given to justifying research on nonmarket goods and services. Studies should go beyond the more traditional financial analyses and introduce social and environmental costs and benefits. Improved evaluation methods are needed for all areas of forestry research.

Develop improved methods for managing forestry research. Research efforts are needed to develop improved methods for establishing research priorities, especially methods that incorporate multiple criteria for judging research concerns of various groups interested in forestry research. Research is also needed to develop improved methods for allocating funds, personnel, and other research resources among different areas of research. Studies of existing forestry research organizations are needed to identify potential barriers to improved performance. Such studies should involve other social science disciplines, including psychology and sociology.

Improve understanding of how forestry research affects and is affected by nonforestry research and foreign research. Studies are needed to document how forestry research and nonforestry research affect each other. Research is needed (micro and macro levels) to document the relationships between U.S. and foreign forestry research activities.

Determine forestry research needs in developing countries and assess roles to be played by the U.S. in meeting such research needs. Studies are needed to identify and set priorities for forestry research needs in developing countries, including identifying the type of research needed and the funding, personnel, training, and facilities required. Research is also needed to determine how forestry research and educational programs and institutions within the U.S. can best support development of forestry research capabilities within developing countries.

Develop a Better Understanding of Technology Development, Dissemination, Adoption, and Utilization

Although forest economists have focused considerable attention on evaluating forestry research in recent years, relatively little attention has been paid to the processes by which technological innovation takes place in a forestry setting. Technology development, adoption, and utilization has been studied in other fields (Gold 1977, 1981; Mansfield 1977); that which have been carried out in forestry is based upon assumptions often unsupported by empirical evidence. Lacking is information about how usable technologies are developed, how they are disseminated to potential users, and what factors influence the adoption and use of available technologies. The linkage between research and technological innovation in forestry is often ineffective (Comptroller General of the United States 1972). More important, lacking is an understanding of the innovation system, an understanding which is critical to identification of barriers to implementation of new technologies and the development of strategies to overcome such barriers.

Forestry and nonforestry research provide new knowledge that is helpful in the development of new technologies. However, formalized research programs are but one source of new knowledge. New technologies also arise from activities occurring outside the domain of formal research programs (Cox 1974). Such informal processes

Allen L. Lundgren

Table 7.1   Technology Flows for Company-Financed Research and Development Expenditures in the Lumber and Wood Products, Furniture, and Paper Mill Products Industries, 1974 ($-million)

Industry Group Originating R&D[a]	Industry Group Using R&D[a]				R&D in Final Consumption	Total R&D Financed
	1	2	3	4		
1	64.2	0.6	0.1	4.9	2.8	72.6
2	0.3	7.8	0.2	20.4	22.4	51.1
3	0.4	0.4	86.4	41.8	73.3	202.3
4	66.2	24.9	119.3	10,167.7	4,012.9	14,391.0
Total R&D used	131.1	33.7	206.0	10,234.8	4,111.4	14,717.0

[a] 1 = Lumber and wood products; 2 = Furniture; 3 = Paper mill products; 4 = Other industries.

Source: Based on data from F. M. Scherer, "Inter-Industry Technology Flows in the United States," *Research Policy*, 1982, 11:232-241.

are poorly documented; their role in the development of new technologies is deserving of much greater attention. Analyses of assigned patents is a step in the right direction (Margl and Ellefson 1987).

Many technological innovations in the forestry sector arise outside the forestry sector. Scherer (1982) found that only 49 percent of company-financed research and development used by the lumber and wood products industry originates within the industry; 51 percent comes from other industries (Table 7.1). In the furniture industry, only 23 percent of company-financed research and development used by the industry came from within the industry, and in the paper mill products industry, only 42 percent comes from within the industry. The importance of external sources of technology to various segments of the forestry sector in technological innovation deserves further study.

Evaluations of technological innovations in forestry should recognize the important role played by international transfer of technologies (Westgate 1986). A review of U.S. interaction with foreign research programs in forestry (Gregersen et al. 1988) concluded that foreign research results are increasingly being incorporated into U.S. technology and products, and that significant foreign research was associated with the development of structural particleboard and containers for tree seedlings (see also Bengston, Gregerson, Haygreen 1988). Styan (1980) points out the key role played by Scandinavia and Japan in the development of new technologies for the paper industry.

New or modified technologies that have been developed must be brought to the attention of intended users via a process of dissemination. Publications are a traditional means of disseminating knowledge. Knowledge about new products and technologies is also passed through news articles, advertisements, and salespersons. Several recent studies of diffusion of innovations in forestry have found that personal contacts with peers is a major source of information about new technologies and one of the most effective means of disseminating information (Muth and Hendee 1980, Roggenbuck and Watson 1980, Nicholls and Prey 1982, Straka, Anderson, and Bullard 1986). Despite a sizable literature on technology transfer in forestry (Bosman 1981, Hale and Rice 1981, Pugh-Roberts Associates, Inc. 1981, Hobbs et al. 1983, Moeller and Seal 1984, Hertel et al. 1985), however, the process by which new technologies are disseminated to potential forestry users is not well understood. Bengston (1985a) concluded that relatively few forestry-specific diffusion studies have been undertaken to date and that coverage has been limited to only a few types of innovations and geographic areas.

Technological innovations are a source of continual change that enable forestry operations and activities to be more effective and efficient in meeting new challenges and opportunities as conditions change (Anderson 1987). However, by all accounts, only a small portion of new technologies are ever adopted and utilized. Midgley (1977, p. 12) reports that only five out of 10 new products ever reach the market and that 92 out of 100 fail to survive in the market beyond one year. There are many potential barriers to successful adoption of new technology (Rogers 1983). For various reasons, organizations often operate below efficiency levels achievable if existing technologies were used (Leibenstein 1966). Reasons include new technologies not compatible with the technologies currently in operation; funds to support adoption of new technologies not available; management unaware of new technologies; employees lack special skills necessary to fully exploit new technologies. Such barriers do exist in the forestry sector (Lingwood 1979). Bengston (1985a) reviews the literature on the diffusion of innovations in forestry and forest products. Studies to identify existing and potential barriers could do much to improve the process of technological innovation in forestry.

A continuous process of modification and adjustment is often necessary before a technological innovation can be fully incorporated into an operating environment. Over a period of about fifteen years, successive versions of FORPLAN (computerized forest planning model used by the national forest system) (Field 1984) were developed in response to problems that arose when the USDA–Forest Service attempted to apply the model in practice (Jones 1987, Weisz 1987). Such adjustments in technology are common throughout the forestry sector.

A variety of additional research has documented the development, adoption, and utilization of new technologies in forestry (Pyne 1981, Gray 1982, Driver et al. 1987). Increasingly apparent are the many factors that influence the success of technological innovations (Cox 1974, Porterfield 1980, Moeller and Shafer 1981, Anderson and Morck 1986). Such factors are far from being fully understood. Research is needed to further identify barriers to the adoption of new technologies and to develop new ways of improving the effectiveness and efficiency of the innovation process.

Two major issues concerning technological innovation processes should be of particular interest to forestry economics research:

- Improve understanding of technological innovation in forestry. Lacking is a thorough understanding of the processes by which new technologies important to forestry are developed, disseminated, adopted, and utilized.
- Improve technological innovation processes in forestry. Lacking is identification of critical barriers that inhibit effective and efficient technological innovation in forestry and development of strategies to overcome such barriers.

Recognizing the above issues, a number of specific research topics should be addressed in order to enhance understanding of technological innovation processes important to forestry. The following are deserving of special attention.

Evaluate factors which influence technological development, dissemination, and adoption in forestry. Research is needed to identify and document processes and factors that influence the development of various types of specific technologies in forestry. The means by which new technologies in forestry are disseminated to potential users should be identified and assessed. Case studies of the extent and rate of adoption of specific technologies should be undertaken.

Evaluate means of improving the effectiveness and efficiency of technological innovation processes in forestry. Research is needed on various kinds of technological innovations in forestry, including innovations in forestry management, timber harvesting, recreation, and forest products industries. Such research should focus on identification of barriers that inhibit technological innovation in forestry and on evaluation of strategies to overcome such barriers.

Develop a Better Understanding of the Impacts of Technological Innovation

Once new technologies are adopted, substantial change can occur in the manner in which forestry-related goods and services are produced, distributed, and used. Such changes can have a direct impact on the forestry sector per se. However, the forestry sector is closely linked to broader social and environmental sectors; technology-induced changes in the forestry sector activities will ultimately have more widespread (direct or indirect) economic, social, and environmental consequences. Both immediate and far-reaching economic, social, and environmental costs and benefits resulting from technological innovations in forestry are in need of greater understanding.

Forestry sector changes induced by technological innovations may affect the kinds and amounts of inputs required for production, distribution, and utilization of particular goods and services. For example, the adoption of new machines may reduce the need for certain types of labor or may change the skill requirements of labor. Technological innovations may also affect the amount, type, timing, and distribution of goods and services produced. Such changes are likely to alter the effectiveness and efficiency of forestry operations and alter the manner in which economic evaluation of technological alternatives are carried out.

Technology does have direct impacts on the operation of forest products industries (Bureau of Labor Statistics 1986). Technology-induced changes in labor productivity have been researched by a number of forest economists (e.g., Ruttan and Callahan 1962, Kaiser 1971, Robinson 1975, Sandoe and Wayman 1977, Buongiorno and Gilless 1980, Stier 1980, Greber and White 1982, Bengston and Strees 1986, Borger and Buongiorno 1985). Most such research has been a one-time effort— the results are often incomplete and in some cases contradictory. Researchers have also carried out a large number of *ex ante* analyses of the anticipated costs and benefits from adopting potential or proposed technological alternatives. Several case studies have attempted to link specific technological innovations in forestry to changes supposedly brought about by their adoption (including changes in cost and revenue structures due to such innovations) (Valfer et al. 1981, Risbrudt and Kaiser 1982, Herrick 1982). In general, however, there has been relatively little long-term focused research on how technological innovations directly impact the effectiveness and efficiency of activities and operations in the forestry sector.

Technology-induced changes in the forestry sector can also result in broader social, economic, and environmental consequences. Such impacts may be immediate or delayed and may be direct or indirect. Impacts of this nature should be recognized by any assessment of benefits and costs associated with the adoption of new technologies. In this respect, income redistribution consequences should be given special attention (Bengston and Gregersen 1988). Comprehensive assessments of important technological innovations, similar to technology assessments recently focused on agriculture (Sundquist et al. 1982a, 1982b), are also lacking in forestry. Several issues regarding the consequences and impacts of technological innovations are of particular concern to forestry economics research:

- Effects of technological innovation on the forestry sector. Lacking is documentation and evaluation of the changes brought about by past technological innovations in the forestry sector, including changes in resource management and in the production, distribution, and utilization of goods and services. There is a need for improved methods of evaluating changes expected from the adoption of proposed technological innovations in the forestry sector, particularly for nonmarket goods and services.
- Effects of technological innovation on society. Lacking are evaluations of technological innovations in the forestry sector, evaluations which include not only the financial effects of technological innovation (e.g., rates of return), but other economic effects, including employment by skill class, income distribution regionally and among different income classes, and social effects (e.g., regional migration and health).
- Effects of technological innovations outside the forestry sector on the forestry sector. Lacking is evaluation of the influence of technological innovations originating outside the forestry sector on operations and activities within the forestry sector, especially evaluation of ways to take such influence into account when evaluating technological innovations in forestry.
- Effects of technological innovation on natural resources and the environment. Lacking are improved methods for evaluating technological innovations in terms of short- and long-term effects on natural resources and the environment.
- Technological innovation for sustainable development. Lacking are approaches for evaluating technological innovations in terms of potential effect on and contribution to long-term sustainable development, including a sustainable resource and environmental base and sustainable human institutions.

Recognizing the above issues, a number of specific research topics should be undertaken in order to enhance understanding of the consequences of technological innovation for forestry and broader social and natural systems. The following are deserving of attention.

Document and evaluate changes resulting from past technological innovations. Evaluate changes resulting from past technological innovations in the management of forest and related resources and in the production, distribution, and utilization of goods and services in the forestry sector (e.g., wood-based products, outdoor recreation, water). Research focused on specific technological innovations should seek to identify specific changes in operations and activities resulting from the adoption of the new technology and should determine the costs and benefits arising from such changes.

Develop improved methods for evaluating changes resulting from adoption of technological innovations, especially those relevant to nonmarket forest outputs. Research should be undertaken to test alternative technology evaluation methods, especially the potential applicability to forestry of techniques currently applied in nonforestry fields. Innovative approaches to evaluating technological innovation involving nonmarket goods and services should receive special attention. Studies should consider financial (e.g., rate of return) and related economic effects of technological innovation (e.g., employment by skill class, income distribution, regional migration, and health). Improved methods are needed in every major area of forestry to insure that evaluations are appropriate for different situations (e.g., forest and related resource management by various owner groups; type of good or service produced).

Develop improved methods for assessing the influence of technological innovations outside the forestry sector on forestry activities. Research is needed to identify and document the influence on forestry activities of technological innovations developed outside the forestry sector. Methods for accounting for such influence in technological innovation processes should also be developed.

Develop improved methods of evaluating the short- and long-term effects of technological innovations on natural resources and the environment. Research is needed to identify and document the effects over time of different kinds of past technological innovations, especially effects on forest and related resources and on the environment. Research is also needed to develop improved methods for evaluating such resource-environmental effects.

Develop methods for evaluating technological innovations in terms of effects on long-term sustainable development. Research is needed to identify and document factors that influence long-term sustainable development as related to forestry. Studies should also be undertaken to determine the role of technological innovation on human institutions and on activities important to forest resource-based sustainable development. Methods for evaluating the contributions of technological innovations to sustainable development in the forestry sector should be developed.

Summary and Conclusions

Growing in importance is the need to understand technological innovation as a means of improving the effectiveness and efficiency of the forestry sector. Despite significant advances in the study of forestry research and related aspects of technology, far from satisfactory is information that would enable improvements in the management of forestry research and enhancement of abilities to guide technological innovations. Especially important is research that would improve understanding of the role forestry research plays in technological innovation, the process by which technology is developed, disseminated, adopted, and utilized, and the impacts of technological innovation on the forestry sector and on more general economic and environmental systems. The products of such research would go far toward improving the forestry community's ability to develop technologies and guide them toward improvements in the efficiency of various forestry activities.

Literature Cited

Anderson, D. H. and V. L. Morck. 1986. Factors affecting information utilization and change: The case of recreation research and river management in the public sector. Journal of Technology Transfer 10(2):53-70.

Anderson, W. C. 1987. Technical changes that solved the southern pine lumber industry's small-log problem. Forest Products Journal 37(6): 41-45.

Babcock, H. M. 1974. Deciding on priority projects. Forest Products Journal 24(9):52-54.

Bauer, R. A. 1969. Second-order consequences: A methodological essay on the impact of technology. MIT Press. Cambridge, MA

Bengston, D. N. 1984. Economic impacts of structural particleboard research. Forest Science 30(3):685-697.

Bengston, D. N. 1985a. Diffusion of innovations in forestry and forest products: Review of the literature. In: Forestry research evaluation: Current programs, future directions by C. D. Risbrudt and P. J. Jakes (compilers). Gen. Tech. Report NC-104. North Central Forest Experiment Station. USDA Forest Service. St. Paul, MN.

Bengston, D. N. 1985b. Economic evaluation of agricultural research; An assessment. Evaluation Review 9(3):243-262.

Bengston, D. N. and H. M. Gregersen. 1988. Income redistribution impacts of technical change: A framework for assessment and empirical evidence. In: Management of technology I by T. M. Khalil, B. A. Bayraktar, and J. A. Edosomwan (eds). Interscience Enterprises Ltd. Geneva, Switzerland.

Bengston, D. N., H. M. Gregersen and J. Haygreen. 1988. Seesawing across the forty-ninth parallel: The international diffusion of a wood-based technology. Journal of Forest History 32(2):82-88.

Bengston, D. N., H. M. Gregersen, and A. L. Lundgren. 1987. Planning, management, and evaluation of forestry research: An annotated bibliography with special reference to developing countries. Staff Paper Series No. 59. Department of Forest Resources. University of Minnesota. St. Paul, MN.

Bengston, D. N. and P. J. Jakes. 1987. Economic evaluation of forestry research: An overview of recent efforts. In: Valuation of forestry research by C. D. Risbrudt (compiler). Proceedings of Working Party S4.05-05. XVIII World Congress. International Union of Forest Research Organizations. Ljubljana, Yugoslavia.

Bengston, D. N. and A. Strees. 1986. Intermediate inputs and the estimation of technical change: The lumber and wood products industry. Forest Science 32(4):1078-1085.

Bengston, D. N. et al. 1988. Forestry research capacity in the Asia-Pacific region. Environment and Policy Institute. East-West Center. Honolulu, HI.

Bethune, J. E. and J. L. Clutter. 1969. Allocating funds to timber management research. Forest Science Monograph no. 16.

Borger, Bruno de and J. Buongiorno. 1985. Productivity growth in the paper and paperboard industries: A variable cost function approach. Canadian Journal of Forestry Research 15:1013-1020.

Bosman, D. L. 1981. The management of technology transfer. Wood & Fiber 13(3):196-211.

Buckman, R. E., R. A. Skok, and J.D. Sullivan. 1982. 1980-1990 national program of research for forests and associated rangelands. Gen. Tech. Report WO-32. USDA Forest Service. Washington, DC.

Bullard, S. H. and T. J. Straka. 1986. Role of company sales in funding research and development by major U.S. paper companies. Forest Science 32(4):936-943.

Buongiorno, J. and J. K. Gilless. 1980. Effects of input costs, economies of scale, and technological change on international pulp and paper prices. Forest Science 26(2):261-275.

Bureau of Labor Statistics. 1986. Technology and its impact on labor in four industries: Lumber and wood products/footwear/hydraulic cement/wholesale trade. Bulletin 2263. U.S. Department of Labor. Washington, DC.

Burns, D. P. (compiler). 1986. Evaluation and planning of forestry research. Proceedings, IUFRO S6.06-S6.06.01. Gen. Tech. Report NE-GTR-111. Northeast Forest Experiment Station. USDA Forest Service. Broomall, PA.

Callaham, R. Z. (coordinator). 1981. Criteria for deciding about forestry research programs. Gen. Tech. Rep. WO-29. USDA Forest Service. Washington, DC.

Callaham, R. Z. 1985. Evaluating social benefits of forestry research programs. IEEE Transactions on Engineering Management. EM-32, No. 2. May 1985.

Clawson, M. (ed). 1977. Research in forest economics and policy. Research Paper R-3. Resources for the Future. Washington, DC.

Claxton, H. D. and G. Rensi. 1972. An analytical procedure to assist decision-making in a government research organization. Res. Paper PSW-80. Pacific Southwest Forest & Range Experiment Station. Berkeley, CA.

Comptroller General of the United States. 1972. The Forest Service needs to ensure that the best possible use is made of its research program findings. Report to the Congress. Jan. 6, 1972. Washington, DC.

Comptroller General of the United States. 1981. Congressional action needed to provide a better focus on water-related research activities. Report to the Congress. June 5, 1981. Washington, DC.

Cox, L. A. 1974. Transfer of science and technology in successful innovation. Forest Products Journal 24(9):44-48.

Driver, B. L. et al. 1987. The ROS planning system: evolution, basic concepts, and research needed. Leisure Sciences 9:201-212.

Duerr, W. A. and H. J. Vaux (eds.). 1953. Research in the economics of forestry. Charles Lathrop Pack Forestry Foundation. Washington, DC.

Fege, A. S. 1987. Evaluation of researchers' decisions in short-rotation forestry. Forest Science 33(1):30-42.

Field, R. C. 1984. National forest planning is promoting U.S. Forest Service acceptance of operations research. Interfaces 14:67-76.

Fox, G. 1986. A framework for identifying public research priorities: An application in forestry research. Gen. Tech. Report NC-109. North Central Forest Experiment Station. St. Paul, MN.

Gold, B. (ed.). 1977. Research, technological change, and economic analysis. Lexington Books, D.C. Heath & Co. Lexington, MA.

Gold, B. 1981. Technological diffusion in industry: Research needs and shortcomings. Journal of Industrial Economics 29(3):247-269.

Gray, G. C. 1982. Radio for the fireline; a history of electronic communication in the Forest Service, 1905-1975. FS-369. USDA Forest Service. Washington, DC.

Greber, B. J. and D.E. White. 1982. Technical change and productivity growth in the lumber and wood products industry. Forest Science 28(1):135-147.

Gregersen, H. and A. Lundgren. 1986. An evaluation framework. In: Alternative approaches to forestry research evaluation: An assessment by P. J. Jakes and E. C. Leatherberry (compilers). Gen. Tech. Report NC-110. North Central Forest Experiment Station. USDA Forest Service. St. Paul, MN.

Gregersen, H., J. Haygreen, S. Sindelar, P. Jakes. 1988. U.S. gains from forestry research. Unpublished manuscript. Department of Forest Resources. University of Minnesota. St. Paul, MN.

Hale, R. A. and W. W. Rice. 1981. Technology transfer is successfully practiced by U.S. dry kiln clubs. Forest Products Journal 31(9):14-16.

Haygreen, J., H. Gregersen, I. Holland, and R. Stone. 1986. The economic impact of timber utilization research. Forest Products Journal 36(2):12-20.

Herrick, O. W. 1982. Estimating benefits from whole-tree chipping as a logging innovation in northern U.S. forests. Forest Products Journal 32(11/12):57-60.

Hertel, G. D., S. J. Branham, and K. M. Swain, Sr. (eds). 1985. Technology transfer in integrated forest pest management in the South. Gen. Tech. Report SE-34. Southeastern Forest Experiment Station. USDA Forest Service. Asheville, NC.

Hobbs, S. D., J.C. Gordon, and G. W. Brown. 1983. Research and technology transfer in Southwest Oregon. Journal of Forestry 81(8):534-536.

Hyde, W. F. (ed). 1983. Economic evaluation of investments in forestry research. The Acorn Press. Durham, NC.

Jakes, P. J. and E. C. Leatherberry (compilers). 1986. Alternative approaches to forestry research evaluation: An assessment. Gen. Tech. Report NC-110. North Central Forest Experiment Station. USDA Forest Service. St. Paul, MN.

Jakes, P. J. and A. S. Van Dyne. 1987. Forestry literature: Who's publishing what where? Journal of Forestry 85(9):33-36.

Jones, D. B. 1987. Technical criteria used in the development and implementation of FORPLAN, Version 1. In: FORPLAN: An evaluation of a forest planning tool by T. W. Hoekstra, A. A. Dyer, and D. C. LeMaster (eds). Gen. Tech. Report RM-140. Rocky Mountain Forest and Range Experiment Station. USDA Forest Service. Fort Collins, CO.

Kaiser, H. F. 1971. Productivity gains in forest products industries. Forest Products Journal 21(5):14-16.

Kaufert, F. H. and W. H. Cummings. 1955. Forestry and related research in North America. Society of American Foresters. Washington, DC.

Krugman, S. L. and E. B. Cowling. 1982. Our natural resources: Basic research needs in forestry and renewable natural resources. Report of the National Task Force on Basic Research in Forestry and Renewable Natural Resources. Forest, Wildlife and Range Experiment Station. University of Idaho. Moscow, ID.

Leibenstein, H. 1966. Allocative efficiency vs. "X-efficiency." American Economic Review 56(3):392-415.

Lingwood, D. A. 1979. Producing usable research: The first step in dissemination. American Behavioral Scientist 22(3):339-362.

Lingwood, D. A. and W. C. Morris. 1976. Research into use: A study of the Forest Service research branch. Institute for Social Research, Center for Research on Utilization of Scientific Knowledge. University of Michigan. Ann Arbor, MI.

Lundgren, A. L. 1982. Research productivity from an economic viewpoint. In: Increasing forest productivity. Pub. 82-01. Proceedings of 1981 Convention. Society of American Foresters. Bethesda, MD.

Lundgren, A. L. 1983. Methods for evaluating forestry research: A prospectus. In: Economic evaluation of investments in forestry research by W. F. Hyde (ed.). The Acorn Press. Durham, NC.

Lundgren, A. L. 1986. A brief history of forestry research evaluation in the United States. Evaluation and Planning of Forestry Research, Proceedings by D. P. Burns (compiler). Gen. Tech. Report NE-GTR-111. Northeastern Forest Experiment Station. USDA Forest Service. Broomall, PA.

Mansfield, E. et al. 1977. The production and application of new industrial technology. W. W. Norton & Co. New York, NY.

Margl, R. A. and P. V. Ellefson. 1987. Assigned patents: Technology trends in 15 U.S. wood-based companies. Forest Products Journal 37(1):47-50.

Midgley, D. F. 1977. Innovation and new product marketing. John Wiley & Sons. New York, NY.

Moeller, G. H. and D. T. Seal (eds.). 1984. Technology transfer in forestry. Forestry Commission Bulletin 61. Proceedings of a IUFRO Conference. Her Majesty's Stationery Office. London, UK.

Moeller, G. H. and E. L. Shafer. 1981. Important factors in the forestry innovation process. Journal of Forestry 79(1):30-32.

Muth, R. M. and J.C. Hendee. 1980. Technology transfer and human behavior. Journal of Forestry 78(3):141-144.

Nicholls, T. H. and A. J. Prey. 1982. Providing information: Researchers to practitioners. In: Urban and suburban trees: Pest problems, needs, prospects, and solutions by B. O. Parks et al. (eds). Proceedings of a conference. Departments of Resource Development and Entomology. Michigan State University. E. Lansing, MI.

Porterfield, R. L. 1980. Successful application of research findings in managing Champion timberlands. Forest Products Journal 30(10):16-20.

Pugh-Roberts Associates, Inc. 1981. A dynamic model of technology transfer in a national forest. Pinchot Institute for Conservation Studies. USDA Forest Service. Milford, PA.

Pugh-Roberts Associates, Inc. 1981. Technology transfer within the national forest system. Conclusions and policy recommendations from a dynamic modeling system. Pinchot Institute for Conservation Studies. USDA Forest Service. Milford, PA.

Pyne, S. J. 1981. Fire policy and fire research in the U.S. Forest Service. Journal of Forest History 25(2):64-77.

Renewable Natural Resources Foundation. 1977. A review of forest and rangeland research policies in the United States. Bethesda, MD.

Risbrudt, C. (compiler). 1987. Valuation of forestry research. Proceedings of Working Party S4.05-05. XVIII World Congress. International Union of Forestry Research Organizations. Ljubljana, Yugoslavia.

Risbrudt, C. D. and P. J. Jakes (compilers). 1985. Forestry research evaluation: Current programs, future directions. Gen. Tech. Report NC-104. North Central Forest Experiment Station. USDA Forest Service. St. Paul, MN.

Risbrudt, C. D. and H. F. Kaiser. 1982. Economic analysis of the sawmill improvement program. Forest Products Journal 32(8):25-28.

Robinson, V. L. 1975. An estimate of technological progress in the lumber and wood-products industry. Forest Science 21(2):149-154.

Rogers, E. M. 1983. Diffusion of innovations. 3rd ed. The Free Press. New York, NY.

Roggenbuck, J. W. and A. E. Watson. 1980. Technology transfer: Lateral diffusion of innovation in forest recreation planning and management. Final report for the USDA Forest Service's Southeastern Forest Experiment Station and the Pinchot Institute for Conservation Studies. Department of Forestry. Virginia Polytechnic Institute and State University. Blacksburg, VA.

Rose, D. W. 1983. Benefit-cost evaluation of the Douglas-fir Tussock moth research and development program. Journal of Forestry 81(4):228-231.

Ruttan, V. W. and J. C. Callahan. 1962. Resource inputs and output growth: Comparisons between agriculture and forestry. Forest Science 8(1):68-82.

Sandoe, M. and M. Wayman. 1977. Productivity of capital and labor in the Canadian forest products industry, 1965 to 1972. Canadian Journal of Forest Research 7(1):85-93.

Scherer, F. M. 1982. Inter-industry technology flows in the United States. Research Policy 11:227-245.

Schreyer, R. M. 1974. Individual orientation and research factors influencing scientific effectiveness in the U.S. Forest Service. PhD thesis. University of Michigan. Ann Arbor, MI.

Seldon, B. J. 1985. A nonresidual estimation of welfare gains from public investment in softwood plywood research. Faculty Working Paper, Ohio Economics Studies 85-16. Department of Economics. Ohio State University. Athens, OH.

Stier, J. C. 1980. Estimating the production technology in the U.S. forest products industries. Forest Science 26(3):471-482.

Straka, T. J., W. C. Anderson, and S. H. Bullard. 1986. An economic appraisal of service forester activities in Mississippi. Tech. Bulletin 137. Agriculture and Forestry Experiment Station. Mississippi State University. Mississippi State, MS.

Styan, G.E. 1980. Impact of North America timber supply on innovations in paper technology. Paper Trade Journal 164(10):25-29.

Sundquist, W. B., K. M. Menz, and C. F. Neumeyer. 1982a. A technology assessment of commercial corn production in the United States. Bulletin 546. Agricultural Experiment Station. University of Minnesota. St. Paul, MN.

Sundquist, W. B., K. M. Menz, and C. F. Neumeyer. 1982b. Technology assessment as a framework of analysis for agricultural production technologies. Staff Paper Series P82-5. Department of Agricultural and Applied Economics. University of Minnesota. St. Paul, MN.

Valfer, E. S., M. W. Kirby, and G. Schwarzbart. 1981. Returns on investments in management sciences: Six case studies. Gen. Tech. Report PSW-52. Pacific Southwest Forest & Range Experiment Station. USDA Forest Service. Berkeley, CA.

Weisz, R. 1987. Technical criteria used in the development and implementation of Version 2 of FORPLAN. In: FORPLAN: An evaluation of a forest planning tool by T. W. Hoekstra, A. A. Dyer, and D. C. LeMaster (eds.). Gen. Tech. Report RM-140. Rocky Mountain Forest & Range Experiment Station. USDA Forest Service. Ft. Collins, CO.

Westgate, R. A. 1986. The economics of containerized forest tree seedling research in the United States. Canadian Journal of Forestry Research 16(5):1007-1012.

Whaley, R. S. and E. F. Bell. 1982. Health of Forest Service economics research. Journal of Forestry 80(6):347-349, 364.

8

FORECASTING DEMAND AND SUPPLY OF FOREST RESOURCES, PRODUCTS, AND SERVICES

Clark Row

Forecasting is an especially important activity of the forest economics community. The outlook for supply, demand, and associated levels of consumption and price are frequently critical inputs to the design of public and private forest policies and programs. Since few forecasts of demand or supply are made for their own sake, the scope of forecasting research necessarily includes the application of demand and supply relationships in market and forest sector models, long-range forest planning, valuation of resources and outputs, and development of market and investment strategies. Models capable of forecasting demand and supply (called sector models or equilibrium market models) have made considerable progress in the last few decades. Such success is evident in the conceptual development of models, construction and refinement of models for a variety of uses, and in the impact such models have had on resource assessments, policy studies, and program and investment planning, both by public agencies and industrial organizations. These accomplishments have been international in scope, with substantial coop-eration occurring between individual economists and organizations in Canada, Europe, the U.S., and other developed countries (Kallio, Dykstra, and Binkley 1987).

Demand and supply relationships and the models that use them are by necessity only abstractions and simplifications of markets and industries operating in an incredibly complex and changing world. The application of models to real world situations frequently surfaces conditions which require models to be corrected, refined, or elaborated upon. Problems fostering such adjustments may stem from inadequate or misleading data, errors in specification of relationships, failure to incorporate an interrelationship, or conceptually inadequate approaches. A number of currently operational models have gone through repeated cycles of use, analysis of problems uncovered, further development and testing, and further use.

Demand and supply are themselves broad areas of interest. In recent times, their study has become dependent on the use of sophisticated statistical techniques,

124

especially econometric techniques. In addition, forest sector and forest planning models use a wide range of mathematical economics and operation research techniques. Of special concern to forest economics research is the adaptation of such disciplines to forestry situations, not the emphasis on research needs in these fields per se.

Research involving forecasts of demand and supply has emphasized timber as a forest output. This emphasis has occurred for a number of reasons: timber often has a dominant role as measured by program size and gross financial returns; timber most always is a private good with well established markets and values; and timber is processed in well-defined stages prior to use by consumers. Futhermore, statistics on production, consumption, and prices of timber and timber products are usually—but not always—readily available. Circumstances such as these are immense advantages to the application of economic analyses, advantages not shared by other forest outputs, which even includes other private goods such as range and water.

Context of Demand and Supply Research

Relatively simple projections of timber requirements and harvests were the mainstay of timber assessments from the 1930s and 1940s until the 1980 RPA assessment. Demands for forest products were projected using "requirement" factors based on historical use per unit of activity in end-use sectors (e.g., housing, manufacturing). Projected levels of activity in end-uses were multiplied by requirement factors (usually modified by trends in technologies). Supplies were the upper estimates of feasible amounts of resource outputs likely to be forthcoming (based on historic harvest levels and estimates of changes in forest area). Period-to-period projections were made with the Timber Resource Assessment System (TRAS) stand table system (treating all timber in a region and ownership class as one stand). TRAS's virtues were not conceptual, they were pragmatic—it gave generally credible projections. The gap by which future demands almost always exceeded supplies indicated potential shortages. Such prospective shortages or resource scarcities were viewed as potential causes of retarded economic growth, increased consumer prices for wood products, and ultimately, a decline in living standards. In response, recommendations were made to conserve timber, promote forest management, reduce waste and residues, and improve utilization. Unfortunately, such recommendations were based on analyses that seldom recognized the role of price in dampening demand and increasing supplies (both directly and indirectly and in the short- and long-term); the numerous and complex interrelationships between markets for competing products, stages of processing, and production regions; and of the role of prices and profits in stimulating investments in processing capacity and resource management.

Research involving demand and supply forecasting has advanced considerably within the past 10 to 20 years (Clawson 1977). Current research almost universally involves price-quantity relationships derived from econometric or similar analyses. Current prices (and often lagged and expected prices as well) are major if not dominant variables. Contemporary demand and supply analyses are typically parts of larger modeling or planning studies which employ (directly or indirectly) demand and supply relationships to forecast future consumption and prices. Such research may have one or the other of the following purposes:

- To guide the strategy of producers and consumers of forest products; and especially to determine whether investments in forest management or forest products processing are justified.
- To assist in improving forest and other policies to solve resource, industrial, trade, and other economic problems. And more specifically, to estimate responses to changes in policies or programs in terms of consumption, prices, and investment in improvements in resources or processing capacity, often by regions or countries.

Recent demand-supply models are strongly dependent on assumptions of a free-market and price- and profit-driven economy—reflective of markets as described in elementary economics texts. Though few markets function as effectively as described by economic theory, these assumptions are useful approximations:

- Economic activities are undertaken if they are profitable (i.e., revenues or benefits exceed all costs). However, what is profitable to one firm or agency may not be profitable to another because of different desired rates of return, time horizons, or complementary activities. The degree to which landowners, loggers, or processors are "economically rational" varies.
- Prices are flexible and are a major (though not exclusive) means of allocating resources and distributing products. This allocation occurs primarily by pricing mechanisms which adjust the amounts bought and sold by purchasers and producers. Such a process clears the market; it can be represented by Marshallian "scissor" diagrams. The actual role of price is often complex; its role in clearing markets may be overestimated.
- Though price and consumption time series (representing successive equilibria of market demand and supply adjustments) are observable data, demand and supply relationships must be inferred by econometric or similar analyses. In short- or market-term periods, gaps between demand and supply (or consumption and production) result in a net change in inventories which in turn affects current prices. In medium- and long-term analyses, inventory changes are assumed to cancel out and thus are ignored. Price and consumption data may include large errors from measurement problems or may be influenced by unpredictable factors. Yet to be intensively studied is how business cycles and short-term fluctuations from various causes (adjusted by changes in output volume and inventories) become long-term trends.
- Nonprice influences shift demand and supply functions left and right. They occur continually and (over time) may change the shape of price-response relationships. Even if no short-term fluctuation occurred, precise equilibriums would rarely be obtained even in the longterm. Market systems, however, would continually move toward equilibrium because of changes in demand and supply functions.

Over time, the basic outlook of demand and supply research has experienced subtle but basic changes—from determining whether society is running out of resources to determining how markets for forest outputs operate; how trade affects forestry sectors; how public and private management can be improved (especially to increase multiple-use outputs and environmental protection); and whether specific resource policies and programs should be changed. Many older concerns (and older terms) have either disappeared from the literature or have been transformed into different concepts. Such terms include "shortages," "scarcity," "requirements," "defense" or "national needs," and "resource self sufficiency."

Increasingly complex approaches have been applied to the forecasting of demand and supply. While simple analyses are useful and often essential to acquaint analysts with problems and to guide research designs, experience has shown that simplistic approaches to complex problems have limited usefulness or credibility. However, trend analysis can isolate seasonal business cycles, and other systematic sources of fluctuations. (Few analyses other than those of business-oriented consulting groups use quarterly or shorter time-period data.)

In general, trend analysis of time series using the Box-Jenkins or other more modern statistical methods has found relatively little application in forest-related analyses. (Some disagree [Buongiorno 1987].) In addition, simple single-equation analyses of consumption trends may produce either a demand equation or a supply equation, or some mixture of both. Thus, even rudimentary analyses of supply and demand often become quite statistically complex. Single-equation regression analyses of consumption, however, have been useful for worldwide analyses of forest product consumption and resource adequacy (still using forms of the "gap" methodology). Such analyses generally use cross-section (or time-series/cross-section) data to develop the influence of local prices and average per capita income on per capita product consumption (Martin 1987, ECE/FAO 1985, Buongiorno 1978, Baudin and Lundberg 1987). Used in forecasting, such relationships assume that as per capita income of a country increases, its use of forest products will resemble that of countries having high incomes. Such analyses may unjustifiably assume away the identification problem—whether the relationship is one of demand or of some mixture of demand and supply.

The prevalent mode of analyzing demand and supply is to construct a market equilibrium model. Usual characteristics of such models are

- Inclusion of several interrelated markets and economic industries; each product's market usually associated with the industry or industries that produces it. The relation between markets or sectors may be: (a) products of one sector are further processed by another sector (forestry-related models frequently extend from the forest to consumption of primary processed products); (b) the sectors are competitive (or complementary) in products or raw materials; or (c) markets or sectors/industries may be separate spatially and differ in availability of raw materials or markets, transportation costs, tariff or nontariff barriers, or regional economic conditions that affect demand or processing costs.
- Demand relationships for processed products are often econometric equations, which for raw materials are derived demands estimated by technical functions

or econometric analysis. Export (part of demand) and import (part of supply) markets may be modeled internally as separate demand and supply areas or may be specified by exogenous factors or relationships.

- Supply is usually defined as a production system (inputs ——————→ (process)—————→ outputs), specified by a set of technologically based coefficients or by econometric relationships involving input requirements and prices as variables.
- A method for determining each period's equilibrium is used (e.g., reactive programming, linear programming, optimal control theory, or general equilibrium modeling system). A frequent rationale is to maximize consumer and producer surpluses. After each period, the exogenous demand factors are updated, the resource changes are simulated, and the processing capacities are adjusted. The system is then solved for a new set of equilibriums.

Multi-sector market equilibrium models have many advantages as vehicles for incorporating demand and supply relationships. They allow more realistically for interaction of demand and supply functions, can incorporate various types of relationships and structures for demand and particularly supply, and most important, can permit use of models for tasks other than forecasting (e.g., estimating responses to public policies, technical innovation, or major foreseen industrial developments).

Recent Development of Equilibrium Market Models

In the last decade, equilibrium market or sector models incorporating demand and supply relationships have achieved both substantial sophistication and practical usefulness. Prominent examples include the following.

Timber Assessment Market Model (TAMM). Based on a variety of previous models, the TAMM model developed into a model which played a major role in improving the 1980 and 1985 USDA–Forest Service timber assessment (Adams and Haynes 1980, Haynes and Adams 1985). TAMM has a number of important features, including coverage of U.S. softwood resources and industries, regional coverage of the nation (divides continental U.S. into six demand and eight supply regions), and is solved for equilibria by an adaption of reactive programming. For purposes of preparing the 1990 U.S. forest and range assessment and for evaluating the South's fourth forest, the following improvements were made in TAMM:

- Replacement of "requirement" type demand functions with price-sensitive econometric equations, and development of price-sensitive demand relationships.
- Replacement of aggregate stand-table resource projection routines with the Timber Resource Investment Model (TRIM) simulation models (using age, site, stocking density, management classes).
- Splitting of stumpage supply/demand markets into sawtimber and pulpwood markets in all regions, and improvements in handling feedback of residues from one industry as raw materials for others.

• Development of improved estimates of supply capacity change for processed products, and inclusion of better estimates of improved technology and utilization on supply functions or modules.
• Inclusion of North American pulp and paper industries as sectors, as well as disaggregation of Canadian and other international trade.

PAPYRUS and other world paper models. Several generations of models have developed increasingly useful representations of the multinational pulp and paper sector (Buongiorno and Gilless 1987; Lange et al. 1987). The most recent model describes markets for 12 commodities, with 13 demand regions and 11 supply regions (all but three in North America). Solved as mathematical programming problems, the models use econometric demand functions of paper by grade and country. Domestic supply is represented by technical input-output representation of pulp and paper technologies and foreign supplies by econometric equations. Supply is limited by regional capacities which change according to past capacity utilization and industrial profitability (measured by the shadow price for an extra unit of capacity). Use of recycled paper is based on price, within a constraint based on a fraction of the previous year's consumption.

Global Trade Model (GTM). A very ambitious worldwide forest sector model, the Global Trade Model was developed by a team of forest economists involved in a five-year project located at the International Institute for Applied Systems Analysis (IIASA) in Laxenburg, Austria. IIASA is a highly respected research organization funded by the academies of science of some 15 countries, including the U.S. and the USSR (Kallio, Dykstra and Binkley 1987). The large and complex GTM has worldwide coverage, representing markets for 16 major softwood and hardwood products in 18 demand and supply regions of the world. The GTM incorporates advanced modeling concepts from many forest sector models around the world (including TAMM and PAPYRUS). The model is "solved" by nonlinear optimization, using numerous assumptions of rates of economic growth in various world regions. The GTM required, and the IIASA project developed, a very large (and problem-rife) data base covering world forest resources, demand factors, industries, trade flows, and transportation costs. Demand relations are based on historical demand moved forward by elasticities for per-capita income change (an exogenous variable) and price (generated endogenously). Supply is represented by industry sector supply functions that incorporate input-output technology typical in each country. New technology is introduced by having three levels of technology in each industry, with the most modern capacity replacing year-by-year existing capacity and the least modern capacity being phased out. Trade flows are based partly on price and transportation costs and partly on continuation of existing trade patterns.

Models by international organizations. Both the United Nations Food and Agriculture Organization (Rome, Italy) and regional United Nation's commissions make assessments of timber situations. A recent assessment is the European Timber Trends and Prospects study (ETTS-IV) (ECE/FAO 1985). Each ETTS has forecast the timber and forest products situation for Europe and has been used extensively for policy development. ETTS-IV is not a formal equilibrium market model, though it uses demand and supply relationships to forecast many decades into the future (Prins 1987). In ETTS-IV, econometric demand functions are based on price and international cross-section income elasticities for paper and paperboard and on end-use elasticities for lumber and panel products (similar elasticities were also used by IIASA's GTM model). Projected timber supplies are each country's estimates of timber supply by species group and size based on analyses of forest area, inventories, growth, current harvests and removals, and processing capacity. These supplies were matched with computed estimates of the supply necessary to meet anticipated end-use consumption of various wood products. The gaps were resolved by expert judgement.

Models of other developed countries. Eminent market model development has been accomplished in Europe, Canada, and in other developed countries. Some example models are

- Swedish national models (two) have unique ways of handling innovation, new technology implementation, and marketing (especially to major foreign markets in western Europe) (Nilsson 1983, Lonnsted 1986).
- Finnish models, including a systems dynamics model, a dynamic linear programming model, and a Finnish application of the IIASA GTM model, have successively been used to study forest policies in Finland (Seppala and Seppala 1987).
- Central European model has been fashioned as an application of the IIASA model to a relatively small region (Kornai and Schwarzbauer 1987).
- Canadian models, developed by a group headquartered at the University of British Columbia, have emphasized techniques for addressing problems of development of logging and processing at the extensive margins of Canadian timber resources.
- Japanese model features complex econometric demand and supply functions, including relationships among domestic industries, imports, and exports within a single market (Nomura 1983).

Models by private consulting organizations. Several industrial forestry consulting organizations have developed supply-demand models. Examples include Resource Information Systems, Inc. (formerly a part of Data Resources, Inc.) of the U.S. and Jaakko Poyry International of Finland (Young 1987, Veltkamp et al. 1983, Uutela 1983). The analysis systems developed by these consulting organizations generally emphasize medium-term projections of great usefulness to industrial clients.

Additional supply-demand models. A number of other individuals and organizations are actively involved in the development of supply-demand models. Among them are:

- World Assessment Market Model (WAMM) with structure and solution algorithm similar to TAMM (Brooks 1987). WAMM provides forest product trade projections for the USDA–Forest Service's 1990 timber assessment study. The model emphasizes Pacific Rim trade.

- Timber Supply Model (TSM) (developed by Resources for the Future, Inc.) is designed to examine the adequacy of world long-run timber supplies (Sedjo and Lyons 1987, 1988). The TSM focuses on the development processes for three types of forest areas—old growth, second growth and managed forest, and exotic plantations (e.g., Brazil, Chile, New Zealand, South Africa, and Spain). The model uses a discrete time optimal technique to forecast harvests, inventory change, and management inputs in seven major timber regions.

Although substantial progress has been made in forecasting demand and supply of forest resources and forest products, the forestry community's ability to foresee future supply and demand conditions remains clouded. Such has serious implications for the development of a variety of public and private policies and programs. In order to improve ability to forecast demand and supply situations, research should be focused in the following strategic directions.

- Develop improved means of forecasting demand and supply of timber and timber products.
- Acquire improved understanding of the influence of technological changes and competitive materials on forecasts of timber demand and supply.
- Develop improved methods of forecasting economic and physical supplies of timber.
- Acquire improved understanding of transportation and related trading costs on prospective demand and supply of timber products.
- Develop better demand and supply forecasts for geographically small areas.
- Develop improved means of forecasting demand and supply of nontimber forest products.
- Develop effective means of applying supply and demand forecasts to important forest policy issues.

Develop Improved Means of Forecasting
Demand and Supply of Timber and Timber Products

Despite the enormous outpouring of demand-supply and forest sector models in recent years, considerable research is needed to solve a variety of problems. Of special concern is further efforts to develop more effective techniques and modeling methods. Among topics in need of attention are the following.

Evaluate market efficiency assumptions relevant to the forecasting of demand and supply. Almost all demand and supply models use price-quantity relationships to equilibrate divergent trends in supply and demand. However, the assumption of relatively high market efficiency (or absence of market failures) may not be valid to the extent required by some modeling techniques. Research should be focused on the following concerns:

- Does inadequate price and related information result in lagged or incomplete market adjustments? Do other institutional factors and subsidies act in a similar manner?

- Can models be developed to accommodate different profit, time horizon, or other guidelines among and within landowner classes? Largely unknown is how such factors vary within major classes of landowners or industries. The significance of this unknown to model development has not been fully explored.
- How can irrationality be introduced to models of market systems? Is it necessary that landowners, loggers and processors always be assumed to make decisions that are rational according to established economic criteria?
- Can long-term models be better adapted to short- or market-term periods, with gaps between demand and supply (or consumption and production) resulting in net change in inventories, and in turn affecting current prices? How do business cycle and short-term fluctuations from various causes become long-term trends through adjustments in output volumes?

Evaluate means of improving the supply of available data required for model development and forecasting. Research is needed to determine more effective means of developing, collecting, and adjusting basic statistical information concerning many aspects of demand and supply. Refinements in concepts and theory are often difficult to quantify because data are lacking or inaccurate. For example, analysts desire but lack access to greater product detail (Adams 1987). Similarly, decisions on geographic regions are often constrained by the availability of regional information (Brooks 1987). In constructing the IIASA GTM model, major inconsistencies in trade data had to be corrected (or at least made consistent) before analyses could be performed (Kornai 1985).

The primary statistical sources for model building and forecasting are scattered— a situation which further compounds the data availability problem. For example, physical resource statistics are provided by the USDA–Forest Service's forest survey; production and forest industry data are issued by trade associations; housing and construction statistics are provided by USDC–Census Bureau; stumpage prices are gathered and published by some states and private services; and forest product prices are provided by USDL–Bureau of Labor Statistics. Problems with sampling, coverage detail, and promptness often limit the usefulness of such statistical bases. Compounding the problem, the efforts of nearly all statistical gathering programs have declined via either direct reductions in budgets or by investments levels that have failed to keep pace with inflation.

Generally lacking (only partially covered—or infrequently collected or not collected at all), has been statistical information on activities such as land use changes, forest management expenditures and programs, harvesting and processing costs, utilization of forest products by industry, and regional or local stumpage prices. Forestry or land management activities are not directly covered by a nationwide economic census such as the Census of Agriculture or the Census of Manufacturers. Stumpage prices are particularly inadequate. A growing number of states publish prices collected by various means. Unfortunately, the data are often statistically questionable. USDA–Forest Service sale bid data are often of doubtful usefulness. In the South, Timber-Mart South summarizes data provided by private consultants. The system has not proven economical elsewhere.

Areas of data availability that are particularly in need of research include:

- How can information on forest management activities and expenditures be more systematically and accurately obtained?
- How can data on use of forest products by end use be obtained with or without expensive detailed surveys?
- Can stumpage price information be gathered and analyzed through cooperative programs of government agencies and other sellers of timber (as in Timber-Mart South)?
- How can different types of partial information be reconciled, and irregularities be identified and corrected?

Assess means of developing and validating structural models. Numerous questions arise during the course of developing models; answers are often based on judgments involving limited conceptual foundations. Furthermore, experience with promising solutions used by other models has been limited. An overall summary of existing models would be useful. Among topics in need of research are:

- For regions of varying size and type, what scope of model or analysis should be attempted? How should solid wood products and pulp and paper be integrated, or at least related? Can timber supply modeling incorporate multiple-use outputs and environmental concerns?
- What guidelines for disaggregation of geographic areas are appropriate? What criteria should guide analysis of areas not requiring great detail (e.g., softwoods in the Northeast and in the Lake States)?
- How can particular problems of exports and imports (e.g., tariffs, exchange rates, non-tariff barriers) be incorporated into models?
- Despite numerous methods for dynamically projecting models, little consensus exists on what methods best fit different circumstances. On what criteria should such decisions be made? Are there improved methods used in other fields that could be adapted?
- How can models best be used to indicate the impact of various projected scenarios on net social benefits and the distribution of such benefits by classes of people or economic interests?
- Rigorous methods of model validation are often not feasible and validation is often ad hoc and judgmental (McKillop 1987). Often the acceptance of model results is basically a matter of credibility of projected scenarios. Can better and more useful, if not more rigorous, methods of validating models be developed?

Evaluate means of determining prospective demands for processed forest products. Most models represent demand with sophisticated, highly complex econometric equations. Such representations are not without problems. Research topics concerning future demand for processed forest products include

- Would measuring the effects of prices of competitive products improve model response to changed economic and technological conditions? Measuring the effects of other product prices (cross-elasticities) is often made difficult by inadequate price statistics and by correlation of the substitute's price with the product price.

- How can the influence of nonprice factors be better recognized in demand forecasting (e.g., styles of homes and furniture, consumer-preferred species of wood)?
- Can better projections be developed for activity in the end-uses of wood such as home building and furniture manufacture? Can they be based on improved information such as demographic trends, housing inventory and disposition, and consumer ability to pay? (The once-active USDA–Forest Service modeling of long-term housing demand has been discontinued.)
- Should the continual updating of demand functions include longer and longer time series? Long series cover greater diversity of the conditions, but some data in long series may have little relevant information (e.g., information early in the series becomes so outdated that it has little content). Similarly, how can the specification of lags in price, income, and other relationships be improved and the long-term effects of shocks be better recognized?
- What are the effects of changing tariffs and increasingly volatile exchange rates on foreign demand for domestic products and U.S. demand for foreign products? Analysis of these factors presents unusual problems (Ckou and Buongiorno 1983, Adams et al. 1986).
- How can nonindustrial demands for timber be more accurately estimated and analyzed (e.g., firewood) (Hardie and Hassan 1986, Laarman 1984)?

Evaluate opportunities for applying futures and group decisionmaking techniques to problems of forecasting supply and demand. Forecasting the supply and demand of forest products and resources is a complicated science. No explicit forecasting technique embodies all that is known about a phenomenon under question. Consideration of factors beyond the model are often of utmost importance to the making of a plausible forecast. As such, the often unarticulated wisdom of experience may have forecasting virtues that are beyond those exhibited by an explicitly defined mathematical model of a market system. Not uncommon is the adjustment of the latter's results in light of what modelers and decision makers view as a more plausible forecast. All such adjustments go under the heading of judgmental. The research question is whether fully or partially judgmental forecasting improves accuracy by bringing expert opinion and intuition into the forecasting process—or whether it detracts from accuracy by introducing the errors of unexamined, perhaps incoherent, intuition. Needed is more careful evaluation of a variety of judgmental techniques as applied to the forecasting of forest product and forest resource conditions. Techniques in need of special attention are nominal group techniques and Delphi techniques applied within a futures context (Ascher 1978, Delbecq et al. 1975).

Acquire Improved Understanding of the Influence of Technological Change and Competitive Materials on Forecasts of Timber Demand and Supply

The advent of new technologies is a major consideration in demand forecasting. An example is that of estimating the pending occurrence of substitutes and the nature of their impact. Though econometric (and most other) techniques often assume trends in the patterns of the past will continue, the history of product use points to numerous cases of relatively large material use changes occurring in short periods of times, followed by perhaps decades of relative stability. The change from one-inch boards to plywood and then to oriented strand board and waferboard is such an example. Researchable topics in this strategic area include

- Are there ways of recognizing technological events that will change product use before such events become evident in statistics of use? Can surveys of research provide early warning systems? Should better research evaluation techniques be used for such approaches?
- Are there repeated patterns of market penetration that can be used to estimate the rapidity and final extent of market penetration early in the process? What are the important factors?

Evaluate technological impacts on derived demands for raw materials. Representations of demands for process inputs differ from those of a model's final outputs. The important variables are not exogenous demand factors (e.g., population, gross national product), but are the scale of the industry or activity using the product. Often the simplest and only feasible representation is a fixed coefficient; for example, 0.9 MBF of sawtimber is required for each MBF of lumber produced. Such fixed ratios are a major feature of standard input-output analysis. Input coefficients often change over time because of gradual or relatively sudden technical innovations or because of changed quality standards or trends in either output product or raw material. Among topics deserving research attention are

- What is the most effective way of analyzing changes in input requirements over time? What role is played by variable rates of technical innovation and by changing quality standards for either output product or raw material? Prices of raw materials often do not significantly affect requirements, especially if technical substitutes do not exist.
- How can the limits by which industries can change the mix of inputs be better analyzed? At any one time, narrow limits of input mixes may be fixed by technical factors (e.g., available equipment). But over time, research has demonstrated that factor proportions may change substantially (Steir 1980, 1985, Merrifield and Singleton 1986, Nautiyal 1985). Under what conditions is the relative price of each raw material important (Abt 1987a)?
- How can demands for residues or recycled materials (often quite price sensitive) be analyzed? They are often used only if price-competitive with virgin materials. Even then, they are often technically limited.

Assess supply relations in the production of processed forest products, especially technological influences. Market supply relations or functions are often more important to market equilibrium models than to old-style timber assessment models. Older assessment research considered demand to be determined by technical end-use factors and price elasticity, while supply was considered as a technically determined cost of production—up to the limit of capacity. Thereafter, supply was considered to be infinitely inelastic (no more could be produced). More recent supply representations, however, include the cost of major inputs, the rates of use, and a measure of all other costs and profit. For lumber, such costs usually include the price of sawtimber, a fixed or variable utilization factor, and a term for all other costs. If statistically significant, variation in labor wages and energy cost must be specifically considered. In the 15 years since the oil embargo, there has been sufficient change in energy cost to statistically test its influence.

Recent supply representations (particularly GTM and European models) have recognized that industries have several levels of technology and different cost structures. An older technology generally has less efficient raw material and labor use—high variable costs, but low fixed costs since capital is largely depreciated. A newer technology often has relatively low variable raw material and labor costs, but high fixed capital (or depreciation) cost (Ince et al. 1987). When output is at low levels, production is concentrated in plants with current technology—older plants are often idle and no new technology plants come on line. As output of the industry rises, older idle plants (not yet abandoned) are brought in and new plants gradually come forth. When output declines (or settles at new levels), new plants continue production while older plants are gradually abandoned.

Important research questions concerning the supply of processed forest products are the following:

- Can techniques developed in production function research be used to develop better cost estimates and to judge the competitive strengths of regions?
- What is the relationship of factor costs to capacity expansion in various regions? Can they be better used to anticipate building of new mills or expansion of older plants? How do the production processes of new capacity differ from those of older capacity? When do plants become obsolete and go out of production?
- What is the tradeoff between econometric models of supply and engineering cost specifications? How can they be fruitfully combined?
- Are improvements in recycling and energy utilization significant? How should they be explicitly handled?

Develop Improved Methods of Forecasting
Economic and Physical Supplies of Timber

Supply of inputs from natural resources (e.g., sawtimber, pulpwood) presents quite different problems for modeling and forecasting than does the modeling of industrial processes. Involved are two sets of biological and three sets of economic phenomena: (1) biological processes of tree growth (including quality improvements) and tree mortality (natural and catastrophic); (2) ecological processes of natural regeneration (after cutting or natural calamity) and change in forest composition or type; (3) economic processes of timber harvest or land conversion (influenced by stand size and quality, accessibility, logging costs); (4) economic processes of investment in timber and resource management to secure higher yields normally not provided by nature; and (5) economic processes of converting forest to nonforest uses (e.g., urban, agriculture), and conversely, converting nonforest uses to forests (e.g., planting idle agriculture land).

Until recently, timber growth (and the response to harvest) was projected using the Timber Resource Assessment System (TRAS), which treated all timber in a region and ownership class as one stand. In recent years, the TRAS model has been replaced, at least for some regions, by the Timber Resource Investment Model (TRIM), which projects the growth of many separate classes of forest area, separated by site quality, stocking, and management intensity (Tedder 1984). Information on forest area is projected jointly by a Markov transition system and the TRIM system (Alig 1986, Alig et al. 1987).

Though new timber supply forecasting methods provide more effective modeling frameworks, their adaption to a region requires development of coefficients and functions that are conceptually consistent with the economics of nonrenewable resources and unique forest management conditions (Hyde 1980). Operational models specific to regions are gradually being developed (Greber and Wisdom 1985, Newman 1987).

Important research questions regarding the forecasting of timber supplies include the following:

- How much and what type of forest land will receive intensive management in the years ahead? Research has long sought to determine the nature of owners willing to invest in forest practices, the type of forest land they prefer to invest in, the returns they expect on their investments, and the nature of tax, cost-share, and technical assistance programs that encourage them to undertake intensive forestry practices. Although limited to specific regions, recent research offers real hope that answers to such questions can be obtained (Royer 1987, Romm et al. 1987, Brooks 1985). How should the results of such research be integrated into models of forestry markets?
- What additional timber volumes will result from various timber management programs? The answer to such a question is dependent on more than simply the number of acres planted or treated—plantations may not be successful, may be harvested prematurely, or may be converted to other uses. To what extent should market models assume private landowners to be economically rational?
- Would it be possible to model landowners in a region, much as recent models do the land classes? The model would track groups of landowners as they bought land, managed land (or failed to manage land), sold land, or bequeathed land to heirs. Such a model might be associated with a land area model.
- What is the influence of type of harvest on the future management of forest land? What are the effects of cutting practices and forest practice regulation on harvest levels and subsequent land management programs of private owners?
- Has replacement of traditional large-scale sawmills with manufacturing facilities utilizing small diameter trees (e.g., flakeboard, stud, and chip-and-saw mills) made some timber resources obsolete, in the sense that mills are no longer available to process large-size timber?

- What is the effect of various environmental controls over private forest management (e.g., restrictions on pesticides, herbicides, sediment runoff, prescribed burning), especially on timber profitability and future timber supply?
- How have changes in tax laws (e.g., elimination of capital gain treatment of timber income) affected profitability and future timber supplies?
- What is the role of interest rates in forest management decisions and timber supply? What are the roles of stumpage price expectations?
- What are the total effects of government subsidy and assistance programs, both directly on recipients and on the actions of neighboring private and industrial landowners?
- What are the effects of transportation facilities or other social overhead on industrial development, forest management, and land ownership patterns?
- What are the total effects of fire, pest, and other protection programs on timber supply?

Acquire Improved Understanding of
Transportation and Related Trading Costs on
Future Demand and Supply of Timber Products

Transportation costs are especially important to timber processing since consumption of timber products is centered in urban and industrial centers while resource centers occur in rural, less populated (often economically undeveloped) regions. In nonspatial models, transport costs can be assumed to be part of supply costs. In spatial models, transport cost rates must be estimated explicitly (Wisdom 1987). Such rates are affected by type of product, distance covered, volume of product, and total trade. Other difficult-to-measure factors affect transport costs. Such factors include ocean-shipping conferences (i.e., international price cartels), availability of port facilities, frequency of shipments, use of full ship cargoes and unit trains, and requirements that shipment occurs via specified routes, modes (e.g., ships, railroads), or facilities. For example, ocean shipments between two U.S. ports must be made in U.S. registered ships.

Though substantial research has been focused on transportation, topics in need of further exploration are

- What influence does deregulation of domestic transport carriers and facilities have on forecasts of timber supply and demand?
- What is the effect of the cycles and fluctuations of ocean transport cost on export and import supply and demand? How will transport costs change as international trade expands?
- What is the best way to represent transportation cost functions in timber supply models and forecasts?

Develop Better Demand and Supply Forecasts
for Geographically Small Areas

Research on prospective demand and supply of timber and wood products is often concerned with large regions, multistate areas, or the nation. Yet much of the potential use of demand and supply information occurs within state and substate areas (e.g., areas within reasonable transport distance of national or state forests). While input-output information is useful for national and region macroeconomic problems (e.g., the impact of program changes on a regional economy), it has limited utility for understanding one or a few sectors of the wood-based industry in a specific local setting. A number of more relevant approaches (or sets of assumptions) can be taken to obtain useful planning information concerning more limited geographical areas (Connaughton and Haynes 1983). Generally unfeasible is the gathering and analysis of extensive local information (the bottom-up approach); thus the use and modification of regional information is usually considered (the top-down approach). A model was recently developed to disaggregate and adjust regional timber demand and supply data for the South and for specific southern states using the top-down approach (Abt 1987b).

Pertinent research topics concerning demand and supply forecasts unique to more limited geographic areas include the following:

- What is the effect of privately produced timber on future public land timber supplies in limited geographic areas, and, more important, what is the effect of publicly produced timber on private owners in geographically small areas?
- What is the effect of roads and other transport facilities on future demand for timber and other forest outputs in geographically small areas?
- What are the dynamics of public and private timber production (e.g., timber sales, timber under contract, harvest rates) within small regional areas? What affect do such conditions have on future timber supplies and demands?
- What are the effects of relatively few processing mills within a geographically small area?

Develop Improved Means of Forecasting Demand and Supply of Nontimber Forest Products

Policies and programs focused on forest resources can be designed to encourage the production of a variety of forest outputs (e.g., recreation, water, timber, wildlife, range). As with timber programs, nontimber forestry programs are often effective only to the extent that prospective demand and supply conditions are known and accounted for in program design. Unfortunately, forecasts and technologies for making such forecasts are not well developed for outputs such as recreation, wildlife, and forest aesthetics. Research should be undertaken to assess the applicability of current forecasting approaches to nontimber resources, and as appropriate, develop new techniques. Once developed, such techniques should be applied under a variety of circumstances.

Develop Effective Means of Applying Supply and Demand Forecasts to Important Forest Policy Issues

Information concerning future supply and demand of forest products can be an important input to the resolution of forestry issues of local, regional, and national importance. Research should be undertaken to determine how demand and supply relationships and forecasts can be effectively applied to a variety of policy problems. For example:

- How might supply and demand models be used in the development of strategies for research and development? For industrial financing decisions? For determining the impact of forestry-related and nonforestry-related innovation in forest management and industrial activities?
- How can supply and demand functions and market models be linked to other evaluation models such as input-output models that determine the impact of forestry programs?
- How can the essential assumptions, structural concepts, data sources and limitations, and results of forecasting models be most effectively communicated in meaningful ways to noneconomists and nonplanners? Improved understanding of modeling activities could remove the aura of the "black box" from forecasting models and could promote the use of demand and supply relationships and models to make strategic and policy decisions.

Summary and Conclusions

Forecasting the supply and demand of forest resources and products is an especially important activity within the forestry community. The often long-time horizons involved in the production of various forest outputs and the nonmarket nature of many forest resources make such an activity especially challenging. Forecasting technologies have improved substantially over the past 20 years; however, much additional research is needed. Future research should focus on the technologies of forecasting, the influence of technological innovations on future consumption levels, the effect of transportation on demand and supply of forest products, the demand and supply of forest products for geographically small areas, the means of forecasting demand and supply of nontimber forest outputs and the application of supply and demand forecasts to important forestry issues. Research in such strategic directions would go far toward more effective design and execution of public and private forestry programs.

Literature Cited

Abt, R. C. 1987a. An analysis of regional factor demand in the U.S. lumber industry. Forest Science 33: 164-173.

Abt, R. C. 1987b. Technical and conceptual issues in sub-regional market modeling. In: Forest sector and trade models by P. Cardellichio, D. Adams and R. Haynes (eds.) College of Forest Resources. University of Washington. Seattle, WA.

Adams, D. M. 1987. Issues in trade modeling: global versus regional models. In: Forest sector and trade models by P. Cardellichio, D. Adams, R. Haynes (eds.) College of Forest Resources. University of Washington. Seattle, WA.

Adams, D. M. and R. W. Haynes 1980. The 1980 softwood timber assessment market model: structure, projections, and policy simulations. Monog. 22. Forest Science. Society of American Foresters. Bethesda, MD.

Adams, D. M. and R.W. Haynes. 1986. Development of the Timber Assessment Market Model for long-range planning in the U.S. Forest Service. Division 4 Proceedings. XVIII World Congress. International Union of Forest Research Organizations. Ljubljana, Yugoslavia.

Alig, R. J. 1986. Econometric analysis of the factors influencing forest acreage trends in the Southeast. Forest Science 32: 119-134.

Alig, R. J., D. Brooks, D. Ingram, and C. Denise. 1987. The role of land use changes in forest sector modeling. In: Forest sector and trade models by P. Cardellichio, D. Adams and R Haynes (eds.) College of Forest Resources. University of Washington. Seattle, WA.

Ascher, W. 1978. Forecasting: An appraisal for policy makers and planners. Johns Hopkins University Press. Baltimore, MD.

Baudin, A. and L. Lundberg. 1987. A world model of the demand for paper and paperboard. Forest Science 33: 185-196.

Brooks, D. J. 1985. Public policy and long-term timber supply in the South. Forest Science 31: 342-357.

Brooks, D. J. 1987. REACTT: An algorithm for solving spatial equilibrium problems. Gen. Tech. Rep. PNW-GTR-204. Pacific Northwest Forest and Range Experiment Station. USDA Forest Service. Portland, OR.

Buongiorno, J. 1978. Income and price elasticities in the world demand for paper and paperboard. Forest Science 24(2): 231-246.

Buongiorno, J. 1987. Statistical approaches in forest sector modeling: philosophical issues. In: Forest sector and trade models by P. Cardellichio, D. Adams and R. Haynes (eds.) College of Forest Resources. University of Washington. Seattle, WA.

Buongiorno, Joseph and J. Keith Gilless. 1987. PAPYRUS: a model of the North American pulp and paper industry. Monog. 28. Forest Science. Society of American Foresters. Bethesda, MD.

Chou, J. and J. Buongiorno. 1983. United States demand for hardwood plywood imports by country of origin. Forest Science 29: 225-227.

Clawson, M. (ed.) 1977. Research in forest economics and forest policy. Research Paper R-3. Resources for the Future. Washington, DC.

Connaughton, K. P. and R. W. Haynes. 1983. An evaluation of three simplified approaches to modeling the regional demand for National Forest stumpage. Forest Science 29: 3-14.

Delbecq, A. L., A. Van de Ven and D. Gustafson. 1975. Group techniques for program planning. Scott, Foresman and Company. Dallas, TX.

ECE/FAO. 1985. European timber trends and prospect to the year 2000 and beyond. Vol I and II. United Nations. New York, NY.

Greber, B. J. and H. W. Wisdom. 1985. A timber market model for analyzing roundwood product interdependencies. Forest Science 31: 164-179.

Hardie, I. W. and A. A. Hassan. 1986. An econometric analysis of residential demand for fuelwood in the United States, 1980-1981. Forest Science 32: 1001-1015.

Haynes, R. W. and D. M. Adams. 1985. Simulations of the effects of alternative assumptions on demand-supply determinants on the timber situation in the United States. Forest Resource Economics Research Staff. USDA Forest Service. Washington, DC.

Hyde, W. F. 1980. Timber supply, land allocation, and economic efficiency. Johns Hopkins University Press. Baltimore, MD.

Ince, P. J., K. Skog, H. Spelter, Henry, I. A. Durbak and J. L. Howard. 1987. Modeling technological change in wood products processing. In: Forest sector and trade models by P. Cardellichio, D. Adams and R. Haynes (eds.). College of Forest Resources. University of Washington. Seattle, WA.

Kallio, M., D. P. Dykstra and C. S. Binkley. 1987. The global forest sector: an analytical perspective. John Wiley Publishers. Chichester, England.

Kornai, G. 1985. Reconciliation of forest products trade data. WP-85-48. International Institute for Applied Systems Analysis Laxenburg, Austria.

Kornai, G. and P. Schwarzbauer. 1987. The Austrian model of production, consumption and trade in forest products (ATM). In: Forest sector and trade models by P. Cardellichio, D. Adams and R. Haynes (eds.). College of Forest Resources. University of Washington. Seattle, WA.

Laarman, J. G. and M. K. Wohlgenant. 1984. Fuelwood consumption: a cross-country comparison. Forest Science 30: 383-394.

Lange, W.J., P. J. Ince, I. A. Durbak and J. L. Howard. 1987. Considerations in the development of a trade model for the pulp and paper sector. In: Forest sector and trade models by P. Cardellichio, D. Adams and R. Haynes (eds). College of Forest Resources. University of Washington. Seattle, WA.

Lonnstedt, L. 1986. A dynamic forest sector model with a Swedish case. Forest Science 32: 377-397.

Martin, R. M. 1987. FAO world outlook for forest products consumptions and supply with emphasis on paper and wood based panel products. In: Forest sector and trade models by P. Cardellichio, D. Adams and R. Haynes (eds.) College of Forest Resources. University of Washington, Seattle, WA.

McKillop, W. 1987. Validation of international trade models. In: Forest sector and trade models by P. Cardellichio, D. Adams and R. Haynes (eds.) College of Forest Resources. University of Washington. Seattle, WA.

Merrifield, D. E. and W.R. Singleton. 1986. A dynamic cost and factor demand analysis for the Pacific Northwest lumber and plywood industries. Forest Science 32: 220-233.

Nautiyal, J. C.and B. K. Singh. 1985. Production structure and derived demand for factor inputs in the Canadian lumber industry. Forest Science 31: 871-881.

Newman, D. H. 1987. An econometric analysis of the southern softwood stumpage market: 1950-1980. Forest Science 33: 932-945.

Nilsson, S. 1983. What we need and what we have in Sweden. In: Forest sector models by Seppala, Row and Morgan (eds.) A B Academic Publishers. Berkhamsted, UK.

Nomura, I. 1983. Long range timber demand/supply prospects in Japan and some problems. In: Forest sector models by Seppala, Row and Morgan (eds.) A B Academic Publishers. Berkhamsted, England.

Prins, K. 1987. European timber trends and prospects to the year 2000 and beyond. In: Forest sector and trade models by P. Cardellichio, D. Adams and R. Haynes (eds.). College of Forest Resources. University of Washington. Seattle, WA.

Romm, J., R. Tuazon and C. Washburn. 1987. Relating forestry investment to the characteristics of nonindustrial private forestland owners in northern California. Forest Science 33: 197-209.

Royer, J. P. 1987. Determinants of reforestation behavior among southern landowners. Forest Science 33: 654-667.

Sedjo, R. A. and K. S. Lyon. 1987. The adequacy of long term timber supply: a worldwide assessment. In: Forest sector and trade models by P. Cardellichio, D. Adams and R. Haynes (eds.) College of Forest Resources. University of Washington. Seattle, WA.

Sedjo, R. A. and K. S. Lyon. 1988. Long-term timber supply: an optimal control approach. (forthcoming) Resources for the Future. Washington, DC.

Seppala, R. and H. Seppala. 1987. Finnish tradition of forest sector models. In: Forest sector and trade models by P. Cardellichio, D. Adams and R. Haynes (eds.) College of Forest Resources. University of Washington. Seattle, WA.

Steir, J. 1980. Technological adaption to resource scarcity in the U.S. lumber industry. Western Journal of Agricultural Economics 5: 165-175.

Steir, J. 1985. Implications of factor substitution, economies of scale, and technological change for the cost of production in the United States pulp and paper industry. Forest Science 31: 803-812.

Uutela, E. 1983. Approaches to the analysis of paper demand and supply in western Europe. In: Forest sector models by Seppala, Row and Morgan (eds.). A B Academic Publishers. Berkhamsted, UK.

VeltKamp, J. J., R. Young and R. Berg. 1983. The Data Resources, Inc., approach to modeling demand in the softwood lumber and plywood and pulp and paper industries. In: Forest sector models by Seppala, Row and Morgan (eds.). A B Academic Publishers. Berkhamsted, UK.

Wisdom, H. W. 1987. Transportation costs in forest sectors models. In: Forest sector and trade models by P. Cardellichio, D. Adams and R. Haynes (eds.) College of Forest Resources. University of Washington. Seattle, WA.

Young, R. L. 1987. The RISI international pulp and paper model. In: Forest sector and trade models by P. Cardellichio, D. Adams and R Haynes (eds.) College of Forest Resources. University of Washington, Seattle, WA.

9

SOCIAL AND ECONOMIC GROWTH OF DEVELOPING NATIONS

Hans M. Gregersen and Jan G. Laarman

Forests and forestry in developing nations are subjects of increasing concern throughout the world (World Resources Institute 1985, Food and Agriculture Organization 1985a). This heightened interest is due in part to mounting evidence that deforestation and forest degradation are linked directly to human suffering in the forms of malnutrition, health problems, lack of wood energy, flood damage, and unemployment. Heightened interest in forests is also the result of distress over the long-term environmental consequences of forest degradation, including adverse climate changes, loss of biological diversity, and shrinking gene pools. Concerns of this nature are indirectly related to current human suffering, yet have serious implications for the long-term survival of mankind.

The developed world has made significant progress in mobilizing efforts to address forestry conditions in developing nations. For example, developed countries and international agencies for forestry development have significantly increased the level of funding available for forestry programs in developing nations (Food and Agriculture Organization 1987a). Concerted efforts have been made to focus such funds on high priority areas (e.g., Tropical Forestry Action Plan) (Food and Agriculture Organization 1985a). In addition, various world leaders have taken an interest in the forestry needs of developing nations. In the summer of 1987, world leaders met to determine means by which the priority forestry needs of developing nations might be better supported (World Resources Institute 1987). Among the subjects discussed was the need for forestry research. In support of the meeting, the Rockefeller Foundation prepared a background paper on research, almost half of which was devoted to policy and social sciences (Rockefeller Foundation 1987).

A focused program of forest economics research can be of substantial value to the forestry needs of the developing world. Review of existing information and examination of previous research reviews suggests that forest economics and policy research would be most beneficial if directed to the following four strategic areas (Romm 1982; World Bank and Food and Agriculture Organization 1981; Gregersen, Draper, and Elz 1988):

- Develop enhanced understanding of forestry programs involving social forestry, watershed management, and nontimber forest outputs.
- Acquire better understanding of environmental implications of forestry developments in developing nations.
- Develop enriched understanding of investment prospects and technology options for industrial forestry in developing nations.
- Develop improved understanding of the effectiveness of international aid for forestry development in developing nations.

Develop Enhanced Understanding of Forestry Programs Involving Social Forestry, Watershed Management, and Nontimber Forest Outputs

The primary focus of international support for forestry development through the mid-1970s was industrial forestry. In recent years, however, a dramatic shift has occurred toward investment in social forestry and watershed management. Nowhere is evidence of such a shift more dramatic than at the World Bank. In 1967-1976, less than 5 percent of the Bank's forestry-related lending was for community and farm forestry, watershed management or conservation, while over 95 percent was for industrial plantations, logging operations, and forest industry development. In 1977-1986, however, fully 60 percent of the Bank's lending involved community and farm forestry, fuelwood, and watershed management (Food and Agriculture Organization 1987a). The early lack of interest in social forestry and watershed management was matched by a corresponding lack of interest in research on such subjects. It was not until the late 1970s that agriculture and forestry researchers began to direct attention to these subjects in the context of developing nations.

Research needed to build the information base for improved social forestry and watershed management programs in a development context has, for the most part, not been undertaken. This is especially true for research involving economics and policy. In the area of agroforestry, for example, some economics research has been undertaken (e.g., ICRAF in Kenya, CATIE in Costa Rica), although the information produced to date is far from that required to make effective decisions about program direction and content. Lack of research has not, however, curtailed rapidly expanding investments in social forestry and watershed management. Such investments are based in part on preliminary evidence from recently completed research and evaluation—evidence which indicates significant positive impacts. Economic rates of return for World Bank projects in social forestry, watershed management, and soil conservation have been estimated to be 15 to 21 percent (World Bank 1984).

Given the scarcity of social forestry and watershed management research currently being focused on developing countries, identification of priority research topics is difficult—nearly all topics are deserving of some attention. Based on extensive reviews, however, the following research topics are suggested (Gregersen, Draper, and Elz 1988; Rockefeller Foundation 1987; Arnold 1987; Gregersen 1987; Repetto 1988).

Expand understanding of forest and tree-related needs and wants of people.
Past failure of outside interventions clearly illustrates what can happen when well-meaning outsiders misinterpret local tree-related needs and wants of people (e.g., early fuelwood projects focused on fast-growing species rather than multipurpose species). As the most numerous economic agents in most rural communities, farmers and their families look at trees within the context of overall farm management. Their need for trees results from very complex circumstances; species preferences may not be readily understood by outsiders (Arnold 1987, Vergara 1984). Wood supply-demand balances also depend on demand by artisans, carpenters, and other traditional wood-using enterprises (including those that burn wood for energy). Understanding such balances is critical to solving many problems associated with deforestation and wood scarcities.

Increase understanding of incentives for tree growing and management. A number of factors create incentives for people to grow and utilize trees. The success of outside interventions to influence tree management and utilization practices depends on how such interventions affect local incentive systems. Thus, understanding such systems is a priority area for social science research—a starting point is knowledge about the economics of tree growing and tree management. Additional effort should be devoted to evaluating the costs and returns associated with different models of farm forestry, common property, and public forest management (e.g., develop a model of the economics, organization, and politics of subsistence forest use within a commercial timber concession). A key requirement is that both market and nonmarket costs and benefits be included and valued from the perspective of different local producers and consumers. Moreover, costs and benefits should be appropriately weighted by relative risks (e.g., loss due to natural causes, tenure insecurity). Such analyses would provide the reference points for comparing alternative incentive mechanisms often suggested as means of increasing net benefits or reducing risk levels.

Most tree growing projects in developing countries include the introduction of various mechanisms to change local incentives. A great deal of research has been done in developed countries to understand the nature and impact of such mechanisms (Royer and Risbrudt 1983). Much less effort has been devoted to empirically researching these mechanisms in developing countries (deCamino 1986, Bombin 1975, Berger 1980). Research should be undertaken to provide better information for planning and evaluating future incentive programs.

Market systems have proven to be a powerful incentive for tree growing and management. While understood in terms of industrial forestry, market systems are poorly understood in the context of smallholder tree growing in developing countries. The complication partially arises because most farmers do not grow trees solely for markets. Farmers are more prone to grow trees for multiple benefits within a farming system; only some trees or tree products are sold on the open market (Arnold 1983). The economics and decision making processes of such complex systems should be researched. Social scientists should work closely with others to evaluate commercial tree growing within a farming systems context.

Increase understanding of employment effects of market-based family activity, especially as relates to local tree growing and to the emergence of an entrepreneurial middle class. Rural unemployment is a serious problem in many developing countries. Since demand for agriculture does not exist in many areas, increasing agricultural employment is rarely a viable solution to the problem. Forest-based activity often is the major source of rural, off-farm employment in many parts of the world (Food and Agriculture Organization 1985c). How can such employment opportunities be improved in terms of both quantity and quality? Does a forest-based cottage industry lead to expansion of the entrepreneurial middle class? Does it provide a smooth transition to more modern industrial activity that should logically be developed in some areas? The issues raised by such questions are deserving of research.

Decisions to favor small-scale enterprises in forest-based sectors require appropriate policies and programs. Doing so can lead to important issues concerning program efficiency and effectiveness (Gregersen 1984). Ironically, very little is known about the benefit-cost ratios of many forestry programs, including forestry (and agroforestry) extension, landowner assistance, credit programs, marketing services, technical assistance, and managerial training. Examination of literature concerning small-scale enterprises suggests that *ex post* evaluations of assistance programs are also very rare (Anderson 1982). Needed is research that conceptualizes the costs and benefits of different assistance modes and empirically tests such frameworks in a variety of circumstances.

Evaluate alternative markets and technologies for sustaining nontimber outputs from tropical and subtropical forests. Prospects for additional employment and enhanced environmental quality could be heightened by expanding nontimber uses of tropical wildlands. In addition to generation of revenue and employment, an objective of interest would be minimal disturbance of ecological systems (Davidson 1985). Example enterprises for accomplishing these multiple objectives include controlled wildlife cropping (Child 1982) and nature tourism (Laarman and Durst 1987). Research should be undertaken to organize and assess existing information regarding the financial success, failure, markets, and general economic viability of such enterprises. Case studies should be undertaken to develop financial profiles and market analyses that will facilitate investment in such enterprises.

Expand understanding of rural land use patterns within a broader watershed management framework. The economic and administrative forces that shape development operate within political boundaries, while the forces of nature affecting land and water resources per se often adhere to watershed or ecological boundaries. In many cases, the boundaries of political and natural systems do not coincide. Since most rural development programs involve the use of land and water resources, failure to properly recognize watershed boundaries can have disastrous effects on rural development. The same occurs if plans ignore political and economic boundaries. Some means must be found to reconcile watershed boundaries and boundaries established for political and economic reasons.

Forest policy and economics research can play a role in unraveling the natural-political boundary dilemma. Unfortunately, present methods for evaluating the economics of watershed management are in their infancy. Needed is cooperative research (e.g., research by economists, political scientists, biologists, physical scientists) that would foster development of useful economic models of forested watersheds (Working Group on Watershed Management and Development 1987; Gregersen, Brooks, Dixon, and Hamilton 1987). Also of some urgency is the need to develop case studies of successful and unsuccessful projects, with emphasis on economic and related factors of importance. Research is also needed on incentive systems and institutional mechanisms for guiding land uses in upland watersheds so they are compatible with downstream objectives concerning environmental quality. Interesting approaches have been developed (e.g., in Japan and Colombia) for ensuring that downstream beneficiaries share in the cost of upstream conservation practices, which are often carried out by owners of small land-holdings (Kumazaki 1982).

Improve economic decision rules and models used to guide construction and resettlement projects located in forests and other wildlands. A number of major projects such as highways, reservoirs and agricultural settlement projects in developing countries have pushed into "empty lands" (Nelson 1973). Such projects often result in displacement of local populations and undesirable land uses (Adams 1985, Schwartzman 1986). Research should be undertaken to develop risk-rating methods for project analysis—methods which more fully assess risk and cost of environmental and social spillovers of forestry projects. In addition, research should be undertaken to define and evaluate financial incentives and penalties keyed to the environmental impacts of a project. The economic effects of different technologies for land clearing also should be explored (Ross and Donovan 1986).

Acquire Better Understanding of Environmental
Implications of Forestry Developments
in Developing Nations

Environmental issues in developing countries are of growing worldwide interest (Guppy 1984, U.S. Congress 1986, Green and Barborak 1987). Among the many environmental concerns at the macro level is tropical deforestation (Moeller 1984, Fonesca 1985, Lundgren 1985, Bowonder 1986, Bushbacher 1986), species extinction (Myers 1984), desertification (Gorse 1985, Anderson 1986) and global climate change (Prance 1986). At the micro level, a like number of ecological issues exist, such as possible negative impact of eucalyptus on site quality and water yield (Food and Agriculture Organization 1985d), definition of criteria for protected areas (Aisbey and Owusu 1982, DeAlwis 1982, Oltremari and Jackson 1985) and economic valuation of wildlife species (Thresher 1981, Decker and Goff 1987). Forest economists have scarcely begun exploration of such topics— partly because of their analytical complexity. Yet pressure for improved environmental management will in all likelihood spread with greater intensity (Norgaard 1985; Blake 1987). Traditional forestry concerns will continue, but they will be joined by major new initiatives and institutions aimed at environmental protection. Within this context, the following topics deserve the attention of forest economics research.

Develop new (or enhance existing) methods for economic valuation of environmental and ecological services. Economics is a framework capable of valuing a variety of goods and services produced by forests and related resources. As such, economics can have much to say about the value of ecosystem preservation, biological diversity, gene conservation, and related interests (Hufschmidt and Hyman 1982; Krutilla and Fisher 1985; Dixon and Hufschmidt 1986; Norgaard 1987). These are objectives of special interest to sections of the humid tropics, where very few species have been cataloged. In temperate regions, the value of preserving natural systems can be many times greater than the value of on-site uses (Peterson and Randall 1984). In the tropics, however, almost nothing is known about values associated with preservation. Research should be undertaken to (1) determine perceptions, attitudes, and preferences which can be used as a base for conceptualizing and estimating willingness to pay for tropical preservation; (2) define and measure option value and existence value of tropical wildlands; (3) determine methods of incorporating environmental degradation into systems of national accounts; and (4) evaluate techniques for incorporating preservation values into project analysis. Such research will require close working relationships between economists and biological scientists.

Develop means of applying economic concepts to physical rates and processes of tropical deforestation. Deforestation and forest degradation are typically measured in physical terms (e.g., hectares per hour or per year). Seldom are deforestation and related concerns identified in terms of benefits and costs. As such, "deforestation crises" have yet to be satisfactorily evaluated from the perspective of change in human welfare (Sedjo and Clawson 1983; Fearnside 1986). Research should be undertaken to develop and test alternative economic models applicable to different kinds of deforestation processes (Burns 1986).

Improve understanding of benefits and costs of wildland protection, especially as related to local residents. Some may argue that poor rural peasants derive no economic benefit from tropical biological reserves or from similar preservation initiatives. As such, they have no economic incentive to protect the designated areas—but may have considerable incentive to eliminate them. The counter to such an argument is that biological reserves can be designed and managed in such a way that they provide local residents with significant employment and income (Halffter 1981, Garrett 1984, Place 1985, Wright 1985, Dobias 1987). Reviews should be undertaken of methods used to assess the flow to local residents of benefits produced by reserved areas. Research should be undertaken to develop and test new methods of assessing such benefits. As appropriate, case studies should be undertaken as a means of further understanding preservation issues. Research of this sort is closely related to research focused on social forestry.

Assess rates at which timber harvesting will shift from natural forests to forest plantations. The notion of plantations as "compensatory forests" is not new. However, very little analysis concerning the substitution of plantation wood for primary and secondary forests has been undertaken. This void exists even though such substitution is a distinct possibility in parts of South America (McGaughey and Gregersen 1983; Sedjo 1983). Research should be undertaken to develop a general analytical framework (costs, yields, utilization, and market demand) for comparing timber obtained from natural forests versus timber obtained from forest plantations. Once developed, such a framework should be estimated and applied to national and regional economies having a significant plantation base. The intent is to improve time-path estimates of primary and secondary forest conversion.

Refine economic tools and models so as to facilitate application to research on biological production functions. A number of ecological issues link forest production with environmental management. Examples are the effect of eucalyptus plantations on soil properties and water yield, the impact of different types of logging on tropical forest succession, and methods of obtaining low-cost natural regeneration in tropical and subtropical forests. The forest economist's contribution to understanding such issues has been limited. Economics research should be undertaken to define and explore such topics, using benefit-cost analysis, welfare analysis, and related techniques appropriate to determining material gain and loss by different economic agents.

Develop Enriched Understanding of Investment Prospects and Technological Options for Industrial Forestry in Developing Nations

Early models of the role of forest industries in economic development have been subject to significant criticism (Westoby 1978, Douglas 1983, Dargavel 1985). Factors not foreseen nor accounted for include changed rural development ideologies, failure of many trickle-down policies, overly optimistic assessment of interindustry linkages, and weaknesses in political institutions such as land tenure. From the perspective of multilateral and bilateral assistance agencies, the new and reformed agenda for forest industries often highlights objectives such as self sufficiency, appropriate scales and technologies, local benefits, and resource sustainability. Themes of this sort are illustrated by a number of forestry papers recently issued by the United Nations Food and Agriculture Organization (Food and Agriculture Organization 1982a, 1986a, 1986b, 1987b). Moreover, the world's timber economy is vastly different from what it was several decades ago. Many developing economies have successfully and dramatically scaled back exports of logs in favor of products (Brion 1983). Industrial forest plantations are now important sources of present and future wood supply (Lanly and Clement 1979; Sedjo 1980). Multinational forest products companies are less sanguine about investments in the tropics (Gregersen and Contreras 1975; McNeil 1981; Bethel 1982). The financing of plantations and industrial enterprises has become a major concern, especially in view of the debt crisis in a number of countries (McGaughey and Gregersen 1983, Gregersen and McGaughey 1988). Given this environment, the following are priorities for future forest economics research.

Conceptualize, develop, and test models which quantify forest industry-based employment and income. Many arguments for and against industrial forest development in developing countries rest more on sentiment (e.g., "small is beautiful") than on empirical evidence gained through systematic research or experience. Issues concerning appropriate technologies and appropriate operational scales command much of the development economist's attention. Yet case studies focused on forestry and forest industries are very limited in number (e.g., Page 1978, Laarman 1982, Amsalem 1983, Koljonen 1983, Salim 1983, Hornick 1984). Large-scale enterprises may be legitimately criticized; small-scale alternatives, however, are not without serious problems (Little 1987). Research should be undertaken to (1) improve methods of accounting for indirect as well as direct employment and income from industrial activities, without constructing expensive input-output models of a nation's economy; (2) develop new approaches for modeling tradeoffs between allocation efficiency and competing social objectives (e.g., local employment); and (3) add economic clarity to engineering concepts of choice of industry, choice of product, and choice of technology.

Improve estimates of supply and demand elasticities, and reexamine supply and demand shifters for tropical timber and wood products. World supply and demand for tropical woods is experiencing dramatic structural change (Wardle and Martin 1986, Boulter 1986, Kallio 1987). This is suggested by Indonesia's cessation of log exports in favor of high-volume plywood production; far-reaching internal adjustments in Japanese economic policy; appearance of China as a large importer; reduced imports of tropical wood based on perceptions of a deforestation crisis; increasing timber volumes derived from forest plantations, particularly in parts of Latin America; and rapid and substantial realignments in currency exchange rates among the world's large exporters and importers. Additional short-term and long-term factors could be cited; even more will emerge in years to come. Research should be undertaken to critique existing models of supply, demand, and trade (Kumar 1985), and to theorize and estimate new models capable of capturing emerging trends.

Assess demand for capital and analyze sources of capital for development of forestry and forest products industries in countries burdened with large foreign debts. The demand for forest products in developing countries is expected to increase substantially in future decades (Food and Agriculture Organization 1982b). Capital needed for investment in forestry and manufacturing plants will be substantial; this has captured the attention of the development community (Food and Agriculture Organization 1987a, Spears 1985, Ganguli 1986, McGaughey 1986, Obura 1986). Yet with large external debts facing many developing countries, the source of such capital is not clear (Laarman 1987). Research should be undertaken to advance analysis beyond capital "requirements" to a market-based definition of demand for capital at different prices. This market-based concept should be used to develop models that are capable of projecting capital supply and demand under a wide variety of scenarios about future economic growth throughout the world and its regions. Some policy prescriptions have already been suggested; privatization of state-run and parastate enterprises, reduction of subsidies and increased tax collection (Laarman 1986), for example. In what ways, and to what degree, will such prescriptions affect the financing of forestry and forest industries? What new institutional arrangements will be needed (e.g., debt-for-equity swaps)? What economic guidance will be needed to inform and possibly redirect subsidy policies for reforestation and afforestation?

Conduct cross-country analyses of government timber concessions and leases, especially regarding the division of economic rent. In most developing countries, publicly owned forests are the primary source of timber. Such timber is purchased by private buyers and wood processors through a wide variety of institutional arrangements, at a wide spread in nominal prices, and with a wide range of consequences for generation of public revenue (Schmithusen 1976, Food and

Agriculture Organization 1983). With few exceptions, rent-sharing between governments and buyers has not been adequately researched (exceptions are Page 1976, Ruzicka 1979, Repetto 1988). Yet the subject has important implications for timber supply modeling, deforestation issues, and efficiency issues. Research should be undertaken to systematically quantify the division of timberland rent across regions and countries in the developing world. Inferences should be drawn from these observations, especially with respect to forest investment and disinvestment.

Develop economic models of technology search and transfer, marketing systems, and capital flows expected from the growth of global information exchange. International links among computers, telephones, and other communication devices lend increasing reality to the term "global village." Markets for stocks, bonds, and some commodities already have been revolutionized by global trading—all facilitated by rapid and low-cost information transactions. Research should be undertaken to identify how low-cost information technologies impact the cost structure and marketing methods of forest industries in the developing world. In addition, determination should be made of how information technologies will affect international comparative advantage in forestry and forest industries.

Continue to develop and validate economic models of inventory techniques, harvest cuts, regeneration decisions, silvicultural prescriptions, and forest management plans for different types of tropical forests. The economist interacts with other disciplinary specialists to develop management tools applicable to tropical forests. Many enterprises and agencies cannot effectively carry out accounting, capital budgeting, and optimization activities since they lack even the most rudimentary cash flow models. Lacking such models, enterprises have difficulty making recommendations to improve management activities (e.g., introduction of log grading, utilization of secondary species, grouping of mixed species by properties). The diffusion and decreasing cost of microcomputers and spreadsheet software will facilitate the application of financial analysis. Research should be undertaken to assess supply of, and demand for, cash flow models. Efforts should also be focused on determining the feasibility of filling the observed gaps with appropriate computers, software, training, and related institutional actions.

Develop Improved Understanding of the Effectiveness of International Aid for Forestry Development in Developing Nations

International aid to developing nations is often subject to substantial criticism. Opponents argue that such aid distorts economic incentives, fosters dependency on outside capital and advisors, concentrates economic power, and leads to adoption of senseless economic policies by government (Bauer 1971; Krauss 1983). In contrast, proponents of aid argue that the present level of development assistance is pitifully small and is provided to developing countries for all of the wrong reasons. Aid for forestry is a specific case. Issues concern not only positions of philosophy but also direction of forestry programs. While little can be done to change the former, economists can play an important role with respect to the latter. Key questions deserving of analysis include: What is the changing level and composition of development assistance to forestry? What factors limit absorptive capacity for additional funds, both in the recipient countries and from the managerial viewpoint of the donors? The first question requires description and projection. The second demands institutional analysis and identification of weak links.

A number of questions concerning the effectiveness of aid in a forestry context can be posed. What are the costs and benefits of promoting increased integration of donor efforts? What policies are most viable for integrating aid for forestry with aid for other sectors? What can be done to promote cofinancing, such as between an international agency and a private corporation or nongovernmental organization? What institutional arrangements will facilitate learning from past experience (e.g., monitoring and evaluation systems)? The challenge to the researcher is to move from general recommendations to models and frameworks that provide a high plane of conceptual refinement and analytical rigor.

Also important in the context of international aid is the building of human capital through education, research, and extension. Very little has been done to link such a theme with forestry and forestry professionals—particularly in developing countries. Needed is a more concrete understanding of "capacity" in the context of research and training, especially the factors that determine such capacity (Bengston, Gregersen, Lundgren, and Hamilton 1988; Bengston, Gregersen, and Lundgren 1987). What economic and social models of human capital are most pertinent for the forestry sectors of low-income countries? What are the returns to such investments? These research issues will continue to challenge us into the indefinite future.

Summary and Conclusions

Trees and forests are often of immeasurable importance to developing countries of the world. To be of value, however, effective and efficient institutions, programs, and policies must be designed and focused on such resources. Forest economics and policy researchers can contribute much to such activities. To be most effective, forest economics research should be designed to improve understanding of social forestry, watershed management, and nontimber forest outputs; enhance ability to effectively address environmental consequences of forestry development; heighten skill in guiding development of industrial forestry enterprises; and improve effectiveness of international aid for forestry development. Guided by such strategic directions, forest economics research can contribute much to the economic and social well-being of developing nations.

Literature Cited

Adams, P. 1985. In the name of progress: The underside of foreign aid. Doubleday Publishers. Toronto, Canada.

Aisbey, E. O. A. and J. G. K. Owusu. 1982. The case for high-forest national parks in Ghana. Environmental Conservation 9(4):293-304.

Amsalem, M. A. 1983. Technology choice in developing countries: The textile and pulp and paper industries. MIT Press. Cambridge, MA.

Anderson, D. 1982. Small industry in developing countries. Staff Working Papers No. 518. World Bank. Washington, DC.

Anderson, D. 1986. Declining tree stocks in African countries. World Development 14:853-863.

Arnold, J. E. M. 1983. Economic considerations in agroforestry projects. Agroforestry Systems 1:299-311.

Arnold, J. E. M. 1987. Research needs and opportunities in tropical forestry: Policy research. Background paper. Rockefeller Foundation. New York, NY.

Bauer, P. T. 1971. Dissent on development. Weidenfeld and Nicolson Publishers. London, UK.

Bengston, D., H. Gregersen and A. Lundgren. 1987. Planning, management and evaluation of forestry research: An annotated bibliography with special reference to developing countries. Staff Paper No. 59. Department of Forest Resources. University of Minnesota. St. Paul, MN.

Bengston, D., H. Gregersen, A. Lundgren, and L. Hamilton. 1988. Forestry research capacity in the Asia-Pacific region: An evaluation model and preliminary assessment. Unpublished report. Department of Forest Resources. University of Minnesota. St. Paul, MN.

Berger, R. 1980. The Brazilian fiscal incentive act's influence on reforestation activity in Sao Paulo state. Ph.D. dissertation. Michigan State University. E. Lansing, MI.

Bethel, J. S., et al. 1982. The role of U.S. multinational corporations in commercial forestry operations in the tropics. Report to the U.S. Department of State. College of Forest Resources. University of Washington. Seattle, WA.

Blake, R. O. 1987. An environmentalist looks at the world's forests. In: 1987 convention proceedings. Society of American Foresters. Bethesda, MD.

Bombin, L. M. 1975. Incentivos Económicos Forestales en América Latina. Coloquio sobre Modernización de la Administración Pública en el Sector Forestal de América Latina. Food and Agriculture Organization. United Nations. Rome, Italy.

Boulter, W. K. 1986. Global supply-demand outlook for industrial roundwood. Forestry Chronicle 62:306-313.

Bowonder, B. 1986. Deforestation in developing countries. Journal of Environmental Systems 15:171-192.

Brion, H. P. 1983. Potentials and requirements of increasing the degree of wood processing in developing countries of Asia and the Pacific. UNIDO Sectoral Working Paper Series no. 5. UNIDO/IS.395. New York, NY.

Burns, D. 1986. Runway and treadmill deforestation. IUCN/IIED Tropical Forest Policy Paper No. 2. IUCN. Washington, DC.

Bushbacher, R. J. 1986. Tropical deforestation and pasture development. BioScience 36(1):22-28.

Child, G. 1982. Managing wildlife for people in Zimbabwe. In: Proceedings of world congress on national parks and protected areas by J. A. McNeely and K. R. Miller (eds.). October 11-22, 1982. Bali, Indonesia.

Dargavel, J., M. Hobley and S. Kengen. 1985. Forestry of development and underdevelopment of forestry. In: Forestry: Success or failure in developing countries by J. Dargavel and G. Simpson (eds.). CRES Working Paper 1985/20. Centre for Resource and Environmental Studies. Australian National University. Canberra, Australia.

Davidson, J. 1985. Economic use of tropical moist forests while maintaining biological, physical and social values. Paper No. 9. Commission on Ecology. International Union for Conservation of Nature and Natural Resources. Gland, Switzerland.

DeAlwis, L. 1982. River basin development and protected areas in Sri Lanka. In: Proceedings of world congress on national parks and protected areas by J. A. McNeely and K. R. Miller (eds.). October 11-22, 1982. Bali, Indonesia.

deCamino, R. 1986. Oncentivos Para Participacion de la Comunidad en programas de conservacion. FAO Conservation Guide No. 12. Food and Agriculture Organization. United Nations. Rome, Italy.

Decker, D. J. and G. R. Goff (eds.). 1987. Valuing wildlife: Economic and social perspectives. Westview Press. Boulder, CO.

Dixon, J. A. and M. M. Hufschmidt, eds. 1986. Economic valuation techniques for the environment, a case study workbook. Johns Hopkins University Press. Baltimore, MD.

Dobias, R. T. 1987. A demonstration project for integrating park conservation with rural development at Khao Yai National Park, Thailand. Parks 12(1):17-18.

Douglas, J. 1983. A reappraisal of forestry development in developing countries. Martinus Nyhoff/Dr. W. Junk Publishers. The Hague, Netherlands.

Fearnside, P. M. 1986. Human carrying capacity of the Brazilian rain forest. Columbia University Press. New York, NY.

Food and Agriculture Organization. 1982a. Appropriate technology in forestry. FAO Forestry Paper No. 31. United Nations. Rome, Italy.

Food and Agriculture Organization. 1982b. World forest products, demand and supply 1990 and 2000. FAO Forestry Paper No. 29. United Nations. Rome, Italy.

Food and Agriculture Organization. 1983. Forest revenue systems in developing countries. FAO Forestry Paper No. 43. United Nations. Rome, Italy.

Food and Agriculture Organization. 1985a. Tropical forestry action plan. Committee on Forest Development in the Tropics. United Nations. Rome, Italy.

Food and Agriculture Organization. 1985b. Tree growing by rural people. FAO Forestry Paper No. 64. United Nations. Rome, Italy.

Food and Agriculture Organization. 1985c. The contribution of small-scale forest-based processing enterprises to rural non-farm employment and income in selected developing countries. Paper FAO:MISC/85/4. United Nations. Rome, Italy.

Food and Agriculture Organization. 1985d. The ecological effects of eucalyptus. FAO Forestry Paper No. 59. United Nations. Rome, Italy.

Food and Agriculture Organization. 1986a. Small-scale forest-based processing enterprises: Summary of key problems. Secretariat Paper for the Expert Consultation on Rural Employment in Forestry-based Processing Enterprises. Document FAO: REFE/86/1.1, October, 1986. United Nations. Rome, Italy.

Food and Agriculture Organization. 1986b. Appropriate forest industries. FAO Forestry Paper No. 68. United Nations. Rome, Italy.

Food and Agriculture Organization. 1987a. Mobilizing funds for world forestry development. Paper CL 91/16. Prepared for the Ninety-first Session of the FAO Council. United Nations. Rome, Italy.

Food and Agriculture Organization. 1987b. Small-scale forest-based processing enterprises. FAO Forestry Paper No. 79. United Nations. Rome, Italy.

Fonseca, G. A. B. da. 1985. The vanishing Brazilian Atlantic forest. Biological Conservation 34:17-34.

Ganguli, B. N. 1986. Trends in financing forestry: The experience of the Asian Development Bank. In: Investment in forestry: The needs and opportunities by S. G. Glover, J. T. Arnot and C. E. Brown (eds.). Proceedings of 12th Commonwealth Forestry Conference. Pacific Forestry Centre. Victoria, British Columbia.

Garrett, K. 1984. The relationship between adjacent lands and protected areas: Issues of concern for the protected area manager. In: Proceedings of world congress on national parks and protected areas by J. A. McNeely and K. R. Miller (eds.). Bali, Indonesia.

Gorse, J. 1985. Desertification in the Sahelian and Sudanian zones of West Africa. Unasylva 37(4)(150):2-18.

Green, G. C. and J. Barborak. 1987. Conservation for development: Success stories from Central America. Commonwealth Forestry Review 66(1):91-102.

Gregersen, H. 1984. Incentives for forestation: A comparative assessment. In: Strategies and designs for afforestation and tree planting by K. F. Wiersum (ed.). Puduc Publishers. Waggeningen, Netherlands.

Gregersen, H. 1987. Social science research needs for tropical forest management. Unpublished report. Rockefeller Foundation. New York, NY.

Gregersen, H., K. Brooks, J. Dixon, and L. Hamilton. 1987. Guidelines for Economic Appraisal of Watershed Management Projects. FAO Conservation Guide 16. United Nations. Rome, Italy.

Gregersen, H. and A. Contreras. 1975. U.S. investment in the forest-based sector in Latin America: Problems and potentials. Johns Hopkins University Press. Washington, DC.

Gregersen, H., S. Draper and D. Elz (eds.). 1988. People and trees: The role of social forestry in sustainable development. Economic Development Institute. World Bank. Washington, DC.

Gregersen, H. and S. McGaughey. 1988. Investment policies and financing mechanisms for sustainable forestry development. InterAmerican Development Bank. Washington, DC.

Guppy, N. 1984. Tropical deforestation: A global view. Foreign Affairs 62:928-965.

Halffter, G. 1981. The Mapimi Biosphere Reserve: A local participation in conservation and development. Ambio 10(2-3):93-96.

Hornick, J. R., J. I. Zerbe and J. L. Whitmore. 1984. Jari's successes. Journal of Forestry 82:663-667.

Hufschmidt, M. M. and E. L. Hyman. 1982. Economic approaches to natural resource and environmental quality analysis. Tycooly International Publisher Limited. Dublin, Ireland.

Kallio, M., D. Dykstra and C. Binkley. 1987. The Global Forest Sector. Wiley and Sons Publishers. New York, NY.

Koljonen, K. 1983. Inter-industry linkages of forestry and forest industry sectors in the Tanzanian economy. Silva Fennica 17:273-278.

Krauss, M. B. 1983. Development without aid. New Press. New York, NY.

Krutilla, J. V. and A. C. Fisher. 1985. The economics of natural environments: Studies in the valuation of commodity and amenity resources. Resources for the Future. Washington, DC.

Kumar, R. 1985. Supply and demand models in forestry—a survey and suggested model. Journal of World Forest Resource Management 1:133-150.

Kumazaki, M. 1982. Sharing financial responsibility with water users for improvement of forested watersheds. A historical review of the Japanese experience. In: The current state of Japanese forestry by R. Handa (ed.) (Vol. II). XVII World Congress. International Union of Forest Research Organizations. Kyoto, Japan.

Laarman, J. G. 1982. Labor intensity and sawmill scale in a labor-surplus economy. Forest Science 28(1):79-91.

Laarman, J. G. 1986. A perspective on private enterprise and development aid for forestry. Commonwealth Forestry Review 65(4):315-320.

Laarman, J. G. 1987. The economic outlook for forestry in Latin America: A hazardous period for projections. Management of the Forests of Tropical America. Department of Forestry. North Carolina State University. Raleigh, NC.

Laarman, J. G. and P. B. Durst. 1987. Nature travel in the tropics. Journal of Forestry 85(5):43-46.

Lanly, J. P. and J. Clement. 1979. Present and future natural forest and plantation areas in the tropics. Unasylva 31(123):12-20.

Little, I. M. D. 1987. Small manufacturing enterprises in developing countries. World Bank Economic Review 1(2):203-235.

Lundgren, B. 1985. Global deforestation: Its causes and suggested remedies. In: Managing Global Issues: Reasons for Encouragement. Proceedings of the Club of Rome Conference (1984). Helsinki, Finland.

McGaughey, S. E. 1986. International financing for forestry. Unasylva 38(1)(151):2-11.

McGaughey, S. E. and H. M. Gregersen. 1983. Forest-based development in Latin America. InterAmerican Development Bank. Washington, DC.

McNeil, D. L. 1981. Tropical forest industries: A transnational corporation view. Commonwealth Forestry Review 60(2):105-112.

Moeller, B. B. 1984. Is the Brazilian Amazon being destroyed? Journal of Forestry 82:472-475.

Myers, N. 1984. The primary source: Tropical forests and our future. W. W. Norton Publishers. New York, NY.

Nelson, M. 1973. The development of tropical lands: Policy issues in Latin America. Johns Hopkins University Press. Baltimore, MD.

Norgaard, R. B. 1985. Environmental economics: An evolutionary critique and a plea for pluralism. Journal of Environmental Economics and Management 12:382-394.

Norgaard, R. B. 1987. Economics as mechanics and the demise of biological diversity. Ecological Modelling 38:107-121.

Obura, W. B. O. 1986. The role of the African Development Bank. Commonwealth Forestry Review 64:212-213.

Oltremari, J. V. and R. G. Jackson. 1985. Chile's national parks: Present and future. Parks 10(2):1-4.

Page, J. M., Jr. 1978. Economies of scale, income distribution, and small-enterprise promotion in Ghana's timber industry. Food Research Institute Studies 16(3):159-182.

Page, J. M., Jr., S. R. Pearson and H. E. Leland. 1976. Capturing economic rent from Ghanaian timber. Food Research Institute Studies 15(1):25-51.

Peterson, G. L. and A. Randall. 1984. Valuation of wildland resource benefits. Westview Press. Boulder, CO.

Place, S. E. 1985. Ecological conservation and rural development in Latin America: The case of Tortuguero, Costa Rica. In: Proceedings of the Conference of Latin American Geographers by L. M. Pulsipher (ed.). Louisiana State University Press. Baton Rouge, LA.

Prance, G. T., ed. 1986. Tropical rain forests and the world atmosphere. AAAS Selected Symposium 101. Westview Press. Boulder, CO.

Repetto, R. 1988. The forest for the trees: Public policies and the misuse of forest resources. World Resources Institute. Washington, DC.

Rockefeller Foundation. 1987. Research needs and opportunities for improved management of tropical forests. Background paper for Strategy Meeting on Tropical Forests (Bellagio, Italy). New York, NY.

Romm, J. 1982. A Research Agenda for Social Forestry. International Tree Crops Journal, 2(1982):25-59.

Ross, M. S. and D. G. Donovan. 1986. Land clearing in the humid tropics. IUCN/IIED Tropical Forest Policy Paper No. 1. IUCN. New York, NY.

Royer, J. and C. Risbrudt (eds.). 1983. Nonindustrial private forests: A review of economic and policy studies. School of Forestry and Environmental Studies. Duke University. Durham, NC.

Ruzicka, I. 1979. Rent appropriation in Indonesian logging: East Kalimantan, 1972/3-1976/7. Bulletin of Indonesian Economic Studies 15(2):45-74.

Salim, A. R. 1983. Employment in the primary wood-based industries of Peninsular Malaysia. Malaysian Forester 46:20-25.

Schmithusen, F. 1976. Forest utilization contracts on public lands in the humid tropics. Paper FAO:MISC/76/6. Food and Agriculture Organization. United Nations. Rome, Italy.

Schwartzman, S. 1986. Bankrolling disasters: International development banks and the global environment. Sierra Club. San Francisco, CA.

Sedjo, R. A. 1980. Forest plantations in Brazil and their possible effects on world pulp markets. Journal of Forestry 78(1):702-705.

Sedjo, R. A. 1983. The comparative economics of plantation forestry: A global assessment. Resources for the Future. Washington, DC.

Sedjo, R. A. and M. Clawson. 1983. How serious is tropical deforestation? Journal of Forestry 81:792-794.

Spears, J. 1985. Role of development banks in forestry financing. In: Proceedings 1985 National Convention. Society of American Foresters. Bethesda, MD.

Thresher, P. 1981. The present value of an Amboseli lion. World Animal Review 40:30-33.

U.S. Congress. 1986. Protecting tropical forests in developing countries. Ninety-Ninth Congress, 2d Session. House Report 99-476. Washington, DC.

Vergara, N. 1984. Agroforestry systems under community forestry: concepts, classification and use in the humid tropics. Food and Agriculture Organization. United Nations. Rome, Italy.

Wardle, P. and M. Martin. 1986. World outlook for forest products. In: Proceedings of Division 4. XVIII World Congress. International Union of Forest Research Organizations. Ljubljana, Yugoslavia.

Westoby, J. C. 1962. Forest industries in the attack on economic underdevelopment. Unasylva 16(4):160-201.

Westoby, J. C. 1978. Forest industries for socioeconomic development. Eighth World Forestry Congress. Jakarta, Indonesia.

Working Group on Watershed Management and Development. 1987. Sustainable growth and development: The role of watershed management. Department of Forest Resources. University of Minnesota. St. Paul, MN.

World Bank. 1984. Annual Report on FY84 Bank and IDA lending for Agriculture and Rural Development. Washington, DC.

World Resources Institute. 1985. Tropical forests: A call for action. Washington, DC.

World Resources Institute. 1987. Tropical Forestry Action Plan: Recent Developments. Unpublished report. Washington, DC.

World Bank and Food and Agriculture Organization. 1981. Forestry research needs in developing countries: A time for reappraisal? XVII World Congress. International Union of Forest Research Organizations. Kyoto, Japan.

Wright, M. R. 1985. Kuna Yala: Indigenous biosphere reserve in the making? Parks 10(3):25-27.

10

INTERNATIONAL TRADE IN FOREST AND RELATED PRODUCTS

Thomas R. Waggener

International trade in forest products has grown considerably as a subject of interest to the nation's forestry community. The essence of forest products trade has been stated by Leigh (1971) as activities designed "to bring the forest resources of the world to those who wish to have them and in the form in which they would be most useful." While biologically somewhat ubiquitous, the economic value of worldwide forest resources is distributed rather poorly between suppliers of wood products and interested consumers. Some of the world's inhabitants enjoy a relative abundance of wood, while other less fortunate consumers must look to other parts of the world for a supply of wood that is appropriate in both form and variety. Again, Leigh (1971) provides a concise summary of this global dilemma:

> Wood is universally abundant and there is scarcely any country in the world in which trees do not grow, but some countries do not have sufficient wood for their own needs whilst others have more than they need. Additionally, the kind of tree varies very much, depending on geographical position and various other factors, which means that an interchange of supplies becomes necessary.

Forest economics research applicable to international trade in forest and related products is, in one sense, only a broadening of the agenda appropriate to forest resource economics research in general. The distinction is that the product (or service) of concern involves two or more nationally sovereign political jurisdictions. To be effective, the application of economic concepts, analytical methods, and research processes must reflect the significance of this international discontinuity. A topic closely related to international trade is economic and social development of lesser developed countries. Experience has demonstrated, however, that a separate program of economic research on the international trade aspects of forestry is both feasible and necessary.

Review of Information and Previous
Research Recommendations

Information and Research Through the 1970s

A number of reports and reviews of forest economics and related research have directed attention to international trade in forest products. One of the earliest assessments was compiled and published by the Social Science Research Council (1935). Bounded by contemporary issues of the time, the assessment emphasized concern over timber depletion and the ability of the nation's forests to provide an adequate supply of timber. International trade was addressed as part of a general concern over forest resource inventories; "studies of the forest resources of several foreign countries have been made . . . in connection with investigations of markets for American lumber" (Social Science Research Council 1935, p. 11). In a marketing and distribution context, the assessment cited several documents containing historic compilations of trade statistics (Bureau of Statistics 1909, Hough 1880). Also introduced was an extensive listing of "surveys of present and potential markets abroad" (Social Science Research Council 1935, p. 30).

The U.S. Department of Commerce and the USDA–Forest Service also directed early attention to international markets in forest products. In the early 1900s, the U.S. Department of Commerce prepared a significant report on the U.S. lumber industry (Bureau of Corporations 1913-1914), while in 1920 the USDA–Forest Service reported (to the U.S. Senate) on the status of various forestry issues, including timber depletion, lumber prices, lumber exports, and concentration of timber ownership (Forest Service 1920). The USDA–Forest Service report is illustrative of early concern over international trade as a factor supposedly contributing to shortages in the nation's supply of domestic timber. Among the report's conclusions are

> The export trade in lumber does not have a serious bearing upon timber depletion from the standpoint of quantity, but does have an important bearing upon the duration of our limited supply of highgrade timber, particularly of hardwoods. . . . The problem presented by lumber exports is not serious from the standpoint of quantity. It may prove serious from the standpoint of quality. Scarcity of high quality products essential to our ship and car building and many other industries is the first and one of the most serious effects of timber depletion. . . . Our fundamental national policy, however, should be for timber growth rather than the regulation of timber use (Forest Service 1920, pp. 4, 58 and 59).

Trade with Canada and heightened interest in Western Europe's reconstruction following World War I, dominated the USDA–Forest Service's 1920 analysis. Forest products trade issues centered on concern over depletion of high quality U.S. hardwood forests and on additional price and supply competition that would have to be met by the U.S. forest products industry as a result of heightened trade activities.

One of the earliest systematic recognitions of forest economics research directed at international trade was contained in *Research in the Economics of Forestry*

(Duerr and Vaux 1953). The document included a brief section identifying international trade in timber products as a legitimate field for research (Elchibegoff 1953). Interestingly, the section was a secondary subtopic under research concerning the location of economic activity, which in turn was a topic under research concerning the "forest economy at large." A companion bibliography (identifying research published during the period 1940-1947) contained a modest section devoted to foreign trade (48 references cited) (U.S. Department of Agriculture 1950). A 1955 supplement to the bibliography (covering 1948-1952) contained an additional 15 citations on the subject (U.S. Department of Agriculture 1955). A similar bibliography, published by the USDA–Forest Service in 1941, cited 33 entries under the heading "Foreign Trade" (Nelson 1941).

International trade in forest products was given additional attention by a 1977 Resources for the Future symposium (Clawson 1977). For the symposium, Zivnuska (1977) prepared a major review of research involving international forest economics. Zivnuska noted the traditional bias against a broader international context for forest economics research when he stated, "There is a common tendency to develop studies in forest economics within the limits of the domestic economy, with international aspects of problems being either ignored or treated as a separable issue, subject to only limited investigation and handled as an addendum or adjustment to the main analysis" (Zivnuska 1977, p. 435). Zivnuska argued the need to incorporate an international dimension in forest economics research, and he urged that international inquiry not be developed as a distinct and separate field of analysis. Zivnuska recognized, however, that some economic research questions involving trade are distinctively international in character and should be treated as such. Zivnuska also suggested that international tourism (often dependent on the availability of forest resources) be viewed as a "forest products export," and that the economic demand for tourism-type recreation (including local economic impacts) be considered a valid topic for economic research (Zivnuska 1977, pp. 456-460). Others at the symposium also noted (briefly) a need for research focused on international trade. Clawson argued that foreign trade in wood products directly impacts the production of wood by U.S. forests and that substantially more should be known about the demand for wood in other parts of the world (Clawson 1977). Similarly, Castle argued that "for much research, the international dimension should be an integral part of the problem; a separate research program on international forestry may not be the best approach" (Castle 1977, p. 12).

As can be seem from the aforementioned reviews, economic research focused on international trade in forest products had grown significantly as a subject of interest through the late 1970s. By 1980, trade was generally acknowledged as a subject deserving placement on the agenda of economic researchers. A number of generalizations can be drawn from these reviews. Consider the following.

International perspective. Nearly all prior reviews of forest economics research implied that *all* such research be broadened to include an international perspective. The forests of the U.S. (or North America) can no longer be viewed in isolation. Whether defined in terms of supply (resources) or demand (markets), contemporary forestry issues often reflect conditions occurring in a dynamic and increasingly integrated global economy. Excluding important global factors unnecessarily limits the usefulness of much economic research.

Implications for U.S. forestry. Prior reviews of forest economics research have also stressed the need to consider international trade as an explicit factor impacting the domestic affairs of industrial forestry enterprises. The trade-related implications for forest management, harvest levels, and trade-offs with multiple-use values are deserving of explicit attention.

Aggregate world demand and supply sources. Acknowledged by prior reviews is a U.S. forestry sector that operates in concert with a global timber economy— a condition which implies a need for greater recognition and analysis of aggregate global demand and supply. The productive capacity of the North American forest must be assessed in the context of alternative global supply opportunities; research in such a context must reflect comparative economic as well as biological productivity. Deserving special recognition are alternatives to domestic production capability, including man-made forests (e.g., intensively managed, fast-growing plantations).

Changing demands for wood. Prior reviews of forest economics research also acknowledge the need to carefully assess the nature of forest products required by specific end users. Product requirements (e.g., grades, standards) in worldwide markets are often significantly different from country to country. As such, dis-aggregation should be introduced into economic analyses in order to adequately reflect these varying demands. Research is also needed on the price responsiveness of demand for specific products (demand elasticity).

International trade policies. Recognized by past reviews is the need to ac-knowledge both national and international policies impacting transactions in forest products. Of special importance are various trade barriers (tariff and nontariff) imposed by both producer and consumer nations. The interdependence of the United States and Canada with respect to the supply and consumption of wood products necessitates a special effort be made to integrate economic analysis of North American forest economies.

U.S. comparative advantage. Previous reviews have rejected the notion that the U.S. wood-based economy is an independent, self-sufficient sector operating within a global economy. To the extent that the U.S. consumes forest products supplied by other nations, U.S. consumers benefit from an enlarged supply of lower priced wood products. To the extent that the U.S. exports wood-based products, the nation's forest products industry enjoys the benefits of enlarged markets for the products they manufacture, that is, greater returns to investments in timber growing and timber processing activities. To sustain a viable forestry sector, however, the U.S. must produce and deliver forest products on a cost-competitive basis in both domestic and foreign markets. Analysis of comparative advantages are viewed as critical to more widespread understanding of operations in a global timber economy. Issues concerning production costs, technology, and distribution should be framed in comparison to overall global situations.

International tourism. Trade in forest products is traditionally considered in terms of wood product commodity flows. Hence, forest economics research has largely focused on market interactions involving the exchange of physical products— not services. Past reviews of forest economics research have begun to recognize international tourism as an activity based (in part) on the attractiveness of forests and related natural resources. The determinants of such trade in recreational services has largely gone unexplored; so too the mechanisms for conducting and subsequently expanding research focused on international tourism.

Descriptive Information and Analytical Research Since 1980

As described above, a modest number of systematic reviews of forest economics research agendas were accomplished prior to the late 1970s. What progress—if any—has been made since then, especially since the 1977 research review conducted by Resources for the Future? More pointedly, what has occurred from the perspective of economics research focused on international trade in forest products? Some highlights of recent activities in descriptive and analytical research are summarized as follows.

The availability of information describing and documenting international trade in forest products has expanded greatly since the late 1970s. Such can be viewed as a significant accomplishment over early reports which described U.S. trade flows patterns (Pratt 1918), European trade activities (Oxholm 1932) and Pacific forest products trade (Elchibegoff 1949). Resources for the Future, for example, documented post–World War II trends in U.S. forest products trade (Sedjo and Radcliffe 1978), while the USDA–Forest Service compiled and reported forest products trade information nationally (Forest Service 1987b) and regionally (Pacific Northwest) (Forest Service 1987a) in conjunction with the preparation of more general reports on production and consumption of U.S. forest products. Inclusion of an international trade dimension in renewable resource assessments (Forest Service 1980, 1982) and reviews of specific market opportunities have also increased nationwide recognition of international trade in forest products (Darr and Lindell 1980).

Workshops and symposia have also contributed to the identification of important research efforts concerning international trade. In 1980, a Resources for the Future symposium concentrated on a number of trade-related topics, including trade barriers and log export restrictions (Sedjo 1981). Similarly, the University of Washington's Center for International Trade in Forest Products (CINTRAFOR) sponsored several international symposia dealing with global developments in wood products (Bethel 1984, Schreuder 1986, Johnson and Smith 1988), and most recently, focused a symposium on trade in pulp and paper products (Schreuder 1988). Trends and developments in North American forest products markets and trade were also subject of a 1986 conference sponsored by the Forest Products Research Society (1987).

Reporting of statistical trends for U.S. wood products trade has also accelerated in recent years. For example, the Foreign Agricultural Service of the U.S. Department of Agriculture now issues (quarterly) brief descriptions of product and market trends (Foreign Agricultural Service 1987), and prepares annual country attache reports for wood products. Internationally, the United Nations Food and Agriculture Organization continues to report statistics concerning global timber harvests, production of forest products, consumption of wood-based forest products, and trade in forest products (Food and Agriculture Organization 1987).

Similar to descriptive studies of trade flows and markets, a wide variety of "trade issue" studies have been undertaken. Economic analysis has often provided the primary analytical framework for such research. Issue studies have included, for example, analysis of log exports and trade restrictions (Josephson 1975), trade barriers (Bourke 1986), exchange rates (Adams et al. 1986), timber inventory (Williams 1987), softwood lumber trade, and transportation and distribution channels (Wisdom 1987). Economic analyses of the competitiveness of the U.S. forest products industry, within a global context, have also been undertaken (International Trade Administration 1984; Office of Technology Assessment 1984), as have assessments of the manner in which wood-based firms organize themselves to undertake international activities (Bilek and Ellefson 1987, Ellefson and Stone 1984).

Although not economics research per se, general descriptions of international trade systems and guides to trade activities have been prepared and published in recent years, with their intent being to increase understanding of international trade activities. Leigh (1971) is an early example which focused on lumber trade with the United Kingdom. More recently, the USDA–Foreign Agriculture Service published a U.S. guide to exporting solid wood products (Foreign Agriculture Service 1986).

International tourism, in the context of forests and natural resources, has received limited attention in terms of descriptive information. While some efforts have been undertaken by sociology and interpretative researchers, studies involving economics have been very few. One of the earliest efforts entailed study of international travel and its relation to domestic outdoor recreation (Outdoor Recreation Resources Review Commission 1962). Gray (1970) helped establish a logical linkage between international travel and international trade, although the natural resource implications (including forestry) of such linkages were not made. In the 1970s, cultural factors influencing international travel were assessed (Gess 1972 relative to Japanese travel, for example). In the same period, international travel statistics were increasingly compiled (U.S. Travel Service 1978). A major 1979 George Washington University symposium addressed a number of tourism topics, including economic issues relevant to forest-based recreation and tourism. For example, Manning (1979) addressed the topic of international tourism and the U.S. national park system. Again, a link to natural resources (including forests) was only obliquely introduced.

In addition to the reporting of descriptive information concerning international trade in forest products, significant analytical research has been undertaken since the late 1970s. In a forthright manner, Zivnuska (1977) argued for additional emphasis on a strong analytical framework for research, one which would integrate international trade flows into comprehensive economic analyses of forestry issues. Referring only to the all-too-common analyses of domestic forestry issues, Zivnuska stated, "There is a need to develop research within a framework which directly includes the international dimensions as an integral element." (Zivnuska 1977, p. 437). In addressing analyses of the North American timber economy, Zivnuska concluded that the situation, then, is that of a single forest product supplying region with two sectors, separated by a political boundary, which serves two major forms of markets, a domestic North American market and an off-shore world export market. . . . it is impossible to develop an adequate analysis of the timber and forest products outlook for either country in separation from the other. Despite this, no such integrated study of the region as a whole has yet been undertaken (Zivnuska 1977, pp. 444, 445).

Forest economics research has responded—at least in part—to the suggestions offered by Zivnuska.

Resource assessments have increasingly taken a more global perspective. For example, USDA–Forest Service assessments (undertaken in response to the Forest and Rangeland Renewable Resources Planning Act of 1974) have given explicit attention to international trade (Forest Service 1980, 1982). Darr (1981) summarized such efforts, while Zivnuska (1981) critically reviewed the theoretical and analytical approaches used to prepare the assessments. A traditional "consumption gap" approach and the lack of a consistent economic equilibrium methodology were identified as the most serious limitations.

Changes in technologies for manufacturing forest products and the relationship of such changes to economic productivity and trade have also been subject of forest economics research. Timber utilization research and related economic impacts were assessed by Haygreen et al. (1986). The relationship between wood processing technologies and competitiveness in a global market received considerable attention from a 1986 Forest Products Research Society (1987) conference. It was also the subject of discussion at a 1987 International Trade Symposium sponsored by CINTRAFOR (Johnson and Smith 1988). The inclusion of technological change in analytical modeling of market dynamics was specifically addressed in a 1987 review of global trade models (Ince et al. 1987).

Research to develop appropriate analytical methods for analyzing international resources and markets has also progressed since the late 1970s (Adams and Haynes 1980; Brooks 1987; Cardellichio, Adams, and Haynes 1987). Global and regional resource assessments have utilized improved analytical and economic methods which have been designed to accommodate economic interdependence in forest products markets and trade (Sedjo and Lyons 1987, Food and Agricultural Organization 1986a and 1986b, Prins 1987). Perhaps the most ambitious modeling effort has been the forest sector project of the International Institute for Applied Systems Analysis (IIASA) (Kallio et al. 1987). As an international cooperative effort, its objective was to model the "economic system that increasingly links the world's forests through interactive trade and by the transnational or even global effects of environmental degradation," and to evaluate "possible structural changes in the forest sector that might develop as a consequence of policy decisions." (Dykstra and Kallio 1987, p. 4). The project resulted in the development of the Global Trade Model (GTM), a model which sought to permit analysis of forest products trade in a worldwide setting.

The Global Trade Model provided significant experience in conceptually formulating complex global forestry and trade systems, structuring analytically the worldwide forest economy, developing data bases sufficient for implementation of trade models, and formulating policy constraints and economic forecasts to simulate future trends. As such, the model represents a landmark "first generation" means of assessing international trade systems in a global environment. It has been installed at the University of Washington (CINTRAFOR) for further evaluation and testing in a North American context (Cardellichio and Adams 1987). Related work has been conducted by other North American and foreign researchers. The University of Washington has also sponsored efforts to further evaluate alternative methods and models (Cardellichio, Adams, and Haynes 1987).

Analytical economic research directed at international tourism and forest-based services has been very limited relative to that focused on international trade in wood products. Wander and Van Erden (1979) explored analytical approaches to estimating the demand for international tourism, although direct ties to forest-based tourism were not made. Similarly, Hawkins (1979) investigated the regional economic impacts of international tourism on U.S. outdoor recreation. A more systematic treatment of international tourism (oriented to natural resource-based recreation in the national park system) was undertaken by Joerger (1979). The latter included identification of related economic studies and the development of a framework for investigating economic and policy issues.

Emerging International Trade Issues

International trade in forest products has prompted some to identify emerging international trade issues in need of forest economics research (Adams 1987, Kaiser 1987, Sedjo 1987). Recognizing the suggestions of these individuals and others, the following areas are offered as topics requiring significantly greater research attention in the years ahead.

- *Trends and determinants of global demand for forest products.* Concern should be directed to disaggregation of product demand so as to reflect significant differences in terms of production and end use requirements. Differentiation should be made between mature markets and emerging markets. Within the context of specific consumer markets, determinants of demand (shifters) should be identified and analytically incorporated.
- *Availability of raw materials for international trade.* Attention should be directed to comparative economics evaluations (in terms of harvesting, processing, and distribution). National and regional advantages in production and distinctions between economic and physical productivity should be identified. Comparative economic productivity of natural managed forests versus plantations should be assessed.
- *Cost structure and productive capacity of regional and national forest industry.* Recognition should be given to raw material, labor, and capital costs as well as profitability of delivered products (including final transportation and distribution costs).
- *North American trade interdependence.* Attention should be directed to U.S. and Canadian trade activities by product, especially the significance of trade for production, regional implications of trade activities, consumer dependence

on trade, complementary versus competitive trade activities, and consumer welfare implications of trade relationships.

- *Value added production.* Recognition should be made of potential comparative advantages for enhancing profitability through complete processing of various products for end market requirements (quality, different product standards) and for total value recovery per unit of input.
- *Technological change in forest products industry.* Of concern should be labor-saving technology (per unit of output), wood saving technology (less wood input per unit output), wood production technology (intensive management, handling, recovery), demand creating technology (efficiencies in utilization, new product technology, product substitution), and economic incentives for technological change.
- *Macroeconomic impacts on international trade in forest products.* Consideration should be given to exogenous variables affecting demand for wood-based products and producer profits (e.g., interest rates, transportation costs, exchange rates, taxes, and tariff barriers).
- *Market access.* Attention should be given to trade barriers (tariff and nontariff) influencing market access, economic competitiveness relative to alternative suppliers, product quality and standards (end use requirements), and economic risks involved in international trade (credit, financing, product acceptance, insurance, inspections, and certification).
- *Market knowledge and intelligence.* Attention should be given to trends and determinants of markets by products, destinations, end use requirements, standards and grades, and producer knowledge of specific markets and channels of distribution.
- *Public policy implications of international trade.* Of concern should be balance of payments, complementary and competitive markets; impacts of export-based trade restrictions; trade barriers and retaliation; regional economic impacts of international trade; direct and indirect benefits of trade; contribution of trade to regional and national economies; and employment and income impacts.
- *Forest-based international tourism.* Attention should be devoted to the relationship of forest and natural areas to demand for international travel, and the impact of international travel on regional economies.

Considerable progress has been achieved in structuring economic research focused on broad issues of international trade in forest products. Review of previous research and assessment of current issues, however, indicates that a considerably enhanced research effort is both feasible and necessary. The agenda is broad, reflecting both the relative lack of specific research in the past and the rapidly expanding global integration of forest products markets. In many respects, strategic research directions should parallel traditional domestically oriented forest economics research programs. Such would be natural, reflecting an overall geographic expansion of the scope of forest economics research to account for international interactions. Some research priorities, however, should focus specifically on the unique nature of international trade and the economic issues which are not normally considered critical to single-country analyses.

Seven strategic directions are recommended for future research involving international trade in forest products. Within each direction, many individual research projects are possible. The scope and definition of such projects is obviously dependent upon the interests and capabilities of concerned research organizations. Over time, new issues will emerge while others may decline in importance. Nevertheless, implementation of the following strategic directions should collectively advance research of international trade in forest and related products. The strategic directions are

- Develop improved understanding of wood and fiber supplies in a worldwide context.
- Develop improved understanding of technical and economic conditions important to comparative advantages in wood product production.
- Develop improved understanding of solid wood and wood fiber consumption patterns.
- Develop cost-effective analytical systems for assessing worldwide trade in forest products.
- Develop analytical ability to address policy issues important to international trade in forest products.
- Develop effective means of facilitating trade and marketing of forest products.
- Develop improved understanding of forest-based international tourism.

Develop Improved Understanding of Wood and Fiber Supplies in a Worldwide Context

Production of forest products has historically reflected the availability of timber from natural forests. Such were composed of unmanaged natural stands from which timber was extracted, transported, and processed locally for marketing either domestically or through trade. Through time, however, managed forests (including intensively managed plantations) have emerged as significant contributors to the world's timber supply. However, meeting timber demands of the future will require more than an adequate physical supply. Comparative advantage in a global context will require that timber be efficiently managed, harvested, transported, and processed into products which can be delivered competitively to viable markets. Investigations of regional supply opportunities, stages of production, and emerging competition for nontraditional supply sources are critical to understanding future global trade patterns. Information about timber supply characteristics (e.g., species, size and quality, growth rates, utilization standards, engineering-processing characteristics) will be required to supplement traditional concerns for gross timber volume. Research topics of the following nature should be addressed.

Evaluate North American softwood's comparative advantage in major world markets. Analyze technical and economic competitiveness of the U.S. in terms of meeting end use requirements of off-shore foreign markets.

Assess enhanced marketing opportunities for North American hardwoods. Examine market opportunities for U.S. hardwoods in major world markets and new product niches for underutilized species and grades of timber.

Examine aggregate supply implications of global plantations. Explore global developments in fast-growing plantations (softwood and hardwoods) with emphasis on product mix and markets targeted by competitive supplier-nations. Determine impacts of these products and markets on U.S. competitiveness.

Assess Asian hardwood, especially substitution potentials and changing raw material supplies. Analyze market strategies of major Asian hardwood producers and associated trade restrictions and market adjustment opportunities. Evaluate industrial restructuring of existing processors of Asian hardwoods and potential substitution of U.S. softwoods.

Develop Improved Understanding of Technical and Economic Conditions Important to Comparative Advantages in Wood Product Production

The availability of adequate raw material has been a dominant factor in the location of forest products facilities. Wood fiber's low value-to-weight ratio has typically required timber processing within or in close proximity to forests so as to reduce waste and residue and to increase value added prior to long distance shipments to markets. Close proximity to physically available raw material, however, is no longer a sufficient condition to assure economic feasibility of production. Total production costs relative to market opportunities are critical to determination of comparative advantage. Changes in cost impacting technology, transportation, exchange rates, raw material mixes, capital and labor, and public policies can all have a dramatic impact on the competitive structure of forest products enterprises—and regional comparative advantage. Topics of the following nature should be addressed by research.

Assess North American industrial restructuring, especially in terms of capacity and investments to serve foreign markets. Evaluate industrial capacity and determine investment levels required to serve foreign markets. Define technical and economic barriers facing adjustment to changing markets.

Appraise global suppliers of softwood, especially regional competition in timber growing and industrialization. Evaluate emerging supply sources and expansion of processing capacity.

Develop Improved Understanding of Solid Wood and Wood Fiber Consumption Patterns

Consumption of wood products is a complex reflection of numerous market conditions (e.g., availability, price, quality, tastes, preferences [including cultural]). Some markets are relatively mature (well-established patterns of end uses) and are impacted mainly by economic and demographic factors leading to demand shifts. Yet others have yet to mature, exhibiting rapid change in the use of wood. Acceptability of wood for specific end uses, together with availability and price, is frequently quite variable. An understanding of the determinants of demand, together with demand shifters, in specific markets is required to identify significant market niches and market expansion potentials. As such, research topics of the following type should be addressed.

Assess future levels of wood product consumption by U.S. and Canadian markets. Evaluate demand shifters for U.S. and Canadian markets, especially in terms of regional specialization and comparative advantage. Determine balance of trade conditions for North American wood product markets.

Evaluate determinants of and trends in Asian markets for wood products. Assess species substitution and self-sufficiency in the context of logs versus processed wood. Determine implications of restrictions on domestic processing and exporting.

Assess implications of maturing Western European markets under conditions of increased competition. Determine North American opportunities for expanding European trade. Evaluate the ability of the U.S. to compete in European markets, given the requirements and product standards imposed by such markets.

Develop Cost-Effective Analytical Systems for Assessing Worldwide Trade in Forest Products

The evaluation of international trade flows and market trends has rested largely on partial equilibrium analysis. The "rest of the world" has been implicitly assumed to be either constant or unimportant. At the extreme, bilateral trade flows have been projected on the basis of trends in consumption or trade without explicit recognition of the interdependence of a rapidly evolving global market for forest products. It is imperative that greater analytical capability be developed for evaluating this interdependence and for improving projections of trends based on underlying economic and political factors. Simulation of alternative market developments will enhance the ability to evaluate political and economic changes in a more complete and realistic manner. Determination of appropriate levels of specification and complexity in analytical systems will be critical in light of the relatively high cost of analytical research. Consideration should be given to the following research topics.

Develop more effective market-determined models of global markets for forest products. Determine and account for demand-pull factors governing global aggregate demand for forest products. Account for the impact of transportation and related barriers to trade, and clarify the responsiveness of regional and bilateral trade flows to changes in market demand.

Develop more effective supply-determined models of global markets for forest products. Determine and account for supply constraining conditions governing consumption of wood products. Give special consideration to exporting ability as influenced by forest land base, timber inventories, size and quality mix, and harvest decisions. Develop analytical ability to assess competition with domestic consumption and pressures for export controls on raw materials. Determine means of relating implications of trade to old growth timber supplies and shifts to second growth harvests.

Develop regional and national models capable of generating aggregate and partial analyses. Develop analytical ability to examine global versus regional systems and to disaggregate by products and markets. Assess effectiveness of bilateral versus general equilibrium systems. Generate ability to determine elasticities for local market demand and regional supply, and cross elasticities for species and product substitution.

Develop Analytical Ability to Address Policy Issues Important to International Trade in Forest Products

Closely related to the development of cost-effective analytical systems for trade analysis is the need to improve the ability to address policy issues and alternative policy scenarios, especially in terms of baseline economic assumptions and forecasts. Such analysis must be capable of incorporating relevant political and exogenous considerations in addition to conventional concern over economic efficiency. Related work on local (regional) distributional impacts (direct, indirect, and induced) resulting from international trade is critically needed in the context of both fairness of trade policies and equity of trade policies (e.g., log versus lumber exports). Issues such as the export of jobs or loss of value added must be addressed in a relevant policy context if economic analysis is to be useful to policy formation. Topics of the following nature are deserving of research attention.

Evaluate global forest product system responses to changes in macroeconomic conditions. Assess international transmission of economic instabilities and the role of trade barriers and tariffs. Evaluate implications of changing exchange rates, uncertainty, interest rates, construction cycles, and shifts in demand.

Evaluate effectiveness of policy instruments used to intervene in forest products trade. Determine the effectiveness of alternative trade policies and strategies and the distributional consequences of trade incentives and regulations.

Analyze regional consequences of trade in forest products. Identify who gains and who loses as a result of international trade activities from a regional perspective (within a country). Assess the regional direct and indirect benefits of trade in forest products and the value added and employment impacts of trade expansion. Evaluate domestic processing in terms of gains for whom.

Develop Effective Means of Facilitating Trade and Marketing of Forest Products

The U.S. forest products industry has historically been oriented to domestic markets. Knowledge of production and supply issues greatly exceeds knowledge of markets and trade. Domestic marketing is largely based on volume transactions of standardized commodities, with end uses well defined in terms of wood applications, standards, grades, and technology. International markets are much more complex, transactions more difficult, and product end uses quite different in terms of product requirements. These factors essentially constitute a market barrier for the inexperienced producer. Various programs and incentives have been implemented to enhance trade, expand markets, and facilitate transactions. Market promotion, trade financing, and information services are generally considered as governmentally justified initiatives to assist forest products trade. Economic analysis of the cost effectiveness of such efforts is very limited. The role, scope, and efficiency of marketing and trade facilitation is itself in need of research. Consider the following research topics.

Assess market access and barriers to trade in forest products. Evaluate national and regional trade activities and economic development objectives in terms of fair and unfair business practices. Evaluate the relative effectiveness of incentives in the context of subsidy effects of national programs and policies. Assess the vulnerability of forest products trade when used as a bargaining chip in trade retaliation actions.

Assess product standards applied by international markets. Evaluate U.S. versus international end market requirements (e.g., product quality). Evaluate consequences of grading standards, conversion factors, and engineering and design for enhanced wood applications. Evaluate implications of the metric system applied in the global marketplace for wood.

Analyze marketing opportunities and information requirements for market identification. Assess marketing information loops and review information needs required to participate in foreign markets (e.g., understanding of wood applications, durability, strength, and economy). Evaluate producer information requirements about foreign markets (e.g., end use requirements, distribution channels, reliability, financing).

Develop Improved Understanding of Forest-Based International Tourism

Wood-based products are relatively easy to identify; their physical movement from point of origin to point of consumption is easy to ascertain. They can be measured with substantial ease as they flow across international boundaries. Recreational and amenity benefits provided by forests, however, occur on site with visitors creating their own value through personal activities or experiences. The provision of such services to an international visitor is an export activity similar in nature to the sale of commodity products. Accounting procedures designed to track international exchange of services have yet to be fully developed; those in existence have not been adapted or applied to forest-based tourism. Identification of some key attributes (demand shifters such as taste and preference) important to international visitors in the context of the U.S. national park system have received some attention. Trends in visitation levels, seasonal patterns of travel, and joint trip purposes have also been broadly studied for outdoor recreation. Application of recreation demand theory to international tourism would assist in documenting forest-based tourism as an international export service. Research is also needed on the local and regional economic impacts of tourism (distributional analysis). Such could address community concerns regarding land use trade-offs and provision of local benefits. Consider the following research topics.

Assess the demand for forest-based international tourism. Determine the significance of forests and related natural resources as means of stimulating international tourism. Assess joint purpose visitation and the role of forests in such visits (e.g., forests versus Disneyland). Evaluate packaged travel as a means of increasing forest-based tourism activities (e.g., education of travel agents, awareness of group visitors).

Evaluate the economic efficiency of and benefits accruing from hosting international tourists. Assess the regional and local economic impacts of tourism, including direct and indirect linkages to tourism, returns to land management, and incremental differences in domestic versus international tourism impacts. Assess externalities associated with international forest-based tourism.

Assess the macroeconomic implications of international forest-based tourism. Analyze balance of payments in services provided and determine the role of international tourism in the stabilization of rural resource economics. Evaluate means of enhancing visitor satisfactions (e.g., cultural and interpretative services as means of stimulating demand).

Summary and Conclusions

International trade has grown significantly as a social and political issue in recent years. For the forestry community at large and for the forest economics research community specifically, international trade in forest and related products implies a number of important and far reaching consequences. Policy and programs, however, must give international trade explicit recognition as a component of a highly complex global system involving forest land use, timber management, timber harvesting, industrial wood-based structure, and consumer demand for a variety of wood-based products. International trade is also growing in importance as a means of bringing the nations of the world together in a positive social, cultural, and economic fashion. Trade in forest products has a very definite role to play in the development of such linkages. Trade in forest products is not, however, insulated from significant problems, be they technical, economic, or political in nature. The existence of such problems and the importance attached to their solutions are ample reasons for directing significant investments toward researching the economics of trade in forest and related products.

Strategies guiding economics research pertaining to international trade in forest products can assume a number of directions. Among the more important would be research that improves—in an international setting—our understanding of wood and fiber supplies; technical and economic conditions important to comparative advantages in wood product production; solid wood and wood fiber consumption patterns; analytical systems for assessing trade; analytical ability to evaluate important policy issues; facilitation of trade and forest products marketing; and forest-based international tourism. Forest resource economics research directed to such areas could do much to improve the effectiveness with which the United States participates in worldwide trade of forest and related products.

Literature Cited

Adams, D. 1987. Research plan: Center for International Trade in Forest Products. Work Plan 1987-89 Biennium (Appendix V). College of Forest Resources. University of Washington, Seattle, WA.

Adams, D. and R. Haynes. 1980. The 1980 softwood timber assessment market model: Structure, projections and policy simulations. Mono. 22. Forest Science. Society of American Foresters. Bethesda, MD.

Adams, D., B. McCarl and L. Homayounfarrokh. 1986. The role of exchange rates in Canadian-United States lumber trade. Forest Science 32(4): 973-988.

Bilek, E. M. and P. V. Ellefson. 1987. Organizational arrangements used by U.S. wood-based companies involved in direct foreign investment: An evaluation. Bulletin 576-1987. Agricultural Experiment Station. University of Minnesota. St. Paul, MN.

Bethel, J. S. (ed.). 1984. World Trade in Forest Products. Proceedings of International Symposium on World Trade in Forest Products. CINTRAFOR and College of Forest Resources. University of Washington Press. Seattle, WA.

Bourke, I. J. 1986. Trade in forest products: A study of the barriers faced by the developing countries. Report prepared for the FAO, United Nations. Forest Research Institute. Rotorua, New Zealand.

Brooks, D. J. 1987. Alternative approaches to modeling international trade: The world assessment market model. In: Forest sector and trade models by P. Cardellichio, D. Adams and R. Haynes (eds.). College of Forest Resources. University of Washington. Seattle, WA.

Bureau of Corporations. 1913-1914. The lumber industry (3 vols.). U.S. Department of Commerce and Labor. Washington, DC.

Bureau of Statistics. 1909. Foreign trade of the United States in forest products, 1851-1908. Bureau of Statistics Bulletin No. 51. U.S. Department of Agriculture. Washington, DC.

Cardellichio, P. A., D. M. Adams and R. Haynes (eds.). 1987. Forest sector and trade models. College of Forest Resources. University of Washington. Seattle, WA.

Cardellichio, P. A. and D. M. Adams. 1987. U.S. experience with the global trade model. In: Forest sector and trade models by P. Cardellichio, D. Adams and R. Haynes (eds.). College of Forest Resources. University of Washington. Seattle, WA.

Castle, E. N. 1977. Research needs in forest economics and policy: An interpretative and evaluative summary. In: Research in forest economics and forest policy by M. Clawson (ed.). Research Paper R-3. Resources for the Future. Washington, DC.

Clawson, M. 1977. American forests in a dynamic world. In: Research in forest economics and forest policy by M. Clawson (ed.). Research Paper R-3. Resources for the Future. Washington, DC.

Darr, D. 1981. U.S. export and imports of some major forest products: The next fifty years. In: Issues in U.S. international forest products trade by R. Sedjo (ed.). Research Paper R-23. Resources for the Future. Washington, DC.

Darr, D. and G. Lindell. 1980. Prospects for U.S. trade in timber products. Forest Products Journal 30(2):16-27.

Duerr, W. A. and H. J. Vaux. 1953. Research in the economics of forestry. Charles Lathrop Pack Forestry Foundation. Washington, DC.

Dykstra, D. P. and M. Kallio. 1987. The IIASA base core projections. In: Forest sector and trade models by P. Cardellichio, D. Adams and R. Haynes (eds.). College of Forest Resources. University of Washington. Seattle, WA.

Elchibegoff, I. M. 1949. United States international timber trade in the Pacific Area. Stanford University Press. Stanford, CA.

Elchibegoff, I. M. 1953. International trade in timber products. In: Research in the economics of forestry by W. Duerr and H. Vaux (eds.). Charles Lathrop Pack Forestry Foundation. Washington, DC.

Ellefson, P. V. and R. N. Stone. 1984. U.S. wood-based industry: Industrial organization and performance. Praeger Publishers. New York, NY.

Food and Agriculture Organization. 1986a. Forest products world outlook projections 1985-2000. United Nations. Rome, Italy.

Food and Agriculture Organization. 1986b. European timber trends and prospects to the year 2000 and beyond (2 vol.). ECE/TIM/30. United Nations. New York, NY.

Food and Agriculture Organization. 1987. Yearbook of forest products: Statistics and economic analysis. Forestry Department. United Nations. Rome, Italy.

Foreign Agricultural Service. 1986. A guide to exporting solid wood products. Agriculture Handbook No. 662. Forest Products Division. U.S. Department of Agriculture. Washington, DC.

Foreign Agricultural Service. 1987. Wood products: International trade and foreign markets. Circular Series, WP 1-87 (Quarterly). U.S. Department of Agriculture. Washington, DC.

Forest Products Research Society. 1987. North American wood/fiber supplies and markets: Strategies for managing change. Symposium Proceedings No. 47351. Madison, WI.

Forest Service. 1920. Timber depletion, lumber prices, lumber exports and concentration of timber ownership. Report on Senate Resolution 311. U.S. Department of Agriculture. Washington, DC.

Forest Service. 1980. An assessment of the forest and range land situation in the United States. FS-345. U.S. Department of Agriculture. Washington, DC.

Forest Service. 1982. An analysis of the timber situation in the United States 1952-2030. Forest Resource Report 23. U.S. Department of Agriculture. Washington, DC.

Forest Service. 1987a. U.S. timber production, trade, consumption and price statistics 1950-85. Miscellaneous Publication. U.S. Department of Agriculture. Washington, DC.

Forest Service. 1987b. Production, prices, employment and trade in Northwest Forest Industries. Resource Bulletin (Quarterly). Pacific Northwest Forest and Range Experiment Station. U.S. Department of Agriculture. Portland, OR.

Gess, K. 1972. A study of Japanese travel habits and patterns. Office of Research and Analysis. U.S. Travel Service. U.S. Department of Commerce. Washington, DC.

Gray, P. H. 1970. International travel—International trade. Heath Lexington Publisher. Boston, MA.

Hawkins, D. E. 1979. The impact of international tourism on outdoor recreation in the United States. In: Tourism planning and development issues. Proceedings of International Symposium on Tourism and the Next Decade. George Washington University. Washington, DC.

Haygreen, J., H. Gregersen, I. Holland and R. Stone. 1986. The economic impact of timber utilization research. Forest Products Journal 36(1):1-18.

Hough, F. B. 1980. Report on Forestry (vol. 2). Prepared under the direction of the Commissioner of Agriculture. Washington, DC.

Ince, P. J. et al. 1987. Modeling technological change in wood products processing. In: Forest sector and trade models by P. Cardellichio, D. Adams and R. Haynes (eds.). College of Forest Resources. University of Washington. Seattle, WA.

International Trade Administration. 1984. A competitive assessment of the U.S. solid wood products industry. Office of Forest Products and Domestic Construction. Basic Industries Sector. International Trade Administration. U.S. Department of Commerce. Washington, DC.

Joerger, S. 1979. International tourism and the National Park Service lands: A framework for identifying economic and policy issues. M.S. Thesis. University of Washington. Seattle, WA.

Johnson, J. and W. R. Smith. 1988. Forest products trade: Market trends and technical developments. In: Proceedings of Third International Symposium on World Trade in Forest Products. CINTRAFOR and College of Forest Resources. University of Washington Press. Seattle, WA.

Josephson, H. R. 1975. Some economic aspects of foreign trade in forest products. In: Social sciences in forestry: A book of readings by F. Rumsey and W. Duerr (eds.). W. B. Saunders Co. Philadelphia, PA.

Kaiser, H. F. 1987. A rationale for trade-related research in North America. In: Forest sector and trade models by P. Cardellichio, D. Adams and R. Haynes (eds.). College of Forest Resources. University of Washington, Seattle, WA.

Kallio, M., D. Dykstra and C. Binkley. 1987. The global forest sector: An analytical perspective. John Wiley Publisher. Chichester, England.

Leigh, J. H. 1971. The timber trade—An introduction to commercial aspects. Volume 12. Pergamon Series of Monographs in Furniture and Timber. Pergamon Press. Oxford, England.

Lovett, S. and A. E. Lovett. 1986. The market for softwood lumber and plywood in the People's Republic of China (2 vols). National Forest Products Association. Washington, DC.

Manning, Robert E. 1979. International aspects of national park systems: Focus on tourism. In: Tourism planning and development. Proceedings of International Symposium on Tourism and the Next Decade. George Washington University. Washington, DC.

Nelson, A. Z. 1941. A selected bibliography on the economics of forestry in the United States. Division of Forest Economics. USDA Forest Service. Washington, DC.

Office of Technology Assessment. 1984. Wood use: U.S. competitiveness and technology. OTA-M-224 (2 volumes). U.S. Congress. Washington, DC.

Outdoor Recreation Resources Review Commission. 1962. A look abroad: The effect of foreign travel on domestic outdoor recreation and a brief survey of outdoor recreation in six countries. Study Report 18. Washington, DC.

Oxholm, A. H. 1932. Europe as a market for American timber: A plan for its proper development. Trade and Information Bulletin 786. U.S. Bureau of Foreign and Domestic Commerce. Washington, D.C.

Pratt, E. E. 1918. The export lumber trade of the United States. Miscellaneous Series 57. U.S. Bureau of Foreign and Domestic Commerce. Washington, DC. 117 pp.

Prins, K. 1987. European timber trends and prospects to the year 2000 and beyond. In: Forest sector and trade models by P. Cardellichio, D. Adams, and R. Haynes (eds.). College of Forest Resources. University of Washington. Seattle, WA.

Schreuder, G. F. (ed.). 1986. World trade in forest products II. In: World trade in forest products (proceedings). CINTRAFOR and College of Forest Resources. University of Washington Press. Seattle, WA.

Schreuder, G. F. (ed.). 1988. Global issues and outlook in pulp and paper (proceedings). CINTRAFOR, Washington Pulp and Paper Foundation and College of Forest Resources. University of Washington. Seattle, WA.

Sedjo, R. A. (ed.). 1981. Issues in U.S. international forest products trade (proceedings). Research Paper R-23. Resources for the Future. Washington, DC.

Sedjo, R. A. 1987. The timber and wood products industry in the Pacific: Emerging issues. Discussion Paper for Pacific Celebration, Washington State Bicentennial. Seattle, WA.

Sedjo, R. A. and K. S. Lyons. 1987. The adequacy of long term timber supply: A worldwide assessment. In: Forest sector and trade models by P. Cardellichio, D. Adams and R. Haynes (eds.). College of Forestry. University of Washington. Seattle, WA.

Sedjo, R. A. and S. J. Radcliffe. 1978. Post war trends in U.S. forest products trade. Johns Hopkins University Press. Washington, DC.

Social Science Research Council. 1935. A survey of research in forest economics. Advisory Committee on Social and Economic Research in Agriculture. Report of Subcommittee on Scope and Status of Research in Forest Economics. Washington, DC.

U. S. Department of Agriculture. 1950. Economics of forestry: A bibliography for the United States and Canada, 1940-1947. Library List no. 52. Library. Washington, DC.

U. S. Department of Agriculture. 1955. Economics of forestry: A bibliography for the United States and Canada, 1948-1952. Library List No. 52, Suppl. 1. Library. Washington, DC.

U. S. Travel Service. 1978. A regional analysis of international travel to the United States. U.S. Department of Commerce. Washington, DC.

Wander, S. A. and J. D. Van Erden. 1979. Estimating the demand for international tourism using time series analysis. In: Tourism planning and development. Proceedings of International Symposium on Tourism and the Next Decade. George Washington University. Washington, DC.

Williams, D. H. 1987. The extensive margin of the mature stock of timber in the coastal region of British Columbia. In: Forest sector and trade models by P. Cardellichio, D. Adams and R. Haynes (eds.). College of Forestry. University of Washington. Seattle, WA.

Wisdom, H. W. 1987. Transportation costs in forest sector models. In: Forest sector and trade models by P. Cardellichio, D. Adams and R. Haynes (eds.). College of Forestry. University of Washington. Seattle, WA.

Zivnuska, J. 1977. Research in international forest economics. In: Research in forest economics and forest policy by M. Clawson (ed.). Research Paper R-3. Resources for the Future. Washington, DC.

Zivnuska, J. 1981. Discussion. Discussion comments on paper presented by David Darr. In: Issues in U.S. international forest products trade by R. Sedjo (ed.). Research Paper R-23. Resources for the Future. Washington, DC.

11

WOOD FIBER PRODUCTION

J. Michael Vasievich

Management of forest resources for purposes of timber production has important implications for forest land owners, for wood product manufacturers, and for the ultimate consumers of products made from wood fiber. More pointedly, the availability of economic supplies of timber is critical to the profitability of wood-using industries and to the competitiveness, vigor, and stability of many rural and urban economies. In addition, consumer prices for products manufactured from wood are dependent, in large part, on the availability of ample quantities of timber. Timber management activities also have implications for the attainment of nontimber forest management goals, which are often positively influenced by the manipulation of forest vegetation undertaken for timber production purposes. Although timber production may be a landowner's primary objective, the positive nontimber effects of such interests can seldom be ignored.

The cost of timber production has in most cases consistently increased over the past decade—without a corresponding increase in productivity. Rising costs for labor and capital, increased restrictions on management practices, and reductions in the availability of land suitable for timber production in some regions are the primary causes of such increases. In the long run, production cost increases without associated productivity gains will lead to higher wood costs, reduced profitability for timber producers, and undesirable social and economic consequences. The latter include diminished wood supplies, declines in economic activity represented by wood processing, and higher consumer prices for wood-based products.

Economic questions about the efficient planning and management of timberlands are important to all timberland owners. Some owners use sophisticated timber management planning methods—others manage with general rules of thumb. In each case, knowledge of the factors affecting the profitability of silvicultural practices is required for efficient timber management, regardless of the level at which timber management planning and analysis is undertaken.

Forest products companies must manage company-owned timberland in an efficient manner in order to remain competitive. Land ownership and land management costs of industrial concerns represent substantial capital investments which are deemed by some to be necessary for the assurance of an adequate strategic supply of timber. Not all large forest products firms own timberland. Some obtain their timber from other private sources or from public agencies; their forestry concerns involve an ability to obtain sufficient supplies from such ownerships at acceptable costs. Largely unanswered is the question of whether or not a forest products company should own timberland. Such ownership is usually justified on the grounds of maintaining an assured supply of timber. Needed, however, is a more complete understanding of the costs of timberland ownership and the strategic benefits resulting from timberland ownership.

State and federal agencies are also concerned with timber productivity. In some cases that concern is focused on nonindustrial private lands. This land accounts for two-thirds of the commercial timberland base in the Eastern United States and represents a substantial portion of the nation's industrial supply of timber. Concerned over the adequacy of timber supplies from nonindustrial private lands, a number of public forestry programs have been developed to increase the timber productivity of such ownerships. The effectiveness of these programs is often confounded by the diverse forestry objectives of landowners and the low priority landowners often give to timber production activities (e.g., owners favor recreation, aesthetics, and wildlife).

Economic opportunities to increase timber supplies from nonindustrial private forests are significant, yet the landowners' frequent lack of interest in and commitment to timber production has fostered substantial uncertainty as to their reliability as a source of timber. Inadequate regeneration following harvest is probably the most limiting factor. Concern for timber supplies from nonindustrial private lands are particularly acute in areas where aggregate timber removals are nearly equal to or exceed growth. Timber supply concerns are not as great in areas where growth exceeds removals, although increased harvest rates can quickly change this perspective.

Managers of publicly owned forest land, including the USDA–Forest Service national forest system, also require economic information concerning the efficient management of forests for timber. In some cases, more intensive timber management on the most productive sites can generate added income and can sustain or increase wood supplies. The economic management of timber has become an important issue on much of the nation's publicly owned forest land in recent years. As pressures grow to simultaneously provide increased levels of timber and nontimber benefits, such forest land ownerships become the scene of intense public controversy. In some cases, timber production capacity is significantly offset by forest land exclusively devoted to nontimber uses. Public land managers need guidance as to the most efficient means of jointly producing a wide variety of forestry benefits, including wildlife, recreation, water, and timber. Also needed for efficient land allocations is information about timber opportunity costs.

Forest economics and policy research that is focused on the economics of wood fiber production has progressed significantly in recent years. However, much additional research remains to be accomplished. To be effective, such research should concentrate on the following strategic areas:

• Enrich understanding of the benefits and costs of various silvicultural systems and timber management regimes.

- Improve understanding of the joint production of timber and other products and services produced by forests.
- Improve methods of planning, economic analysis, and decision making germane to timber production.
- Enhance the economic and institutional framework affecting timber production and long-run timber supplies.

Enrich Understanding of the Benefits and Costs of Various Silvicultural Systems and Timber Management Regimes

Economic analyses have been conducted on a wide variety of timber management practices. In spite of such advances, new analyses are needed as economic conditions change, new management practices are developed, or old practices are modified or extended to new areas. New financial comparisons of timber management options are needed as costs and production methods change and timber growers seek ways to increase timber production efficiency.

Current timber production trends favor more intensive management, especially in the South and Pacific Northwest, where softwood supplies appear limited. Producers substitute timber capital for land and may reduce future transportation and harvesting costs through intensive management. In general, shifts toward more intensive management of softwoods have been motivated by projected timber supply shortfalls, higher land costs, and increased emphasis on capital intensive harvesting systems. Also, silvicultural research continues to identify potential gains from new or revised practices. Fundamental financial questions regarding intensive management practices are: Which practices are sufficiently profitable? On which sites should such practices be applied? Analyses should focus on financial comparisons of incremental costs and benefits from alternative management practices.

Research is needed for several intensive stand management activities, including artificial regeneration, intermediate stocking-control treatments, commercial thinning, and timber harvesting. Also needed are economic assessments of tree improvement programs, short-rotation plantations, and forest protection activities. Studies should also be carried out to compare even-aged and uneven-aged timber management strategies for mixed hardwood stands.

Develop improved production functions and economic data bases. The ability to conduct economic analyses of silvicultural practices depends in large measure on the availability of adequate production functions. Quantitative models of timber growth responses to proposed treatments are a prerequisite to determination of profitability. Few growth and yield models are available to predict responses to a wide variety of common timber management practices. Although the development of appropriate production functions may appear to be the domain of specialists in silviculture and mensuration, economists have a role in defining the requirements of such yield models. Models must describe the effects of management on timber yields and should account for factors that affect management costs and timber values.

The shortage of data on management costs and stumpage prices also limits economic analyses of silvicultural practices. Such data is costly and difficult to maintain—but essential to effective analyses. Data should account for variations over time and location and should reflect differences due to economies of scale, timber quality, and treatment difficulty. Information is particularly limited for operations on nonindustrial lands but is less so for large organizations with satisfactory accounting systems.

Undertake financial analyses of regeneration options. Regeneration activities influence timber productivity more than any other timber management practice. Timber producers need to know which methods of regeneration are cost effective and how alternative treatments affect timber production levels.

Artificial regeneration offers the greatest assurance of full stocking, control of competing vegetation, minimal production delays, and maximum wood production. Research is needed most for southern pine plantations in the South, Douglas fir in the Northwest and spruce and pines in the Northeast. Of special concern should be the economics of artificial regeneration of highly productive sites and of sites where natural regeneration methods have failed. Financial questions on artificial regeneration involve the cost effectiveness of site preparation, and tree planting. Although such practices may be common, new techniques, proposed changes in existing practices, and combinations of existing and new practices deserve research attention.

Information on the profitability of natural regeneration practices is limited for most timber types. Research is needed to clarify the costs and productivity of natural regeneration options. Natural regeneration requires fewer direct outlays, but may have greater opportunity costs and negative silvicultural results (e.g., variability in stocking and volume production). Natural regeneration and less intensive silvicultural practices offer potential gains over custodial management of many acres. Information on the most effective natural stand management practices are needed for situations where capital is limited (e.g., where landowners do not favor intensive management for mixed stands).

Production and use of container grown seedlings has greatly expanded in all regions of the nation. This method offers improved control of seedling quality, extension of the planting season, reduced planting costs, and better early survival (particularly on adverse sites). Needed are cost effective methods for production of containerized planting stock and comparison of such methods with bare root planting procedures. Although container grown seedlings are more costly, potential benefits appear great for extension to new areas.

Practices designed to enhance site conditions have also increased significantly in the last decade. Timber managers interested in increased wood production have learned that water budgets, nutrient budgets, and soil structure can be improved by timber management practices, often in an economic fashion. Analyses of practices such as drainage, bedding, fertilization, and subsoil ripping are needed as trends favor more intensified timber management activities.

Evaluate the economic performance of noncommercial stand improvement treatments. Intermediate stand treatments present managers with opportunities for mid-course stand corrections. These treatments are often overlooked as profitable methods for achieving timber production goals. Such management activities may include noncommercial thinning and timber stand improvement (e.g., elimination of undesirable trees, release of stagnated stands, cleaning operations designed to focus growth on higher value crop trees). Available information is inadequate to guide timber managers in the selection and application of corrective treatments. Rehabilitation of high-graded hardwood stands, hardwood control in softwood stands, precommercial thinning of overstocked seedling and sapling stands, and sanitation cuts in damaged stands are all candidates for additional research.

Develop efficient commercial thinning and timber harvesting guidelines. Optimal harvesting is an economic problem in all forested regions. Timber growth and earnings are reduced if stands are harvested too early or too late or if stands remain overstocked. Timber managers require economic guidelines to formulate long-term harvesting strategies and to operationally select harvest chances. Development of optimal harvest strategies generally involves sophisticated methods capable of meeting constrained goals (e.g., linear and dynamic programming, simulation techniques). Optimal strategies depend on production objectives, resource constraints, and on many site, stand, and market conditions. As a result, the development of appropriate harvest schedules requires consideration of many interacting effects.

Although many methods have been developed to solve optimal timber harvest scheduling problems, they are limited in several ways. Improvements are particularly needed to make the methods easier and less costly to use. At present, only large landowners have the computational and data resources needed to solve complicated harvesting scheduling problems. Also needed is incorporation of more realistic objective functions and constraints so as to enable testing of the effects of input errors and assumptions on optimal solutions and to more effectively define and account for uncertain futures.

Timber managers also need guidelines to aid operational thinning and harvesting decisions. Efficient harvest rules are needed for such stand conditions as stagnated or overstocked softwood plantations, high-graded hardwood stands, partial harvests of uneven-aged stands, salvage and regeneration of damaged timber, and conversion of inadequately stocked stands. Productivity and profitability of such stands can often be improved if these stands are appropriately thinned or harvested and regenerated.

Establish guides for the management of economically sound tree improvement programs. Tree improvement programs offer great potential for increasing timber supplies. Tree selection and breeding strategies are costly and time-consuming but can produce trees with preferred traits for growth, form, and pest resistance. As such, economic information is needed to help tree breeders select the most promising and profitable tree breeding strategies. Research that clarifies the potential value of different breeding approaches, the probability of attaining improvement goals, and the costs of producing them could significantly improve tree improvement decisions. Analytical systems are needed to clarify the financial payoffs for different breeding strategies and the probability of success.

Determine returns for intensively managed short-rotation plantations. Production of woody biomass in short-rotation plantations has been subjected to much silvicultural research over the past decade. Research has focused on the culture of fast-growing sycamore, cottonwood, sweetgum, and green ash in the South, on poplars and soft maples in the Northeast, on alder and related species in the West, and on eucalyptus in southern Florida and Hawaii. Biomass grown quickly at an acceptable cost is considered an alternative to fossil fuels and as a raw material for pulping or wood-based chemicals. Competitiveness of short-rotation biomass with conventional fuel and fiber sources remains a particular concern.

Short-rotation biomass production is a relatively new technology, and the few analyses that are available are site-specific assessments based on limited cost and production data. Some research has examined production costs, market development, risks, and the efficiency of specific management practices. Additional research is needed to address the profitability of these plantations since technology continues to evolve and few operational plantations are available for on-site evaluations.

Evaluate protection treatments focused on timber production. Forest protection activities directly impact timber supplies by limiting losses from pests and other natural hazards. Protection of forests involves treatments to reduce stand hazards, to reduce the frequency of damaging events, to detect pest activity, to respond to and treat active infestations, and to salvage damaged timber. Research is needed to determine the payoff from alternative management responses to specific pests or threats (e.g., various levels of detection). Few studies have investigated the economic consequences of integrated pest management strategies. Better information is needed on the costs and returns of management strategies to limit pest impacts before such strategies are applied.

Assess the economics of management options for mixed hardwood stands. Mixed hardwood stands in the Northeast are some of the most complex and valuable forests in the nation. These forests include mixed stands of red oak, cherry, ash, maple, basswood, and other valuable hardwoods. Many stands have suffered from past cutting practices that have left them far below their productive potential. In addition, regeneration success may be impeded by animal or pest predation, allelopathy, and competition. Managers need financial evaluations of alterative methods for rehabilitating and managing such stands with even-aged and uneven-aged silvicultural systems.

Managed hardwood stands on high sites can grow very high-quality, valuable sawtimber. Unfortunately, the financial merits of alternative silvicultural systems applied to such sites (ranging from uneven-aged single tree selection to even-aged clearcutting) are still largely undetermined. Guidelines are needed to help timber managers compare management strategies for different stand conditions. Comparisons of even-aged and uneven-aged systems are particularly needed for hardwood forests because the acceptability of clearcutting and even-aged management is being challenged in many places. Analyses must consider existing stand conditions, yield responses to management, scale economies for management and harvesting costs, timing of treatments, site productivity, and critical harvest diameters. Information is especially needed on optimum stocking levels and harvest frequency for partial harvests. Extension of these analyses to include nontimber considerations is a logical next step.

Improve Understanding of the Joint Production of Timber and Other Products and Services Produced by Forests

Forests are often managed for multiple uses (including use as timber), especially publicly owned forests and forests owned by nonindustrial private interests. Even industrial forest lands are capable of producing nontimber outputs. Unfortunately, for whatever ownership category, there exists very little information to guide decisions involving the joint production of goods and services from forests (e.g., timber, water, wildlife, recreation, range, aesthetics). Full recognition of joint production potentials and associated costs is essential to efficient forest management, avoidance of destructive conflicts over the use of forests, and achievement of maximum net private and social benefits. Economic information is needed which fully defines the production possibilities for different forest types and broad management regimes.

Forest management practices do not uniformly affect the production of all possible forest outputs. For example, each variation in timber management practice results in a different mix of wildlife outputs—some enhance habitat quality for certain species while at the same time diminish habitat quality for others. Ironically, many private landowners believe that timber production is uniformly incompatible with nontimber goals, even though the mechanisms and impacts of timber management practices on nontimber outputs are not well known. Lack of knowledge about output tradeoffs can significantly bias resource management decisions and can curtail effective landowner decisions about the multiple output options available to them. Significant research is needed to develop theoretical and practical information on joint production functions for a multitude of forest management situations.

Define joint economic production functions for multiple forest outputs. Multiple output production functions are needed to guide management decisions where nontimber forestry objectives are important. Development of an appropriate theoretical base is needed to define production possibilities, output expansion paths, and efficiency measures. Research is needed to define feasible production relationships for specific situations, to identify output tradeoffs, and to specify how changes in stand conditions affect the mix of outputs. Many possible combinations of timber and nontimber outputs can be considered for specific forest types or management situations. The most promising candidates for research involve situations where conflicts appear the greatest. Information on timber opportunity costs is a potentially valuable measure of the effects of management alternatives which favor nontimber outputs over timber production.

Assess effects of nontimber resource demands on timber supplies. Timber supplies from public and nonindustrial private lands can be significantly affected when commercial timberland is allocated wholly or in part to the production of nontimber forest outputs. Modification of timber management regimes to satisfy nontimber objectives reduces potential timber productivity and wood supplies by causing shifts in rotation ages, changes in stand structure, or other limitations on stand management. Little is known about the timber supply effects of managing forests for nontimber outputs. Such information is vital to decisions involving proposed shifts in public land management priorities that may limit the supply of industrial wood.

Income from timber management often ranks lower than other ownership objectives in many surveys of nonindustrial landowners. Emerging demands for nontimber outputs can affect landowner management behavior, timber management investments, and long-run wood supplies. A greater understanding is needed of the interactions between public nontimber demands and landowner production behavior and the subsequent effect of such interactions on timber supplies, especially in the context of planning for future industrial wood supplies.

Evaluate the timber supply consequences of practices undertaken to enhance forest wildlife habitat. Wildlife considerations affect timber production decisions more than most nontimber management considerations. Greater recognition of wildlife benefits is needed as the demand for recreational wildlife increases. Most foresters can readily select effective timber practices—integrated timber and wildlife practices are more elusive. Timber producers may expand wildlife outputs by more intensive wildlife management of selected acres. Such may reduce wildlife pressures on highly productive timber sites. Information on the costs and effectiveness of wildlife management practices is needed to guide forest management activities.

Assess the spatial economics of patterns of forests management for multiple outputs. A particularly promising research area involves spatial patterns, tract size, and diversity for optimal production of timber and nontimber outputs. Spatial effects on forest outputs are not well known (most forest production functions consider temporal but not spatial effects on productivity). Efficient production of wildlife, endangered species, recreation, and water may require specific spatial arrangements of forest conditions. Also, research involving stand size, location and spatial arrangements can enhance timber production goals by identifying more efficient land use patterns.

Economic analyses are also needed to identify cost-effective patterns, critical habitat sizes, appropriate mixes of stand conditions, and similar production requirements to meet nontimber goals. A better understanding of such production relationships could improve the allocation of land to meet timber and nontimber goals.

Improve Methods of Planning, Economic Analysis, and Decisionmaking Germane to Timber Production

Economic and financial analyses provide important information necessary to the resolution of forest management problems at several levels. Cost-benefit analyses, for example, are important to stand level decisions. At the forest level, such analyses are critical to the efficient development of forest plans, the effective regulation of forest regimes, and the efficient allocation of land and management resources among competing objectives. At the public policy level, economic analyses provide important information needed to judge whether or not specific policies and programs are effective means of meeting social goals.

Many forest landowners and managers do not use available economic and financial methods to analyze and solve timber management problems. They often lack means for applying such methods—means that include necessary data, technical expertise, or computing resources. In some cases, the gains from more sophisticated analytical methods may not appear to justify the associated costs. Also, available analytical methods may not be easily adaptable to specific management goals and constraints faced by individual landowners. Timber managers need low-cost methods to help guide timber management decisions and subsequently produce timber more efficiently. Such methods include optimal harvesting and forest regulation actions, selection and scheduling of management activities, comparison of silvicultural alternatives, and assessment of future scenarios for strategic planning. In addition, data on forest management options and costs, timber yield responses, nontimber effects, and price expectations are needed for such analyses.

Public policies which address perceived economic timber scarcity are constantly under examination. Research is needed to identify the social benefits and desirability of financial assistance programs, tax policies, technical assistance, and similar efforts aimed at improving timber productivity. The results of such research can guide judgments as to whether such programs are needed, and if so, what changes might be made to improve their effectiveness.

Evaluate financial analysis methods used to guide stand management decisions. Efficient timber management decisions require comparisons of silvicultural alternatives with other management opportunities. Methods for cash-flow analysis are well known. However, better implementation of analytical procedures could help managers evaluate and compare timber options more quickly and easily. Also, managers are more likely to use improved analytical techniques to support their decisions if they are inexpensive, easy to use, and under their control. Most financial analyses are based on fixed estimates of costs, prices, yields, and management schedules. Methods are needed that explicitly recognize and test the effects of inaccurate estimates and the variability of these parameters.

Develop methods to track financial performance of timberland investments. Unlike financial instruments such as stocks and bonds, a financial track record is not commonly available for timber growing. This shortage of financial return information often limits investor interest in timber investment opportunities. Although some information is available on historical stumpage prices, such time series do not account for land and management costs, variation in yields, or other factors that affect profitability. Efforts should be undertaken to develop and maintain a reliable record of timber growing costs and profits.

Institutional investors are one group of investors that could benefit from historical information on the profitability of investments in timber production. Large pension funds, limited partnerships, investment divisions of insurance companies, and other similar groups have invested in timber to some degree. Their interest is often to reduce the riskiness of typical asset portfolios by including timberland as a diversifying element. Unfortunately, little information is available on the scope of current investments and factors that might lead to increased participation by the banking and investment community—a community that could be an increasingly important source of capital for increased timber production. More information is needed to better understand the group and the factors that affect their consideration of timber as a portion of broader portfolios. Research on the investment information requirements of this group (e.g., need for casualty insurance, risk preferences, timberland market imperfections, profitability requirements) could prove especially fruitful.

Develop improved land classification systems involving the economics of timber management. New multi-factor systems of land classification have been developed over the past decade. These systems recognize the many edaphic, ecological, and climatic factors that affect the productivity of plant and animal communities on different sites. Multi-factor ecological classification systems offer managers a way to prescribe management treatments with greater assurances of biological success. Little information is available to link such forest land classification systems with economic measures of forest productivity. Integrated ecological and economic classification systems could provide more effective means of allocating forest land to timber and nontimber uses. Greater recognition of management costs, productivity, and potential returns could further enhance the utility of such systems.

Evaluate economic risks involved in timber management decisions. Timber management is often viewed by managers and investors as highly risky in that financial performance varies greatly. Natural hazards (e.g., fire, insects, diseases, weather) are usually cited as most significant, although market, technological, or institutional factors may impose even greater variance on timber profitability. Perceptions of risk and willingness to accept risk affects management decisions, but insufficient information is available to fully define the degree of economic risk involved in timber management. Research is needed to identify important sources of risk, assign probability values, estimate effects on profitability, and identify methods of reducing risk.

Methods are also needed to identify optimal management strategies under conditions of risk. This requires estimates of the variability of factors that affect returns to timber management. In addition, strategies are needed to reduce the effects of risk on forest management. Such strategies may include changing timber management regimes, spreading risks by diversification and insurance, and reducing the effects of risk through improved loss recovery decisions.

Develop effective means of using forest resource data bases in economic decision making. Reductions in the cost of computer systems have led many large landowners to implement timberland data systems. These systems are highly variable and may include information such as stand and compartment descriptions, continuous forest inventories, and data bases portraying land characteristics. While these systems permit greater accountability, their use for selection of treatments is limited to biological and physical criteria. By linking such systems to economic data (e.g., stand value, value growth rate, treatment or harvest costs, potential returns), more effective decisions based on economic measures could be made. However, better methods are needed for including economic information in these systems and using them for economic decisions. Promising research areas include the development of economic selection criteria for stand management practices and the use of spatial data for management decisions.

Enhance the Economic and Institutional Framework Affecting Timber Production and Long-Run Timber Supplies

A large number of market and institutional factors affect timber management activities and future timber supplies—most are not well understood. For example, timber productivity on nonindustrial lands is greatly influenced by a wide variety of state and federal programs, many of which are designed to address perceived economic timber scarcity or concerns over soil and water conservation. Such programs take a variety of forms, including tax incentives, cost-sharing incentives, technical assistance, and regulatory programs. These programs are typically implemented at the state level, often requiring reforestation after harvest and sometimes imposing limitations on some timber management practices. An overriding concern about programs focused on nonindustrial private forests is their effectiveness and social desirability. In general, the responsiveness of nonindustrial landowners to changing markets and public programs is not well understood. Yet, nonindustrial private timber is a critical source of supply in many locations.

Timber management and long-term supplies are also affected by changes in demands for wood products and technological advances associated with forest management, timber harvesting, and wood product manufacturing. Forest management decisions made today affect future species and product options and, in turn, affect prices and profitability. Information on future wood requirements is essential; it provides needed insights for public program management and decisions concerning actual stand management.

Assess the efficiency of markets for timber. Efficiently operating timber markets are essential to sustained economic timber supplies. Unfortunately, timber markets frequently are burdened by major flaws in their operation. Research is needed on prices and market behavior and on distortions arising from institutional constraints, limited availability of market information, and adverse market structures. Factors that affect market prices for standing timber vary from place to place. Many timber sellers and buyers could benefit from better models designed to improve trading efficiency by more clearly explaining variations in stumpage prices. Availability of price reports and related market information is believed to be an effective mechanism for improving market performance. Research is needed to determine if additional market information can improve the functioning of timber markets.

Some timber markets are very competitive while others have highly concentrated producer or consumer markets. Markets dominated by few timber sellers or buyers do not operate as efficiently as more competitive markets. Research is needed to better understand the performance of markets where competition is limited.

Responsiveness of federal, state, and local public land managers to market changes is also in need of research. The timber investments of such landowners is affected by a variety of factors, including budget limitations, organizational structures, and management objectives. Changes in timber production and harvest approaches as well as changes in market pricing systems merit consideration. Also in need of research are innovative strategies for funding increased levels of timber investment on public lands.

Assess impact of timber product demands, prices, and technological advances on optimum silvicultural regimes. Forest managers base timber production decisions on expectations of future markets for timber products. Future demands influence markets for specific products and, in turn, determine the profitability of timberland investments. Shifts in demands for wood products, due to changes in consumer preferences, export markets, and other factors, influence anticipated markets for stumpage. Similarly, changes in wood processing technologies affect manufacturing costs and wood requirements. In both cases, these cause changes in optimum silvicultural regimes and in the expected profitability of timber management investments. To fully meet this need, economists must define alternative future supply and demand scenarios for appropriate timber products and timber producing regions. Also needed are recommendations for preferred silvicultural strategies which appear best suited to meet future timber needs.

There exists a long history of timber assessments—most influenced by fears of timber shortages. Such assessments reflect opportunities to change future supply expectations by changing investment levels, by reallocating timber production and processing capital, and by changing public and private timber management programs. The question of whether supply responses have actually resulted from past projections is problematic. Such is deserving of research in the spirit of designing assessments that are more likely to have an impact on investment in timber supplies. In doing so, it must be recognized that timber supply and demand projections have relevance at the policy as well as the land management level. Land managers need to know how timber supplies and demands are likely to change for their timbershed so they can effectively allocate timberland to the most productive uses, make industrial plant location and capacity decisions, and affect the efficient implementation of public programs.

Projections are available for aggregate timber products and for broad regions. Projections are well developed for softwoods in the South and West, but are limited for hardwoods in the Northeastern region. In many cases, forest planners and managers need timber supply and demand projections disaggregated to state and substate regions—projections not available for many locations. Research is needed to develop cost-effective methods and projections appropriate for smaller geographic regions.

Evaluate the costs of environmental regulations affecting timber management practices. Significant wood production costs are due to environmental restrictions that limit timber management activities. Such restrictions are often imposed on the use of chemicals, prescribed fire, intensive site preparation, wetland drainage, and harvesting practices. They are typically designed to protect air and water quality, preserve wetlands, reduce erosion, and maintain critical habitat for endangered species. Such regulations are costly in both dollars and in foregone timber volumes. When applied, restrictions should be based on full knowledge of their cost and associated impacts. Insufficient information currently exists for judging these costs, estimating their impact on timber outputs, and assessing the effectiveness of alternative treatments. Some research has been done (e.g., restrictions of 2,4,5-T herbicide use, cost of administering state forest practice regulations), but much more is needed if informed timber management decisions are to be made. In most cases, restrictions apply only to certain forest types or locations. As such, research focusing on the impacts of local or regional situations is needed.

Evaluate the effects of competing land uses and landuse patterns on economic timber supplies. Changes in land use patterns can have an especially adverse affect on long-term timber supplies. Often severely limiting timber supplies are forest land withdrawals for nontimber uses, special nontimber use designations, and conversion of forests to nonforest uses. In addition, many demographic, economic, and institutional factors affect timberland area. Unfortunately, the causes and consequences of land use shifts are not well understood. Improved models are needed to estimate land use changes and their potential impacts on timber supplies.

Land fragmentation from subdivision of timber tracts into smaller units affects production costs because of economies of scale. Such is a special problem where urban populations are expanding into rural areas. As forest tracts become smaller, they become more costly to manage and harvest—to the point where small tracts become unavailable for timber production. Capital-intensive management and harvesting systems are more efficient on larger tracts. The high cost of acquiring land and timber in areas with small average tract sizes has limited industrial expansion in some areas. Research is needed to determine areas where fragmentation is occurring, the economic factors causing reduced tract size, and the effect of reduced tract size on timber supplies.

Assess the timber production response of nonindustrial private landowners to critical economic and policy considerations. Timber from nonindustrial private forest lands is critical to industrial forestry operations in all regions. Unfortunately, timber production from such ownerships is far below its potential. The failure of many nonindustrial private landowners to reforest their timberland following harvest is well documented, even though many have profitable opportunities to do so. With less than complete success, past research has often attempted to model nonindustrial private landowners as profit-maximizing producers. Increases in timber prices elicit increased harvests, but do not seem to lead to increased investments in timber production. In general, economic models are inadequate to explain the timber growing actions of this group. More research is needed to describe the response of nonindustrial private landowners to market changes and to public programs. Rather than consider such landowners as rational producers, the most promising research approaches involve the development of qualitative assessment models to predict behavior and the application of consumption or utility concepts.

A related research need is the development of realistic timber supply goals for nonindustrial private lands. In some areas, public pressure to reduce timber outputs and expand nontimber uses of public lands has focused attention on private lands as the most promising replacement for public timber. Significant unanswered questions include how much additional timber could be produced on private lands, at what cost, and what is the probability that such supplies will actually be forthcoming. This is a particular concern in the Pacific Northwest, the Lake States, and the South.

Summary and Conclusions

Research designed to augment the nation's economic supply of timber is deserving of significant attention in light of the importance of timber to the nation's social and economic well-being. In order to facilitate the availability of long-term timber supplies at acceptable prices, economic research involving the production of wood fiber should be directed toward avenues such as improving the economic understanding of silvicultural systems and timber management regimes, joint production of timber and nontimber forest outputs, institutional factors affecting timber production, and methods of analyzing and planning timber management activities.

Many of the research topics discussed here are applied rather than basic or theoretical. They point to the practical need for adequate data, appropriate methods, and sufficient analyses to effectively plan timber operations and improve on-the-ground management decisions. The need for financial analyses will exist as long as economic conditions and the techniques of forest management continue to change. Economics research can blend silvicultural and decision sciences so as to identify the most economically efficient and biologically appropriate methods of managing natural forest systems. Gains can come about from improvements in the way timber management decisions are made and in the quality of those decisions.

A more theoretical research approach will, however, be required to gain a better understanding of problems in some areas. For example, increases in demands for nontimber forest outputs have added a new dimension to timber production—a dimension that requires explicit consideration of tradeoffs and opportunity costs. Economic approaches on this front are currently inadequate. New theories are needed to define production functions and to understand the implications of timber and nontimber management activities applied in concert.

Research to develop new methods of analysis and planning has been strong over the past decade. Advancements in computer and information technology have greatly expanded options for more quantitative analysis of timber production. However, such methods are costly and require substantial increases in information— all of which does not necessarily translate into more efficient timber management decisions. Improved timber efficiency can only be achieved by development of more cost-effective methods, establishment of economic information in the most appropriate form, and interpretation of that information for better management decisions.

Institutional and economic factors affect the way forests are managed, their profitability and, ultimately, the future supply of timber. Concerns about efficient timber markets, environmental impacts, land use shifts, and other factors that affect the economic climate for forestry are important determinants of timber supplied by private and public forest lands. Research to clarify the impacts of these factors on timber production are important to all levels of planning and management. Similarly, there are significant policy questions about the need for and effectiveness of public programs aimed at increasing timber supplies. The social welfare implications of these programs are likely to come under ever greater scrutiny if state and federal budgets are further constrained.

Research focused on the economics of timber production must recognize that a variety of biological, economic, and political factors influence the profitability of growing timber. Such research is often driven by immediate unsolved problems as well as by subtle but significant changes in technology and in economic and political climates. Over time, many researchable problems have remained, despite significant past efforts to research them. Challenges to research involving the economics of wood fiber production will continue. They are deserving of the research community's continuing attention.

Bibliography

Clawson, M. C. (ed.). 1977. Research in forest economics and forest policy. Resources for the Future. Washington, DC.

Forest Service. 1988. The South's third forest: Alternatives for the future. Resource Report No. 24. U.S. Department of Agriculture. Washington, DC.

Forest Service. 1982. An analysis of the timber situation in the United States: 1952-2030. Resource Report No. 23. U.S. Department of Agriculture. Washington, DC.

Gregersen, H. and G. Stabler. 1977. Economics of forest management: Research needs and priorities. In: Research in forest economics and forest policy by M. C. Clawson (ed.). Resources for the Future. Washington, DC.

Leefers, L. A. , D. T. Cleland and J. B. Hart. 1987. Ecological classification system: Information in economics. In: Proceedings of central hardwood forest conference (IV) by R. L. Hay, F. W. Woods and H. DeSelm (eds.). Knoxville, TN.

McLintock, T. F. 1987. Research priorities for eastern hardwoods. Hardwood Research Council. Memphis, TN.

Phillips, W. E., J. A. Beck and G. W. Lamble. 1986. Forest economics research needs for west-central Canada. Information Report No. NOR-X-281. Canadian Forest Service. Toronto, Ontario, Canada.

Royer, J. P. and F. J. Convery (eds.). 1981. Nonindustrial private forests: Data and information needs. School of Forestry and Environmental Studies. Duke University. Durham, NC.

Royer, J. P. and C. D. Risbrudt (eds.). 1983. Nonindustrial private forests: A review of economic and policy studies. School of Forestry and Environmental Studies. Duke University. Durham, NC.

Science and Education Administration. 1978. National program of research for forests and associated rangelands. Report ARM-H-1. U.S. Department of Agriculture. Washington, DC.

Southern Industrial Forestry Research Council. 1986. Priority research needs. Report No. 4. American Forest Council. Atlanta, GA.

U.S. Department of Commerce. 1984. A competitive assessment of the U.S. solid wood products industry. Office of Forest Products and Domestic Construction. Washington, DC.

Yoho, J. G. (no date). Economics research: Critical needs for industry in the South. Unpublished report. Department of Forestry. Purdue University. West Lafayette, IN.

12

TIMBER HARVESTING

Frederick W. Cubbage

Harvesting economics research offers many opportunities for large improvements in efficiencies and profits of firms of all sizes. Forest economics research has traditionally focused on timber valuation and investment analyses, following the biological orientation of the forestry profession. While the biology and economics of growing trees are important, they have perhaps been studied to the neglect of an equally important aspect of forestry—the economics of harvesting and transporting timber to wood-based manufacturing firms.

The market costs of wood procurement and timber harvesting often exceed the actual costs of growing wood. For example, southern pine pulpwood sold for $15 to $25 per cord as stumpage (standing timber) in 1988, but brought $50 or more per cord when delivered to the mill. Hardwood pulpwood sold for about $5 per cord, and fetched $40 to $50 per cord at the mill in the same year. Sawtimber price differences were less extreme but still substantial. Southern pine sawtimber stumpage sold for about $150 per thousand board feet and brought $200 per thousand at the mill in 1988. Hardwood sawtimber sold for $100 at the mill and $50 on the stump in the same year.

The preceding illustrates the magnitude of efficiency gains that could result from performing useful timber harvesting economics research. The procurement and harvest of wood fiber constitutes a large share, perhaps more than half, of the cost of supplying wood fiber to forest products processing facilities. This importance is compounded by the fact that any efficiency gains achieved in timber harvesting can be realized as soon as applied. Timber growing efficiency improvements, however, must wait for years to be realized, even if they are applied immediately to tree planting or timber stand improvement activities. Additionally, economic improvements in timber harvesting may be more rapidly adopted than forest management improvements since market systems force innovations on timber harvesters more quickly than on forest landowners managing forests for many purposes other than timber production.

Current Timber Harvesting Information and Research

Current research pertaining to timber harvesting, forest engineering, and wood procurement focuses mostly on the development and testing of various harvesting and road-building equipment. It consists of engineering studies that try to develop new equipment for a particular application, improve upon existing equipment, or use existing equipment in new and more efficient harvesting system configurations. Economics enters into such research but is not the principal concern—just as it is not the crux of silvicultural research. Nevertheless, a review of harvesting research and its economic applications is useful. As used here, timber harvesting economics involves the application of economics to a wide range of harvesting activities, including planning, purchasing, cutting, and delivering wood to a forest products facility. The focus is on economic issues of timber harvesting—those relating to market supply, demand, and prices for harvested wood.

The greatest amount of timber harvesting research has focused on development of production functions—underlying input-output relationships. Such research has led to tabular or mathematical relationships between number of employees, amount of equipment used, stand (or tree) conditions, and the time required to harvest timber. Matthews (1942) published early work on time-motion studies of harvesting operations. In the 1960s, the Battelle Institute performed an intensive study of timber harvesting operations in the East. The American Pulpwood Association and many cooperating firms sponsored the extensive Harvesting Research Project (HRP) in the early 1970s, a project which led to summary tables of production rates for southern timber species. Many U.S. researchers continue to perform time-motion studies of ongoing or experimental timber harvesting operations.

Research has also focused on estimation of harvest costs. Most often such research involves a machine rate formula also developed by Matthews (1942). The formula presents a means of calculating the average annual investment of a piece of equipment, including depreciation, interest, taxes, and insurance costs, spread over the period of ownership. Fixed costs per hour to operate equipment are calculated by dividing total yearly costs by scheduled or operating costs per hour. Operating costs consist of the hourly equivalents for operation (fuel and lube), maintenance and repair, and tire or track. Total costs are the sum of fixed and operating costs per hour.

Typical production and cost research to date has used calculated machine rate costs and the amount of timber harvested to calculate an average cost per unit of wood produced. The classical machine rate formula has been reiterated and published by modern authors as well, but remains essentially the same as published by Matthews four decades ago (American Pulpwood Association 1965, Warren 1977, Miyata 1980). Several authors have also researched general machine rate computations for broad classes of equipment over the years (Plummer 1967-1982, Cubbage 1981, Werblow and Cubbage 1985, Dorris and Cubbage 1987). Despite its popularity, the machine rate formula has been criticized for various theoretical shortcomings—including failure to represent actual cash flows, poor (if any) incorporation of tax flows, and little basis for use in equipment replacement decisions (Harou 1980, Butler and Dykstra 1981, Tufts and Mills 1982).

Timber harvesting research has also attempted to improve field methods and techniques for timber harvesting and forest engineering. This area has not focused on immediate economic evaluation of costs and returns, but rather on mechanical or physical means of increasing efficiency. Engineering of new equipment and harvest systems is one large component of timber harvesting research. Developing better, cheaper, or more environmentally safe roads, particularly in the West, has also received a large amount of research effort. Ergonomics—the study of the interactions and relationships of operators and equipment—and safety have become increasingly important research topics. Scandinavian countries have studied such topics for decades. U.S. researchers began investigating ergonomics in the 1980s; the efforts have been fairly modest to date (Smith and Sirois 1982).

Trucking safety for primary forest products has become a very important issue in the 1980s, as have trucking costs. As the wood-based industry shifted from rail to trucking as a means of transportation, truck safety and regulation issues have increased, especially as a result of tree-length log transportation (Cubbage and Greene 1987). Means of increasing truck safety, limiting liability, and preventing adverse state legislation have all become important to cost containment efforts.

Local regulation of logging has also become an important issue in the 1980s. Many countries or municipalities in the West and Northeast are beginning to enact regulatory ordinances. State and national wetlands legislation and nonpoint source water pollution laws also affect logging costs. Research on the cost impacts of such regulations has only recently begun.

Procurement of wood supplies is also an area of concern to timber harvesting. This is actually the broad framework in which timber harvesting takes place. Procurement includes planning, purchasing, and scheduling the timber harvests. To date, very little research has addressed wood procurement per se. Some research (especially in the West) has addressed purchasing of bidding for stumpage.

Current research in timber harvesting also encompasses the development of new analytical methods for evaluating harvesting productivity and costs. Analytical methods include mainframe computer simulation of harvesting operations and costs and microcomputer packages to estimate productivity and costs. The most widely accepted integrated mainframe harvesting analysis package (Stuart 1981) has been GENMAC (Generalized Machine Simulator) and HSS (Harvesting System Simulator). Several microcomputer packages for analyzing harvesting costs have also been developed in the last four years (Hendricks and Curtin 1985, Tufts et al. 1985).

A final and particularly important development in harvesting research is the integration of harvesting economics with forest management practices. Too often, foresters have grown trees assuming that all timber is equally easy to harvest (regardless of tree size or spacing) and that all trees will receive equal stumpage prices. In practice, tree size is a critical variable—even within a particular product class—in determining harvesting costs. Such subjects certainly deserve additional research; little has been performed to date.

Based on the status of current information and research, several strategic directions seem promising for future timber harvesting economics research:

- Improve production function information for timber harvesting activities.
- Improve estimation of timber harvesting costs.
- Develop improved means of economically integrating timber management and timber harvesting activities.

- Develop better understanding of the economic efficiency and effectiveness of new timber harvesting methods.
- Develop more efficient and effective wood procurement systems.
- Develop better understanding of public regulatory programs focused on timber harvesting activities.

Improve Production Function Information
for Timber Harvesting Activities

Production functions are the underlying relationships among inputs of terrain, trees, employees, equipment, capital, managers, and other factors that determine the type and level of output from timber harvesting. To date, most information on production rates for timber harvesting operations has consisted of case studies of individual machines, often performed on only one stand. The results of such studies may be valid for a particular stand or set of operator conditions, but the transferability of such information to situations with differing input factors is moot. Probably the most important need for basic timber harvesting research is the development of broader, more reliable, and more generalizable timber harvesting production functions.

Why have researchers focused on case studies of individual machines or harvesting systems? First, it is obviously easier and cheaper to time an individual machine or system in one or a few locations than to develop a random sample of a variety of operators across a broad geographic region. Second, the statistical results—most usually a regression prediction equation—are generally better for an individual case study than for a broader sample. Terrain, weather, tree size, operator skills, management talents, machine maintenance, and a host of other factors vary so widely among logging operations that generalized, statistically significant equations are difficult to develop. Not all input factors are easily quantified and the variability among operations may be so great it precludes successful statistical estimation.

The most notable modern effort at estimating regional production functions for timber harvesting was the Harvesting Research Project (HRP), sponsored by the American Pulpwood Association and a number of forest products firms operating in the South. The project led to the development of the seminal HRP production tables (Lanford and Haver 1973). The tables were distributed to member companies and participating researchers only; they never were released officially for use by broader interests in timber harvesting. Additionally, they are now outdated and do not include new feller-buncher machines, modern grapple skidders, or other equipment improvements.

Although additional aggregate-level production research is probably not needed, such information—if desired—can be obtained without efforts as large as HRPs. Researchers from different universities, experiment stations, or firms could coordinate their efforts in collecting and analyzing such data. Efforts of this sort could be accomplished under the auspices of regional professional associations (e.g., Council on Forest Engineering, Society of American Foresters), or might be accomplished with university funding through regional projects. In any case, the development of statistically valid regional production functions for a variety of modern machines would be a great improvement over isolated case studies.

Several other possible improvements could be made in harvesting production functions. First, a thorough up-to-date bibliography of harvesting production data (by regions and type of equipment) should be assembled. Second, case studies or regional production functions should be developed in a manner so that different utilization standards could be incorporated into the prediction equations. For example, growth and yield equations usually can predict volume with different top diameters and products, such as sawtimber or pulpwood. Similar developments in harvesting production equations would be more difficult, but possible. Third, researchers should develop new equipment and better methods to make time-motion studies and to develop production equations—such as reliable video recorder methods or widely accepted statistical approaches.

Harvesting productivity research should also facilitate integration of elemental, machine, and system-level production functions. Elemental time studies measure each individual machine movement that occurs in the harvesting process—such as moving to the tree, grasping it, severing it, accumulating it, or dumping it. Machine (or function) level studies measure production rates for the entire cycle made by a piece of equipment in processing a stem or bunch of stems. System productivity focuses on a set of machines that take a tree from the stump to the roadside or mill. Some modern studies have integrated elemental and function-level machine productivity rates, and a few have done so with entire harvesting systems. Production functions developed in this manner will facilitate further economic-engineering studies.

Improve Estimation of Timber Harvesting Costs

Production functions are the bases for harvesting cost estimation. In the last four decades, the sophistication and reliability of production function estimation has increased noticeably; harvesting cost estimation probably has not. As previously stated, the standard means of calculating average harvesting costs—the machine rate method—was developed in the 1940s (or even 1930s) and has seen only minor refinements to date. A priority for harvesting economics research should be development of modern methods of estimating timber harvesting costs, especially methods that incorporate modern economic and financial theory and that reflect actual cash flows experienced by logging firms.

Economists customarily use one of three broad means of estimating long-run performance of firms—economic engineering (synthetic firm), statistical cost, and survivorship. The survivorship method alone is probably not helpful in analyzing costs, but it is useful to compare results that may be obtained by the other two approaches with actual field data on the number and type of firms harvesting timber. In the South, the American Pulpwood Association periodically has conducted surveys of the logging producer workforce (Watson et al. 1977, Weaver et al. 1981, 1982), which have been useful. Similar studies might also be useful in other regions of the country.

The economic-engineering approach to cost estimation is similar to that of the classical machine rate calculations, where input production rates and machine costs, labor rates, and so on are used to calculate costs per unit of output. Almost all economic engineering harvesting cost studies have been based on Matthews's 1942 machine rates. Butler and Dykstra (1981) and Tufts and Mills (1982) have suggested new methods of calculating harvesting costs that incorporate tax and firm cash flow considerations; their suggestions have not been widely adopted by researchers or harvesting practitioners. Firm cash flows and tax positions are important, however. Harvesting analysts may be making serious errors by not performing such detailed cost analyses. Thus, one important direction for harvesting economics research would be development of new cost methods.

Sophisticated financial methods may not be adopted because they are hard to understand, difficult to compare with prior machine rate studies, or quickly become outdated due to changing tax laws. New economic engineering cost methods must overcome these limitations. Sophistication and ease of understanding are often hard to achieve simultaneously, but they both will be required if the classical machine rate is to be supplanted. Probably it will be necessary to develop detailed financial models that calculate machine and harvesting costs yet also incorporate default assumptions (i.e., interest rates, repair costs, tax brackets) that make models easy to use by researchers or practitioners—even if some data are missing. In order to provide a comparison with prior machine rate calculations, data used in any new financial models should be similar to that used in old models, although more detail will be necessary to some extent. Costs calculated by any new models could thus be compared to calculations made with classical machine rates.

Continually incorporating new income tax laws in a model is always difficult. For a harvesting cost model, the best way to approach the problem would be to develop before and after tax calculations. Calculations performed without the effect of income taxes would have similar results throughout time, as have the machine rate calculations. The after tax calculations could be based on the current tax law, with revisions being made as tax laws change.

All suggestions for an elegant, understandable, accurate, and adoptable equipment cost model (or models, since different approaches may be better for different studies) are desirable but are admittedly difficult to accomplish simultaneously. It may, however, be possible to incorporate such considerations in a model using micro-computers and spreadsheet software. The latter would allow researchers to set up "templates" that incorporate desirable financial calculations yet suggest default inputs for users lacking sophistication or data. Such computer models could calculate before and after tax annual cash flows and yearly or hourly average cost (such as with the machine rate, or even the machine rate itself). Most researchers and users have microcomputers. They could quickly adopt new cost calculation methods if procedures and data were provided in an easily accessible format. The advantage of being able to calculate actual cash flows, before and after tax information, and average equipment costs should be compelling to both harvesting researchers and harvesting firms. The equipment costs calculated for all machines, labor and other costs in a system can then be used to calculate harvesting costs using the economic-engineering approach.

Statistical cost estimation is a second approach to harvesting cost research. Statistical cost methods directly estimate costs of producing various levels of output based on the costs of individual firms. Involved are empirical analyses for a large number of firms, the intent being to estimate average costs for entire harvesting functions and regions. The machine rate or any new replacement models consist mostly of deterministic calculations of costs for a piece of equipment or harvesting operation. Economists are not only concerned about costs for individual firms, however; they also have an interest in classes of firms. To date, little research has been published that estimates aggregate costs for broad classes of firms harvesting similar or diverse products. Stier (1982) analyzed the impacts of technology on timber harvesting, and generally found that such changes were labor-saving. Cubbage and Wojtkowski (1988) performed a preliminary aggregate cost study based on the 1979 American Pulpwood Association southern logger survey. Considerably more aggregate cost research needs to be done in order to characterize the economic structure and changing efficiency of the industry. One problem with statistical cost estimation is that data quickly become outdated. Especially useful would be development of an accepted means of updating historical cost data via the application of price indices appropriate for harvesting equipment.

In addition to economic-engineering and statistical cost methods applied to harvest cost estimating, there are several other harvesting cost research needs that deserve attention. Collection and dissemination of machine costs and rates, in whatever form is accepted, should continue to occur as in the past. This should include machine costs, labor costs, social legislation, fringe benefits, insurance rates, interest, tax rates, and other relevant data. Although mundane, this data gathering research is not easy; it is, however, very useful to logging managers or for generalized research analyses. Also desirable would be development of standardized methods for integrating machine cost information into system cost information for case study approaches. Currently, every researcher seems to use a different approach for calculating total harvest system costs. Some discussion and agreement among researchers as to the most effective way to calculate system costs would be helpful. Ongoing research is also needed to evaluate harvesting cost trends. To date, studies have generally found that real logging costs decreased in the 1970s and 1980s (Hassler et al. 1981, Herrick 1982, Sinclair et al. 1985, Cubbage et al. 1988). But this bears ongoing research, with particular attention to developing good estimates of logging costs. The latter are not directly reported in any data source.

A cost-related topic that deserves further research is the development of new computer models for integrated harvesting and cost analyses. Deterministic microcomputer machine cost models and the stochastic mainframe HSS models have been widely accepted by harvesting researchers and practitioners. It may be possible, however, to improve on such models by enhancing modeling of machine interactions, better simulating the stochastic nature of harvesting, and integrating improved cost analyses. Better linkages of geographic information systems (GIS) with harvesting models would also be a logical enhancement for economic research, particularly as more affordable GIS systems become available. Devine and Field (1986) discussed GIS applications in general, and Davis and Reisinger (1988) and Jackson and Brinker (1988) have published research on GIS applications in harvesting. The development of artificial intelligence applications in harvesting also shows promise for future research (Mills and Reisinger 1988).

Also deserving of research attention is additional analysis of logging road costs and how to allocate such costs to other forest management activities. This subject underlies much current debate over below cost timber sales on the national forests; it is relevant to private lands as well. Rideout and Hof (1987a, b) have performed the most notable research on allocation of joint costs, but considerably more work needs to be done.

Develop Improved Means of Economically Integrating Timber Management and Timber Harvesting Activities

Financial returns to timber management investments usually are calculated using publicly available stumpage price information. Such information may (or may not) reflect the value of trees of a particular size or location. In practice, stumpage prices are affected by many factors, including residual values based on product prices, total harvest costs, and the supply of timber. Harvest costs are crucial in determining stumpage costs. They are affected by tree diameter, tree length, volume per acre, space between trees, steepness, and access to the stand. Forest management economics often ignores such considerations by simply maximizing total return at a given volume per acre and product class. For example, a pulpwood stand composed of 300 eight-inch trees would be less costly to harvest than one of equal volume composed of 700 six-inch trees. In the future, harvesting economics researchers should work more closely with silviculturists, biometricians, and forest management researchers to develop integrated models of the effects of tree spacing, size, and other factors on delivered wood costs (LeDoux 1986, Lee 1986, Sessions et al. 1986).

Biomass energy harvests and the effects of more (or less) intensive stand utilization on reducing (or increasing) subsequent stand establishment costs is also a management-harvesting topic in need of research. If prices are high enough, it may be possible to utilize much of the low-quality wood in a forest stand for fuelwood. There has been considerable research into biomass harvesting and economics in the last ten years (i.e., Forest Products Research Society 1975, Vasievich 1984). Analyses should also consider the impacts that increased utilization has on reducing future site preparation costs. Most of the biomass energy harvesting analyses performed to date have included both production rate and cost information, and some have included the impacts of increased utilization on site preparation costs. Various types of biomass harvesting equipment and systems have been examined, including conventional equipment, cable yarders, swathe felling-chipping machines, and wood bales (Stuart et al. 1981, Tennessee Valley Authority 1983, Stokes et al. 1985, Miller et al. 1987). These studies generally estimate the cost of harvesting the entire stand for biomass, or the cost of harvesting residual volumes not valuable as conventional timber products. Such costs are then compared to the price of various alternative fuels in order to determine the economic feasibility of biomass harvesting. Conventional fuel prices have dropped substantially in recent years; energy wood is less attractive than in the early 1980s. However, conventional fuel prices probably will rise in the future, again making research in this area very fruitful.

Intensive utilization of harvesting residuals is also an area in need of research. More intensive utilization reduces site preparation costs. A stand that is harvested with low utilization levels will leave a large amount of residual, undesirable trees on the site. These trees and tops must be killed, cut and slashed, piled and burned, or treated in some other manner to prepare the site for regeneration. Income tax treatment of harvesting versus site preparation costs make this particularly important. Timber harvesting costs can be deducted from the timber sale revenue as part of the sales' expenses. Site preparation costs, however, must be capitalized and deducted at the time of any future harvests. Thus, greater utilization may have substantial income tax advantages (in addition to the ability to use the residues for energy wood). Stokes and Watson (1985) examined the effects of energy harvests on reducing site preparation costs in the south. They did not consider the tax aspects of the harvests. Considerably more research could be performed in this area.

Develop Better Understanding of the Economic Efficiency and Effectiveness of New Timber Harvesting Methods

Improvement of timber harvesting methods can be an especially important area of research. Although principally the domain of harvesting and forest engineering specialists, the subject area does have important economic components. Improved harvesting techniques and equipment must be continually developed in order to harvest timber more efficiently. Improvements are needed most to help meet the harvesting needs of a timber resource that is increasingly composed of smaller trees. Improving ability to harvest small trees on wetland soils and steep sites in an economically and environmentally acceptable manner is sure to increase in importance. Operator comfort and safety are also important research topics. Scandinavian forest researchers have performed and published considerable research on safety and on ergonomics (comfort). North American harvesting publications have focused almost exclusively on safety. Foremost among these have been a continuing series of periodic technical releases published by the American Pulpwood Association. The development and adoption of formal training programs for loggers also would be helpful.

The development of mechanical sawing heads that reduce damage to harvested trees and increase efficiency will also grow in importance as an engineering research topic. The use of mechanical shears and feller-bunches on tracked and rubber tired prime movers has increased substantially in the last two decades. Such shears usually cause butt damage to the most valuable part of the tree. New felling heads that are as efficient but that cause less damage would be a great economic asset.

Highway transport of raw forest products is also an engineering research topic that has economic consequences. Public awareness and governmental scrutiny of trucking has increased steadily. In fact, trucking issues are raised continually, usually with state attempts to restrict means of trucking wood in order to ensure public safety. The problem is related to safe loading and driving practices of individual loggers and to public relations in general. Engineering research should be able to develop new and safer means of transporting raw forest products, particularly for trucking. The challenge would then be to have such methods approved by state legislatures and state departments of transportation.

All of the above engineering areas relate in some manner to the cost-efficiency of harvesting and transporting timber. Research in such areas should seek to increase logging productivity so as to produce an acceptable, low-cost output— all accomplished in a manner that is environmentally sound and uniformly safe for operators and the general public. Economic evaluations and adoption by loggers will be the measure of whether engineering research has met productivity, environmental, and safety goals.

Develop More Efficient and Effective Wood Procurement Systems

Wood procurement involves all components of supplying wood fiber to a mill— in some instances, even growing timber. Planning, purchasing, cutting, transporting, and storing wood fiber are crucial yet vastly under-researched topics. A number of research avenues exist.

New means and models of evaluating wood procurement as part of a forestry business (all phases of obtaining wood from the stump to the mill) is an especially important area in need of research. Included should be investigation of various means of owning land to be used for timber production purposes—by forest products firms, by traditional nonindustrial private forest landowners, by corporate investors, or by some new institutional means. Wood procurement research might also strive to develop means to rationalize the purchase and delivery of primary wood products. Rationalization could consist of cooperation among forest products firms, forest landowners, and loggers—given existing laws and institutional constraints.

Wood procurement research could also develop more effective means of purchasing and selling timber—means that are equitable to both buyers and sellers. Such research could be controversial; empirical data may be difficult to obtain (Shaffer 1987). Harvesting economics could help in analyzing optimal means of sorting primary products at the stump, deck, or concentration yard, the intent being to increase utilization or value added. Similar analyses could investigate the efficiency of various procurement delivery systems (i.e., trucking wood near mills, outlying rail transport, satellite chip yards, and tree length versus shortwood delivery systems). Various mill inventory control systems and strategies should also be investigated. Research could also examine the comparative merits of various wood procurement and marketing systems (i.e., wood dealers, direct purchase, or mixed systems). Marketing of primary forest products has been a traditional forest research topic, but applying modern scientific methods to the field would be a significant step forward.

Develop Better Understanding of Public Regulatory Programs Focused on Timber Harvesting Activities

Timber harvesting activities have been regulated by many western states since the 1940s. In the 1970s, most western states either revised existing state forest practice laws or enacted new laws. Beginning in the late 1970s and continuing through the 1980s, many local governments have established ordinances governing logging practices; at least two states enacted new forest practice laws (Irland 1985, Henly and Ellefson 1986, Cubbage et al. 1986). Almost all such laws and ordinances regulate harvesting in some manner. Research into the extent of harvesting regulation and its impacts on operating efficiency and costs will increase in importance in the future.

Timber harvesting regulations that affect public and private forest landowners are often enacted by disparate state and local governmental units. Keeping track of enacted laws, summarizing their contents, and assessing their impacts is difficult. Legal research designed to clarify and publicize operating rules for loggers has been performed by many authors. Summarizing laws not only helps individuals directly affected by a law's content but also helps in comparisons among jurisdictions and in drafting new legislation. Legal research and publication will clearly need to continue in the future as regulation of forestry and harvesting activities increases.

State forest practice laws and regulations are added to and modified with great frequency. Such changes should be monitored. Oregon, for example, has revised riparian zone regulations; Massachusetts recently enacted state timber cutting regulations; and California frequently revises its state forest practice law. Changes such as these can have a significant effect on timber harvesting activities. Other state legislation that can affect timber harvesting should also be monitored (e.g., zoning, water quality, scenic rivers, and wetlands laws).

A more difficult group of regulations to keep abreast of are logging ordinances established by local governments—counties, townships, or cities. Local governments in the West have enacted ordinances regulating logging, and some are more restrictive than state forest practice laws. Municipalities in the northeast have enacted a plethora of ordinances governing logging, including over 100 in New Jersey alone (Hogan 1983). Connecticut, New York, Massachusetts, and Pennsylvania also have local ordinances aimed at strictly controlling timber harvesting activities. Little explicit local logging regulation has occurred in the South or Midwest, except for Georgia (15 or more county logging ordinances as of 1987) (Cubbage and Raney 1987). Simply becoming aware of and summarizing information about local ordinances—their origins and their impacts—will require a significant research effort.

Research on often conflicting relationships between restrictive local laws and state laws also has importance for timber harvesting activities. Considerable debate often occurs over which governmental unit has precedence. The Pacific Coast states have generally resolved this question in favor of state regulation; other areas have not clarified authority. In 1982, Massachusetts enacted a forest cutting practices law, partially in hopes that one state law administered by a single agency (Department of Forests and Parks) would be preferable to loggers operating under a multitude of often differing local ordinances. The merits of local versus state ordinances addressing timber harvesting activities will be increasingly important in the years ahead.

Another area of legal-descriptive research having relevance to harvesting activities (in fact, all forestry activities) is the degree of allowable regulation, or "taking." Forest practice regulation has been held constitutional by state and federal courts

(Cubbage and Siegel 1985), but recent U.S. Supreme Court decisions have suggested some basis for challenge of regulatory ordinances. Some decisions on "inverse condemnation" have held that, although constitutional, local governmental bodies may be required to pay for the loss of use suffered by private parties due to regulation (Berg 1987).

Adequately monitoring the proliferating forest practice and logging regulations will require more scientists than are likely to be available. Yet research is also needed to determine the economic and administrative impacts of timber harvesting regulations. How much do the regulations cost local landowners to implement in terms of cash flow or foregone income? What are the public agency administrative costs of the laws? What are the measurable biophysical improvements attributable to the laws? How may laws be changed or modified to achieve the best obtainable benefits compared to the public and private costs incurred? Only limited research has addressed these crucial questions (Henly and Ellefson 1986).

A related area of impact research is that of determining why communities enact regulatory laws and ordinances. Political science theory could provide insights about the process that leads to logging regulation, and perhaps actions the forestry community may take to forestall or ameliorate such regulation. Relating the amount of regulation to input from citizens, interest groups, government officials, foresters, and other participants in the policy process would help in understanding and responding to regulatory issues. To date, only Salazar (1985) has performed any significant research along these lines.

Forestry regulation research may also go beyond description and analysis to suggest appropriate means of regulation. The Society of American Foresters (SAF) considered development of a model state forest practice law in the 1970s but instead adopted suggested criteria that should be considered if a law were being contemplated (Society of American Foresters 1978). Some state Society of American Forester organizations have recently considered developing model local logging ordinances. Based on legal research, impact assessments, and professional opinion, forest economists certainly have a role to play in describing the economic consequences of such model laws.

Summary and Conclusions

Forest economics research addressing timber harvesting and related activities may well prove most effective if focused on six strategically important areas. Most fundamental is development of more accurate production functions for use in further economic analyses of timber harvesting activities. Improved research on harvesting costs is a second area of needed economics research, and integrating harvesting economics with timber production economics is a third strategic direction. Development of new harvesting techniques for field use comprises a fourth economic research direction that has economic components. And broadening the definition of harvesting economics leads to identification of two additional strategic areas, namely, wood procurement and public regulation of timber harvesting activities.

A vast amount of research could be focused on economics of timber harvesting. In fact, timber harvesting economics probably warrants more attention than many other areas of forest economics research. Much forest economics research has focused on planting, growing, and managing of trees. Most harvesting research has tended to focus on production rates and engineering case studies—broader economic questions are often ignored. In total, harvesting economics and wood procurement research probably has fewer scientist years devoted to it than most timber management research. Yet procurement, harvesting, and transportation functions may account for up to one-half of the total cost of delivering wood fiber to a wood processing facility. Additionally, any savings in such costs can be realized immediately; there is no need to wait for decades as is the case with timber growing investments.

Overall, harvesting and procurement economics offers a wealth of possible research topics and great potential for large economic payoffs, yet it suffers from a paucity of funding and research scientists. One must logically conclude that harvesting economics research should be increased. Given current public budget austerity, such seems highly unlikely—at least in the immediate future. As such, any increase in harvesting economics research must come from reallocation of existing forest economics or harvesting research programs. Some reallocation might be worth the effort, since harvesting and procurement economics research potentially has greater marginal returns than timber management research. If the challenge of providing low cost wood fiber to primary processing facilities is to be met, scientists, research administrators, and users of research should make a more distinct and convincing case for research devoted to the economics of timber harvesting.

Literature Cited

American Pulpwood Association. 1965. Machine rate calculation. Technical Release 65-R-32. Washington, DC.

Butler, D. A. and D. P. Dykstra. 1981. Logging equipment replacement: A quantitative approach. Forest Science 27(1):2-12.

Berg, S. 1987. Wetlands, nonpoint source pollution, smoke management update. American Paper Institute and National Forest Products Association. Washington, DC.

Cubbage, F. W. 1981. Machine rate calculations and productivity rate tables for harvesting southern pine. Staff Paper Series 24. Department of Forest Resources. University of Minnesota. St. Paul, MN.

Cubbage, F. W. and W. D. Greene. 1987. Logging regulation: The emphasis is on trucking. In: Proceedings of Southwide Forest Operations Conference. Cooperative Extension Service and School of Forest Resources. University of Georgia. Athens, GA.

Cubbage, F. W. and K. P. Raney. 1987. County logging and tree protection ordinances in Georgia. Southern Journal of Applied Forestry 11(1):76-82.

Cubbage, F. W. and W. C. Siegel. 1985. The law regulating private forestry. Journal of Forestry 83(9):538-545.

Cubbage, F. W., W. C. Siegel and K. P. Raney. 1986. Logging, politics and regulation in the east: Status and prospects. In: Proceedings of 8th Annual Meeting, Council on Forest Engineering.

Cubbage, F. W., B. J. Stokes, and J. E. Granskog. 1988. Trends in southern forest harvesting equipment and logging costs. Forest Products Journal 38(2): 6-10.

Cubbage, F. W. and P. A. Wojtkowski. 1988. Analysis of factors affecting southern pulpwood harvesting costs. In: Proceedings of 1987 Annual Meeting, Council on Forest Engineering. High Technology in Forest Engineering. Syracuse, NY.

Davis, C. and T. Reisinger. 1988. Harvest planning using GIS. In: Proceedings of 1987 Annual Meeting, Council on Forest Engineering. High Technology in Forest Engineering. Syracuse, NY.

Devine, H. A. and R. C. Field. 1986. The Gist of GIS. Journal of Forestry 84(8):17-22.

Dorris, J. and F. Cubbage. 1987. 1986 Timber harvesting costs. Technical Release 87-R-13. American Pulpwood Association. Washington, DC.

Forest Products Research Society. 1975. Wood Residue as an energy source. Proceedings No. P-75-13. Madison, WI.

Harou, P. A. 1980. How to decide when to replace high cost capital equipment. World Wood 21(4):26.

Hassler, C. C., S. A. Sinclair, and D. J. Ferguson. 1981. Trends in pulpwood logging cost during the 1970s. Forest Products Journal 31(9):53-58.

Hendricks, G. L. and D. T. Curtin. 1985. Description of a microcomputer-based timber harvesting model. Paper No. 85-1610. American Society of Agricultural Engineers. Chicago, IL.

Henly, R. and P. V. Ellefson. 1986. State forest practice regulation in the U.S.: Administration, cost, and accomplishment. Station Bulletin AD-SB-3011. Agricultural Experiment Station. University of Minnesota. St. Paul, MN.

Herrick, O. 1982. Estimating benefits from whole-tree chipping as a logging innovation in northern U.S. forests. Forest Products Journal 32(11/12):57-60.

Hogan, E. 1983. Catch 22 in New Jersey. American Tree Farmer 32(4):10, 16.

Irland, L. 1985. Logging and water quality: State regulation in New England. Journal of Soil and Water Conservation 40(1):98-102.

Jackson, B. and R. Brinker. 1988. An evaluation of the northeast Louisiana pulpwood procurement environment using a geographic information system. In: Proceedings of 1987 Annual Meeting, Council on Forest Engineering. High Technology in Forest Engineering. Syracuse, NY.

Lanford, B. L. and G. F. Haver. 1973. Analysis of Production Study Data for the South. Harvesting Research Project.

LeDoux, C. B. 1986. MANAGE: A computer program to estimate costs and benefits associated with eastern hardwood management. General Technical Report NE-GTR-112. Northeastern Forest Experiment Station. USDA Forest Service. Broomall, PA.

Lee, R. G. 1986. Integrating tree growing and wood utilization. In: Proceedings, Juvenile Wood: What Does It Mean To Forest Management and Forest Products? Proceedings 47309. Forest Products Research Society. Madison, WI.

Matthews, D. M. 1942. Cost Control in the Logging Industry. McGraw-Hill Publishers New York, NY.

Miller, D. E., T. J. Straka, B. J. Stokes and W. F. Watson. 1987. Productivity and cost of conventional understory biomass harvesting systems. Forest Products Journal 37(5):39-43.

Mills, W. L. and T. G. Reisinger. 1988. Expert systems for management decisions. In: Proceedings, Current Challenges to Traditional Wood Procurement Practices. Forest Products Research Society. Madison, WI.

Miyata, E. S. 1980. Determining fixed and operating costs of logging equipment. General Technical Report NC-55. North Central Forest Experiment Station. USDA Forest Service. St. Paul, MN.

Plummer, G. L. 1967-1982. Harvesting Developments Machine Cost and Rates. Georgia Kraft Corporation. Atlanta, GA.

Rideout, D. and J. Hof. 1987a. Cost sharing in multiple use forestry: A game theoretic approach. Forest Science 33:81-88.

Rideout, D. and J. Hof. 1987b. Allocating joint costs in applied forestry. Western Journal of Applied Forestry 2:45-48.

Salazar, D. J. 1985. Political processes and forest practice legislation. Ph.D. Dissertation. University of Washington. Seattle, WA.

Shaffer, R. M. 1987. Bidding strategies in a competitive market. In: Proceedings, Southwide Forest Operations Conference. Cooperative Extension Service and School of Forest Resources. University of Georgia. Athens, GA.

Sessions, J., J. W. Mann and D. Brodie. 1986. Technical advances in wood utilization, quality premiums for stumpage, and profitability of forest investment. In: Proceedings, Assessing Timberland Investment Opportunities. Proceedings 47436. Forest Products Research Society. Madison, WI.

Sinclair, S. A., C. C. Hassler, K. Bolsted and E. Kallio. 1985. Characteristics of independent loggers: Productivity, employees, profitability. Forest Products Journal 35(5):19-24.

Smith, L. A. and D. L. Sirois. 1982. Ergonomic research: Review and needs in southern forest harvesting. Forest Products Journal 32(4):44-49.

Society of American Foresters. 1978. Criteria for a competent state forest practice act. Position statement. Bethesda, MD.

Stier, J. C. 1982. Changes in the technology of harvesting timber in the United States: Some implications for labor. Agricultural Systems 9:255-266.

Stokes, B. J. and W. F. Watson. 1985. Integration of biomass harvesting and site preparation. Paper presented at seventh annual forest biomass workshop. Gainesville, FL.

Stokes, B. J., W. F. Watson, and I. W. Savelle. 1985. Alternate biomass harvesting systems using conventional equipment. In: Proceedings of the 1984 Southern Forest Biomass Workshop, Sixth Annual Meeting of the Southern Forest Biomass Working Group. Southeastern Forest Experiment Station. USDA Forest Service. Asheville, NC.

Stuart, W. B. 1981. Harvesting analysis technique: A computer simulation system for timber harvesting. Forest Products Journal 31(11):45-53.

Stuart, W. B., C. D. Porter, T. A. Walbridge, and R. G. Oderwald. 1981. Economics of modifying harvesting systems to recover energy wood. Forest Products Journal 31(8):37-42.

Tennessee Valley Authority. 1983. An evaluation of logging systems designed to recover harvesting residues for energy. Technical Note B-50. Division of Land and Forest Resources. Tennessee Valley Authority. Norris, TN.

Tufts, R. A., B. L. Lanford, W. D. Greene, and J. O. Burrows. 1985. Auburn Harvesting Analyzer. The Compiler 3(2):14-15.

Tufts, R. A. and W. L. Mills, Jr. 1982. Financial analysis of equipment replacement. Forest Products Journal 32(10):45-52.

Vasievich, J. M. 1984. Economics of biomass—a new found resource in the South. In: Proceedings of the 1983 Southern Forest Biomass Workshop. Fifth Annual Meeting of the Southern Forest Biomass Working Group. Southeastern Forest Experiment Station. USDA Forest Service. Asheville, NC.

Warren, B. J. 1977. Analyzing logging equipment costs. In: Logging Cost and Production Analysis. Timber Harvesting Report No. 4. LSU/MSU Logging and Forestry Operations Center. Long Beach, MS.

Watson, W. F., J. M. Kucera, R. K. Matthews, and R. A. Kluender. 1977. Pulpwood producer census, Southwest Technical Division of the American Pulpwood Association. Agricultural and Forest Experiment Station. Mississippi State University. Mississippi State, MS.

Weaver, G. H., R. A. Kluender, W. F. Watson, W. Reynolds, and R. K. Matthes. 1981. 1979 pulpwood producer census, Southwest and Southeast Technical Divisions of the

American Pulpwood Association. Agricultural and Forest Experiment Station. Mississippi State University. Mississippi State, MS.

Weaver, G. H., R. A. Izlar, R. A. Kluender, W. F. Watson, and R. K. Matthes. 1982. 1980 survey of high volume independent wood producers in the South. Information Bulletin No. 23. Agricultural and Forest Experiment Station. Mississippi State University. Mississippi State, MS.

Werblow, D. A. and F. W. Cubbage. 1986. Forest harvesting equipment ownership and operating costs in 1984. Southern Journal of Applied Forestry 10(1):10-15.

13

PRODUCTION AND VALUATION OF FOREST AND WILDLAND RECREATION

George L. Peterson and Thomas C. Brown

Outdoor recreation is an important use of the nation's forest and wildlands for millions of Americans. In the face of rapid growth in population, urbanization, and technological development, Americans are increasingly realizing the benefits of recreational activities which encourage direct interaction with natural environments. The broad spectrum of recreational experiences available from forest and wildland resources is conducive to provision of a variety of social benefits, including rest from the stresses of daily life, greater family solidarity, increased worker productivity, greater self-confidence, and improved physical fitness (Forest Service 1981).

Along with individual and social benefits of outdoor recreation come a number of problems of importance to recreational users and managers. Many such problems are deserving of research attention, especially the attention of persons trained in the field of forest resource economics. Economics research focused on the recreational use of forest and wildlands would be most helpful to recreationists and managers if focused on the following strategic areas:

- Develop more suitable means of defining and measuring forest and wildland recreation products.
- Acquire improved knowledge of markets for forest and wildland recreation products.
- Develop better perceptions of production and supply of forest and wildland recreation products.
- Acquire improved knowledge of demand and economic valuation of forest and wildland recreation.
- Develop more effective means of pricing, financing, and rationing forest and wildland recreation.
- Acquire better understanding of forest and wildland recreation as a means of encouraging economic development.

- Secure a better understanding of off-site and non-use values of forest and wildland recreation.
- Develop means of enhancing the usefulness of recreation economics in policy development processes.
- Ensure development of recreational data bases and foster ability to apply information about recreation economics.

The subject of economics research concerning forest and wildland recreation is especially rich; a subject which, for complete coverage, requires more space than is available here. As such, the above strategic directions will be only briefly described as will research topics within each strategic area. Specific research studies or methods for their accomplishment will not be presented. No attempt has been made to organize strategic directions or research topics in order of priority.

Develop More Suitable Means of Defining and Measuring Forest and Wildland Recreation Products

Outdoor recreation products are not well understood (King and Hof 1985, Driver 1985). A major cause of this problem is the diversity of objectives and decisions that surround recreation products. Managers and researchers often fail to recognize that different objectives and different decisions focus on different products.

At one extreme the product may simply be access to sites or facilities. A person passing through a gate is one unit of the product. At the other extreme it may be a subjective and unobservable recreation experience. Between these two extremes are such things as trips or activities, which are forms of observable behavior.

The heterogeneous and composite nature of most recreation outings aggravates the definitional problem (Harrington 1987, King and Hof 1985). People who take recreational trips often seek economies of joint production by visiting several destinations, thereby obtaining more net benefit than if they visited each destination independently. At a given destination, people usually participate in several activities that produce composite experiences. There is no such thing, for example, as a fishing trip where fishing is the only activity experienced. Complicating the matter more, each site, trip, activity, or experience may be unique. Recreation does not occur through mass production of identical units that are perfect substitutes for each other.

Many definitions exist because there are many different recreation products— each of which is relevant to a different policy question. Needed are decisions about which questions focus on which products. Products then need to be defined in operational terms before product quantity, quality, and price can be measured (Churchman and Ratoosh 1959). It is important that researchers and managers use standard product definitions and that these definitions agree with the definitions used by consumers in their recreation behavior.

Given precise definitions, however, measurement methods still require extensive development and refinement. The question, "How much of the good?" requires not only clear and precise definition of the good in operational terms, but also standard units and methods of measurement. At present, the units used to measure recreation quantity are a collection of things like recreation visitor days (RVDs), recreation activity days, wildlife user days (WFUDs), persons at one time (PAOTs), trips, visits, acres of land, units of a given facility (e.g., number of campsites or picnic tables, miles of trail). Misunderstandings and misapplications are common. Adding to these problems with quantitative measurement, measurement of recreation quality is primitive and largely unsuccessful.

Thus, one of the most important and fundamental strategic directions for research is to improve the state of the art of definition and measurement of forest and wildland recreation products. Research topics within this strategic direction include the following:

- Develop good definitions of forest and wildland recreation products that consider quality differences, joint production and consumption, and product heterogeneity.
  - Analyze planning and management decisions to identify the different recreation products. Decide which products are of economic concern, and evaluate the adequacy of existing definitions and measurements. Identify and separate products supplied and controlled by landowners and operators from products supplied by consumers via household production.
  - Where needed, define recreation products in terms that allow magnitude to be identified and measured quantitatively.
- Develop standard units, methods, and procedures for measuring or estimating recreation outputs, including quantity and quality.

Acquire Improved Knowledge of Markets for Forest and Wildland Recreation Products

Outdoor recreation markets are complex, imperfect, and poorly understood (Hof and Kaiser 1983, Hof and Loomis 1983). In part the cause is product fuzziness and heterogeneity as described above. Four additional factors, however, also contribute to the problem:

- Inefficient decisions by product suppliers or regulating agencies may cause market distortion. For example, the price of recreation opportunity on public land tends to be below cost because of the high cost of fee collection, political traditions, legal constraints, or unknown marginal costs.
- Spatial lumpiness of the recreation market, and the need to travel to locations where recreation takes place, cause transportation cost to be a significant component of price. As a result, different people face different prices, and different sites (although identical in all other respects) face different demand curves.
- Outdoor recreation tends to have the characteristics of a congestible public good. There is a tendency toward non-rivalry in consumption, but quality also tends to decrease with increasing congestion.

- Outdoor recreation sites often coincide with unique natural features that are responsible for site attractiveness and for site designation. These unique features create strong product differentiation which, in combination with spatial lumpiness, leads to monopolistic conditions.

Thus, an outdoor recreation market is one of production and consumption of strongly differentiated public or quasi-public goods under imperfect competition in a distorted spatial market. The theories and paradigms needed to understand such markets are available but not adequately applied to recreation. Several schools of knowledge in economics need to come together in an integrated theory of recreation markets. The theory of nonpriced goods is well developed in recreation economics (Randall 1984a), but the theory of spatial markets, as developed and applied in transportation, urban economics, analytical geography, and regional science (Anas 1982, Mills 1980, von Thunen 1826, Losch 1954, Alonso 1964) has yet to be widely applied to recreation economics. These bodies of knowledge need integration in applied recreation economics. Also needed is more knowledge about product differentiation and imperfect competition.

Recreation resources and opportunities are also involved in long-run market questions that have not received sufficient attention. The supply of scenic and recreation resources is an important influence on quality of life and attractiveness of land for residential and industrial use (Abelson 1979, Flattau 1985, Pope 1985). These resources, therefore, exert an influence on land markets and land values. For example, part of the value product of investment in public forest and wildland may be rent captured by adjacent private landowners. Complete evaluation of the productivity of recreation-related investments requires identification and measurement of such rents. With adequate documentation, they are external benefits that the government might, for example, internalize through taxation.

Many recreation goods and services are produced and consumed in parallel public and private markets. For example, the private sector produces opportunities and charges competitive prices for camping and hunting on private land. The public sector also produces such opportunities on public land without charge or at fees determined administratively or legislatively. Information about the cost and price structure of the private market may be useful for guiding resource allocation and management decisions in the private sector (U.S. Department of Agriculture 1987). However, knowledge of the interaction between public and private recreation markets and comparability of products is not adequate.

Overcoming these and other problems requires research that will improve description, explanation, and modeling of recreation markets and of markets affected by recreation resources. Research topics within this strategic area include:

- Combine and apply location economics, monopolistic market theory, and the economics of public goods to develop an integrated theory and model of outdoor recreation markets.
- Use this theoretical framework to develop models of imperfect equilibrium and simulated competitive equilibrium for specific forest and wildland recreation markets. Such models will allow analytical evaluation of alternative policies

for intervention in imperfect markets, determination of efficient prices and quantities for resource allocation, and estimation of welfare changes resulting from policy changes.

- Apply the theories and methods of land economics and urban economics to explain and model land market relationships between recreation resources and private land values.
- Identify and analyze private recreation markets that offer recreation products and services comparable to those offered on public land. Evaluate the competitiveness of such markets and analyze the interaction between public and private sectors.

Develop Better Perceptions of Production and Supply of Forest and Wildland Recreation Products

The theory of recreation production and supply is poorly organized (Harrington 1987). Lack of clear product definition, jointness in production and consumption, severe product heterogeneity, spatial lumpiness of recreation markets, and lack of rigorous theoretical attention have left the supply-side of recreation economics in a primitive state. There even seems to be disagreement about what is meant by the supply-side of recreation.

Recreation supply is a complex chain of inputs and outputs. The chain begins at one end with production of facilities or equipment and management of natural areas. At the other end, consumers of recreation combine sites and facilities with travel, time, energy, activity, equipment, skills, and other input factors in household production of recreation experience (Becker 1965, Morishima 1959). One person's production is another person's demand. Is the supplier of recreation the owner of the land where the activity occurs, the operator of facilities or concessions on the land, the supplier of equipment used in the activity, the operator of transportation services that provide access to recreation sites, or the recreationist who combines these things to produce recreation experience? It is not even clear that consumers define recreation products in terms that are compatible with the production process. As in automobile transportation, the supply side controls facilities, capacities, and opportunities, but the demand side controls the quantity and type of output.

Even with agreement on product definition and the point of market reference by which to define supply and demand, there still is a great void of knowledge about production and cost functions for many such points. For example, public agencies produce forest and wildland recreation opportunities in a hierarchical framework of multiple public objectives (Hof and Pickens 1986, Peterson 1985). Many purposes jointly motivate and serve public investments that contribute to recreation opportunity (Hof et al. 1985, Rideout and Hof 1987). Roads built for timber harvest may improve access for recreation, and timber harvest may enhance or impair wildlife habitat, which, in turn, affects the quality of recreation opportunity. Allocation of joint costs among different products is not justifiable without arbitrary assumptions (Herfindahl and Kneese 1974, Bowes and Krutilla 1979, Duerr 1960, Hof et al. 1985). Furthermore, although the theory of marginal cost in joint production is clear, we do not understand the joint production processes well enough to empirically describe joint marginal cost functions.

Another important area of concern in the production of recreation opportunity is the cost imposed by congestion and user impact. To be complete, a marginal cost function needs to include these congestion and impact costs (Rosenthal et al. 1984, Freeman 1979, Freeman and Haveman 1977). Their inclusion is, however, beyond the state of the art. Supply-side analysis of costs is largely limited to estimation of budget or direct expenditure functions, not true cost functions.

If the product is recreation experience, the supply side is even more complex and obscure. The supplier is the recreationist who jointly produces multiple-destination trips and composite experiences at each destination. Some of the inputs to such a production process are non-priced and may even be unobservable subjective commodities (Harrington 1987). Thus, defining the recreationist as the supplier of recreation experience, as in household production, does not solve the supply-side problem and is not particularly useful to suppliers of recreation opportunity. Recreation resource managers control some but not all the inputs to production of recreation experience.

However, the household production process is important to understand, if only because it reveals the complexity of the recreation supply problem and its entanglement with recreation demand. Recreation resource planners and managers want to know the demand and supply prices for various forest products in order to estimate the present net value of alternative forest management prescriptions. Managers and planners ask for information about the value of specific recreation activities. However, because of joint production of activities by the recreationist on the demand side of the manager's problem, separation of demand prices by activity requires arbitrary assumptions. One way to solve the problem is to reframe the question in terms of the supply of opportunity for independently consumed demand packages, such as visits to a site. While such an approach may solve the joint production problem, it substitutes a new and equally severe problem—a very large number of unique demand packages that vary in magnitude and quality.

These and other problems offer some interesting challenges. There is no satisfactory measure of recreation supply, relative scarcity, or effective price (Harrington 1987). It is thus not possible to obtain a good inventory of recreation supply; evaluating adequacy of existing resources or planning effectively for future needs become insurmountable tasks. Absence of reliable marginal cost information remains an obstacle to efficient pricing, even without the other obstacles (Rosenthal et al. 1984). Difficulty with identification and estimation of supply and demand prices hampers benefit cost analysis. The combined effect of these problems is like flying through clouds without adequate instruments.

Thus, improvement of knowledge about production and supply of recreation requires major research efforts in topical areas such as the following:

- Identify the specific products and points of reference in the recreation production process where there is a need for better supply information.
- For these products and points of reference, develop a theory of recreation production and supply in imperfect markets that is suitable for specification

and estimation of recreation supply functions and production functions, and for integration with imperfect market demand theory and application to benefit-cost analysis.

- Apply this theory to public and private situations to describe recreation production relationships and processes, including marginal cost functions, joint production relationships, and sensitivity of recreation quality to site management and level of investment.
- Reconcile long- and short-run cost relationships, and develop criteria for deciding when to include long-run costs (e.g., opportunity cost of land allocation) in the marginal cost function. Explore and develop a hierarchical model of resource investment objectives, decisions, and cost allocation.
- Develop and apply measures of recreation and supply, effective price, and relative scarcity suitable for evaluating the adequacy of recreation resources.
- Describe the relationship between level of investment and quantity and quality of opportunity.
- Develop methods for including the costs of congestion and user impact in marginal cost analysis.
- Explore relationships between public and private recreation markets. How does government production of opportunity affect the private recreation sector and vice versa? What, if any, public recreation services can the private sector provide more effectively?
- Identify and measure the costs of alternative ways to provide recreational opportunities to special populations, and expose the joint production and income transfer costs of income redistribution through public provision of recreation services.
- Describe and analyze the cost structure of private recreation suppliers, and evaluate the transferability of cost information from the private to the public sector for comparable products.

Acquire Improved Knowledge of Demand and Economic Valuation of Forest and Wildland Recreation

The theory and methods of the demand-side of recreation economics are well developed (Peterson and Randall 1984). However, intriguing questions have begun to challenge the theory, and serious problems confront practical application. Because of these problems and the complexity of the subject, credibility of demand-side valuation methods for nonpriced recreation is weak in some management and policy circles.

On the theoretical side, there is a need to reconcile the behavioral axioms and propositions of microeconomic theory; implications of empirical economic measurements of value; and behavioral propositions of other disciplines, such as psychology (Peterson, Driver, and Gregory 1988; Simon 1985; Brown 1984; Peterson, Driver and Brown [in press]). Serious unresolved differences threaten the credibility of value measurements. For example, economic theory predicts that willingness to pay (WTP) and willingness to accept compensation (WTA) are equal except for income effects, which for most recreation cases are trivial. However, empirical experiments show substantial differences that are inconsistent with economic theory (Fisher et al. 1988, Gregory 1986, Knetsch and Sinden 1984). Some economists argue that inadequate experimental methods cause the empirical discrepancies (Gregory 1986). Some behavioral scientists contend that economic theory is inconsistent with human behavior, and that such differences are real and expected (Tversky and Kahneman 1981).

Even without these challenges to the well-developed theory of economic behavior, other problems inhibit practical application. Jointness in household production of multiple destination trips and composite recreation experiences raises questions about estimation of the activity-specific and destination-specific values needed by the current framework of resource planning and management. Substitution in forest and wildland recreation needs more work (Rosenthal 1985). Understanding of the effect of variation in congestion and other site characteristics on the quality and value of recreation experience is not adequate (Fisher and Krutilla 1972, Freeman and Haveman 1977). Econometricians disagree among themselves about empirical methods. Although the plans that require knowledge of demand values reach into the future, knowledge about the variation of demand and value over time is sparse (Peterson, Loomis, and Sorg 1985; Peterson, Stynes, and Arnold 1985; Peterson 1988).

Given these problems and the virtually boundless complexity of the subject, it is no wonder that credibility suffers in policy circles. Well-placed empirical validation studies might narrow the credibility gap, but good validation studies are few. It is difficult to show that values estimated for imperfect market goods are valid estimates of monetary exchange value commensurate with competitive market prices for private goods.

Lack of high quality and comparable data also hampers demand-side valuation. The required primary data collection is expensive and time-consuming. Most economic valuation studies are either non-comparable and non-generalizable ad hoc research exercises, or too general and comprehensive to be applicable to specific policy questions.

There are also some more fundamental demand questions that transcend the traditional domain of economics. The demand economics problems identified above are largely concerned with specification and estimation of demand functions and the economic benefits of existing recreation products. But traditional economics has little to say about how to find out what people want and why they want what they do, about the beneficial and detrimental consequences of what they choose, about how to design more valuable products, or about how to estimate the economic value of things that people have not experienced. Hicks (1956), Morishima (1959), Becker (1965), and Lancaster (1966) have flirted with some of these questions, but economics simply is not self-sufficient. Thus, in addition to research on the economics of demand, there is also a need for research on basic relationships between recreation and human health and well-being.

To solve these problems, research is needed to improve and apply theories, methods, and information for recreation demand analysis, imperfect market demand-side valuation of recreation resources and products, and assessment of recreation needs. Specific research topics include the following:

- Further develop and apply the theory of recreation demand and household production for imperfect markets and heterogeneous and composite recreation products.
- Improve and validate economic valuation theory. Reconcile theoretic and empirical discrepancies. Conduct basic research on human economic behavior to determine whether microeconomic theory adequately describes that behavior.
- Develop a well-balanced framework for empirical validation of economic value estimates. Do revealed preference and hypothetical market estimates of WTP and WTA for non-priced goods effectively measure monetary exchange values? Evaluate and standardize economic valuation methods such as the travel cost method (TCM) and the contingent valuation method (CVM). How does the validity of different methods compare for different goods and different contexts? What constraints and caveats limit application of the estimates?
- Sensitize economic valuation theories to variations in resource quality, and develop methods for determining the effect of changes in congestion, resource quality, and level of investment on recreation participation and economic value.
- Develop a better understanding of substitution in recreation demand. Develop improved demand function specifications that adequately incorporate substitute prices. Explain the choice process by which people make decisions to substitute, and develop rigorous and practical methods for estimating a substitute price index for recreation.
- Estimate and catalog generalized recreation demand models and valid estimates of marginal and non-marginal recreation demand prices.
- Describe changes in recreation participation and demand over time and develop effective forecasting or contingent forecasting techniques, theories, and models.
- Investigate the problem of joint consumption (joint household production) of recreation activities, composite experiences, and multiple destinations. If possible, develop methods for estimation of values for separate recreation activities and destinations. Explore new strategies for valuation and planning, strategies that estimate and compare total recreation value under alternative management prescriptions.
- Identify and describe special recreation needs (e.g., senior citizens, children, ethnic minorities, inner-city residents) whose recreation preferences and economic values may not be represented adequately in economic studies, market forces, or political processes. Evaluate and compare the distributional consequences of alternative recreation management policies on such groups.
- Evaluate the applicability of private recreation market demand information to comparable goods and services offered on public land. Where applicable, use such private market information to estimate demand functions and economic values for non-priced public opportunities (U.S. Department of Agriculture 1987).

Develop More Effective Means of Pricing, Financing, and Rationing Forest and Wildland Recreation

Pricing outdoor recreation for efficiency, revenue, and rationing is not well understood (Rosenthal et al. 1984). Inefficient fees cause income transfer and wasteful allocation of resources (Musgrave and Musgrave 1973, Loomis 1980, Miller and Higgins 1981, Loomis 1982). A political choice to accept income transfer and inefficient use of resources as the price of social objectives is legitimate, but good choices require knowledge of the costs.

The economic theory of pricing and price rationing of social goods is well developed (Winston 1985, Musgrave and Musgrave 1973), but application to publicly supplied and non-priced recreation is primitive (Rosenthal et al. 1984). Multiple part pricing, as needed for efficiency in hierarchical joint production systems such as recreation, is not well understood or effectively employed (Guldin and Kroon 1987, Peterson 1985). Price rationing is a potentially useful but unused tool for management of congestion in public recreation areas (Hof and Loomis 1986, Loomis 1982).

In the private sector, landowners and facility operators pay competitive prices for the services they provide. They control access to their services, and they must recover costs. In theory, supply and demand drive private prices to efficient levels, and prices regulate supply and ration consumption. However, private and public recreation markets interact. Inefficient decisions in the public sector may distort the market and cause inefficiency in the private sector. Needed is better understanding of the interaction between public and private recreation opportunities. The competitiveness and efficiency of private markets also needs to be evaluated so that pricing information can be transferred to the public sector for comparable services.

The distributional impact of pricing policy is also in need of better understanding (Musgrave and Musgrave 1973; Vaux 1975; Just, Hueth, and Schmitz 1982; Baumol 1980; Driver 1984; Harris and Driver 1987; Driver and Koch 1987; Crompton 1984; Crompton 1982). Prices can be important distributional agents, particularly for local opportunities or in local areas. Public services (e.g., recreation) offer opportunities to achieve income redistribution, as directed by social choice. Such income transfer is a political decision, but good political decisions require technical information about the distributional consequences of alternative choices. Where or whether income transfer occurs and how often is not clear, in part because of the jointness in production of recreation opportunity and the hierarchical allocation of resources to multiple public purposes (Herfindahl and Kneese 1974, Bowes and Krutilla 1979, Duerr 1960, Hof et al. 1985).

A commonly neglected and poorly understood type of income transfer is external benefit to private landowners from public investment in recreation resources. The private landowner who controls access to public land is a recipient of such an external benefit. External benefits also flow to private landowners when nearby public recreation resources enhance land productivity and land value (Abelson 1979, Flattau 1985, Pope 1985, Freeman 1979). The improvement in productivity is a technological externality of public investment and constitutes an income transfer. Subject to political constraints, documentation of such externality gives the public sector the option to capture the transferred value through taxation (Anas 1982).

Thus, it is important for policy makers to understand the impact of prices on efficiency and distribution. Development of effective and equitable pricing, financing, and rationing policies for public forest and wildland recreation requires research on topics such as the following:

- Apply the theory of efficient marginal cost pricing, second best pricing, multiple part pricing, and social merit investment to imperfect market recreation.
- Apply methods for analyzing discriminatory or distributional impacts of resource location, allocation, and pricing to forest and wildland recreation. Compare such impacts under different circumstances (e.g., urban versus nonurban resources).
- Analyze the impact of alternative pricing and rationing policies on the quantity of recreation supplied and consumed. Evaluate the use of fees to ration recreation services and resources. Compare the efficiency of price rationing with other rationing policies.
- Describe subsidies, income transfers, and external diseconomies in recreation markets. Evaluate the effect of alternative price policies on income transfer. Identify those recreation opportunities available at zero marginal cost where the efficient fee is zero.
- Evaluate the potential for exposing external land rent benefits induced by public investments in recreation resources. Identify and evaluate alternative ways to internalize such external benefits.
- Evaluate transferability of private pricing information to recreation opportunities provided by the public sector (U.S. Department of Agriculture 1987).

Acquire Better Understanding of Forest and Wildland Recreation as a Means of Encouraging Economic Development

Recreation is an important industry in many local economies, and information on the economic impact of recreation and tourism is of great interest to state and local governments. Recreation attracts tourism dollars, generates local income and employment, plays an important role in balance of payments within a geographic jurisdiction, and has an important influence on local land and labor markets. Recreation resources (especially major public forest and wildland areas such as wilderness areas, national forests, and national parks) contribute to quality of life in the local area and have a significant long-term influence on residential and industrial location decisions. International tourism also is important for the same reasons at the local and national levels.

Planning at the federal level focuses primarily on internal efficiency and external balance of payments, but the local jurisdiction worries more about its income, employment, and local balance of payments. Analytical methods for economic impact assessment are well developed, and current practice is a straightforward technical exercise where adequate data are available (Alward 1986, Propst 1985, Miller and Blair 1985). Data for recreation are not widely available, however. Development of the coefficients needed for input-output analysis of the recreation sector (as in Impact Analysis for Planning [IMPLAN]) requires an extensive data collection effort (Alward and Lofting 1985). Without adequate data, it is not possible to use available methods to evaluate the importance of various recreation expenditures at local, regional, and national levels. Even with such data, however, current practice in economic impact assessment falls short of the need, because it includes only direct monetary transactions and ignores non-priced and external exchanges and long-term impacts on the land market.

Improvement and application of methods for analyzing economic impact, development, and distribution to forest and wildland recreation require research in areas such as the following:

- Apply impact assessment methods (e.g., IMPLAN) for analyzing and displaying local and regional impacts of recreation investment and management policies.
- Develop and apply methods to include external economies and diseconomies (including non-priced effects) in analysis of economic impact, distribution, and development.
- Improve and apply theories and methods for evaluating the impact of transportation improvements on recreation supply and demand.
- Assess the role of recreation and tourism in local economic development.
- Apply the theories and methods of location and urban economics to develop land market models that include the effects of recreation and recreation resources on land values and land development.
- Assess the role of international tourism on national balance of payments and national economic development.

Secure a Better Understanding of Off-Site and Non-Use Values of Forest and Wildland Recreation

The domain of recreation economics falls almost entirely within the set of activities that require use of and visits to recreation sites or facilities. Even the apparent exception of land market externality by which recreation-related resources affect the value of private land holdings is a form of site-use, albeit in a long-run sense of demand for land with enhanced access to recreation sites. There are real exceptions, however, that are somewhat neglected in the current state of the art. These exceptions include off-site recreation use values and recreation-like non-use values.

Demand for indirect or vicarious enjoyment of recreation resources through use of literature, films, photographs, and paintings generates off-site recreation use value. The substantial industries built around such services are evidence that demand for a recreation resource goes beyond direct on-site use. The relationship between this kind of demand and on-site activity is not clear, however. How much do recreation resources contribute to the value received by the consumers of off-site products, and what is the economic value of that contribution?

Resource-based photographs, paintings, films, and literature are non-rival and non-excludable uses of recreation resources. Fees, therefore, are difficult or impossible to collect, even where the value received is large. Furthermore, if economic efficiency is the objective, the fee should be zero where the marginal cost of such use is zero. Failure to collect a fee, however, is not cause for resource allocation decisions to ignore the value received by off-site users.

Non-use demands for recreation resources fall into three main categories: option demand, existence demand, and bequest demand (Peterson and Sorg 1987; Randall and Stoll 1983; Boyle and Bishop 1987; Randall 1987; Bishop, Boyle, and Welsh 1987). Such concepts are complex and controversial. Briefly stated, option demand is a demand to forego irreversible or long-term commitment of a resource to preserve the option to use it in other ways in the future, even though those other uses are not known now. For example, part of the motive for preserving an endangered species that now serves no known human purpose is the possibility that the species or its genetic information may serve an important human need at some future time.

Willingness to pay to preserve a resource without any intention of personal use is an expression of existence demand. Both ethical and altruistic motives drive this kind of demand (Boyle and Bishop 1987, Fisher and Raucher 1984). An example of an ethical motive is a belief that another species has the right to exist. Given such a belief, humans who have the power to destroy or preserve the species have an ethical responsibility to respect its right to exist. An example of the altruistic motive is a willingness to pay to preserve a resource because it is beneficial to someone else (or to another species) even though the preserver has no anticipation of personal gain. Bequest demand is a special case where future generations of humans are the beneficiaries.

Whether non-use values belong in a discussion of recreation production and valuation is an open question, depending on one's point of view. However, such values are recreation-like, they may involve an anticipation of or potential for recreation, and they often involve resources used for recreation. They are included here because non-use values are an important and neglected component of the justification for preserving the wildland resources and protecting the wildlife species that comprise much of the nation's forest and wildland recreation resource base.

Indirectly or implicitly, non-use values probably help drive the political processes that make resource preservation decisions. Such values do not generally receive adequate formal recognition, however, in the economic information systems that serve participants in the political processes of social choice. The state of the art of economic definition and measurement of these values still is in the developing stage.

Thus, to identify, describe, and measure the off-site and non-use benefits of forest and wildland resources, research should be focused on topics of the following nature:

- Describe, analyze, and assess the off-site use values of recreation resources generated through such media as literature, films, photographs, and paintings.
- Develop rigorous and operational definitions of the non-use benefits of outdoor recreation resources.
- Identify and describe, for each forest and wildland product or resource, the environmental and policy situations in which non-use values are substantial and significant. Evaluate the ability of economic information systems and policy institutions to identify, measure, and respond to off-site and non-use values.
- Evaluate the non-use economic question from policy and research positions. Is it a useful question from a policy point of view? Is it an answerable research question? Is non-use value a separable component of total value? Is it separable from recreation value?
- Evaluate and improve the theory and method of production and valuation of non-use benefits. Is it possible to design and carry out experiments capable of yielding defensible and valid measurements of the monetary value of non-use benefits?

Develop Means of Enhancing the Usefulness of Recreation Economics in Policy Development Processes

Some problems in the application of economic information to production and valuation of recreation lie outside the domain of economics. The answers to such questions require better understanding of the role of economic information systems in social choice (Ellefson 1986), not more technical information about recreation economics.

For example, the economists' rational benefit-cost model and potential Pareto improvement criterion sometimes disagree with the outcome of social choice (Loomis 1987, Bolan 1967). Economic efficiency is only one of many objectives on the pluralistic agenda of political conflict resolution. Viewing benefit cost analysis as a technical information system, rather than as a social choice criterion (Randall 1984b), raises three important research questions: How is economic information used in wildland recreation decisions? Can technical economic questions be separated effectively from political issues (Frey et al. 1984)? And does economic information get communicated to all who need it (Gray and Ellefson 1987)?

More technical research on recreation economics cannot answer these kinds of questions. The need is to improve the usefulness of recreation economics in policy decisions via research on topics such as the following:

- Describe and evaluate the influence of economic information in forest and wildland recreation policy decisions.
- Develop better understanding of management and policy needs for economic information about recreation. Clarify the needed roles and forms of such information in public policy analysis, planning processes, and management

decisions. Clarify the separation between political value judgments and technical judgments of fact.
• Evaluate the adequacy of existing land management and planning procedures. Can these procedures cope effectively with the complex process of recreation production and consumption and with the imperfection of recreation markets?

Ensure Development of Recreational Data Bases and Foster Ability to Apply Information About Recreation Economics

Besides the above identified research needs, two important problems impede the conduct of research on recreation economics and the usefulness of information generated by that research. The first problem is a lack of adequate data bases. The high quality longitudinal and cross-sectional data needed for policy application and for research require better data systems. Second, decision makers who need economic information often lack the training and advice required to use it. Effective application of research requires programs that provide suitable economic training for decision makers.

Programs such as the following will support disciplined research in recreation economics and will enhance effective application of economic information to recreation resource allocation and management:

• Establish and maintain a continuing process for inventory and evaluation of national recreation resources, recreation costs, and recreation outputs. Accumulate valid and reliable data on key supply variables.
• Develop and maintain valid and reliable data and data collection systems for monitoring recreation participation and estimating recreation demand models.
• Develop and maintain data collection systems needed for effective assessment of the economic impacts of investments in recreation resources.
• Evaluate and improve educational programs, both in natural resource management and in recreation economics. Integrate a better understanding of location economics, monopolistic market theory, and the theory of non-priced goods for outdoor recreation into managerial practice and educational programs.

Summary and Conclusions

The recreational use of forest and wildland resources is an important social and economic activity. The means by which such resources are made available for recreational purposes are not without problems—many of which involve the allocation of scarce resources. Such problems warrant the attention of forest economics research. To be most effective, research involving recreation economics should be designed so as to enhance the forestry community's understanding of product definitions and measurement; production and supply; demand and economic valuation; pricing, financing, and rationing; economic development; off-site and non-use values; recreation economics in policy development processes; and information bases which characterize forest and wildland recreation.

There are, of course, broader more fundamental recreation problems that do not fall under the umbrella of economics research. For example, in the public sector, market imperfection and public trust sever the market connections between supply and demand, linkages that typically enable private producers to decide what and how much to produce. Should managers of national forests, for example, focus on people's recreational preferences (providing them what they want) or are there also public trust responsibilities that must be served? If preferences are the guiding criteria, whose preferences? Such questions are the business of politics, not economics. Given the answers, the business of economics is to identify and estimate economic costs and benefits, and to provide the economic information required by the political process to evaluate efficiency and distribution. Properly crafted, economic research can play an important role in this respect.

Literature Cited

Abelson, P. W. 1979. Property prices and the value of amenities. Journal of Environmental Economics and Management 6:11-28.

Alonso, W. 1964. Location and land use. Harvard University Press. Cambridge, MA.

Alward, G. S. 1986. Local and regional impacts of outdoor recreation development. In: Literature review: President's commission on Americans outdoors. President's Commission on Americans Outdoors. Washington, DC.

Alward, G. S. and E. M. Lofting. 1985. Opportunities for analyzing the economic impacts of recreation and tourism using IMPLAN. Paper presented at the Regional Science Association Meeting. Washington, DC.

Anas, A. 1982. Residential location markets and urban transportation. Academic Press. New York, NY.

Baumol, W. J. 1980. Theory of equity in pricing for resource conservation. Journal of Environmental Economics and Management 7:308-320.

Becker, G. S. 1965. A theory of the allocation of time. The Economic Journal 75(299):493-517.

Bishop, R. C., K. J. Boyle and M. P. Welsh. 1987. Toward total economic valuation of Great Lakes fishery resources. Transactions of the American Fisheries Society 116:339-345.

Bolan, R. S. 1967. Emerging views of planning. Journal of the American Institute of Planners, July, pp. 233-245.

Bowes, M. D. and J. V. Krutilla. 1979. Cost allocation in multiple-use management: A comment. Journal of Forestry 77(7):419-420.

Boyle, K. J. and R. C. Bishop. 1987. Valuing wildlife in benefit-cost analysis: A case study involving endangered species. Water Resources Research 23(5):943-950.

Brown, T. C. 1984. The concept of value in resource allocation. Land Economics 60(3):231-246.

Churchman, C. W. and P. Ratoosh (eds.). 1959. Measurement: Definitions and theories. Wiley and Sons Publishers. New York, NY.

Crompton, J. L. 1982. Psychological dimensions of pricing leisure services. Recreation Research Review. October.

Crompton, J. L. 1984. How to establish a price for park and recreation services. Trends: User fees and charges 21(4):12-21.

Driver, B. L. 1984. Should the user pay? American Forests. March.

Driver, B. L. 1985. Specifying what is produced by management of wildlife by public agencies. Leisure Sciences 7(3):281-295.

Driver, B. L. and N. E. Koch. 1987. Cross-cultural trends in user fees charged at national outdoor recreation areas. In: Economic value analysis of multiple-use forestry by H. F. Kaiser and P. Brown (eds.). Proceedings of XVIII World Congress. International Union of Forest Research Organizations. Ljubljana, Yugoslavia.

Duerr, W. A. 1960. Fundamentals of forestry economics. McGraw-Hill Publishers. New York, NY.

Ellefson, P. V. 1986. Decisions about public forestry programs: The role of policy analysts and analytical tools. Journal of Resource Management and Optimization 4(1):65-77.

Fisher, A., G. H. McClelland, and W. D. Schulze. 1988. In: Valuation of public amenity resources: An integration of economics and psychology by G. L. Peterson, B. L. Driver and R. Gregory (eds.). Venture Press. University Park, PA.

Fisher, A. C. and J. V. Krutilla. 1972. Determination of optimal capacity of resource-based recreation facilities. In: Natural environments: Studies in theoretical and applied analysis by J. V. Krutilla (ed.). Johns Hopkins University Press. Baltimore, MD.

Fisher, A. and R. Raucher. 1984. Intrinsic benefits of improved water quality: Conceptual and empirical perspectives. Advances in Applied Micro-Economics 3:37-66.

Flattau, E. 1985. Dollars and sense and scenic beauty. Chicago Tribune. Chicago, IL. September 21.

Forest Service. 1981. An assessment of the forest and rangeland situation in the United States. Forest Resource Report No. 22. U.S. Department of Agriculture. Washington, DC.

Freeman, A. M. III. 1979. The benefits of environmental improvement. Resources for the Future. Washington, DC.

Freeman, A. M. III and R. H. Haveman. 1977. Congestion, quality deterioration and heterogeneous tastes. Journal of Public Economics 8:225-232.

Frey, B. S., W. W. Pommerehne, F. Schneider and G. Gilbert. 1984. The American Economic Review 74(5):986-994.

Gray, G. J. and P. V. Ellefson. 1987. Statewide forest resource planning programs: An evaluation of program administration and effectiveness. Station Bulletin 582-1987. Agricultural Experiment Station. University of Minnesota. St. Paul, MN.

Gregory, R. 1986. Interpreting measures of economic loss: Evidence from contingent valuation and experimental studies. Journal of Environmental Economics and Management 13:325-337.

Guldin, R. W. and H. J. Kroon. 1987. Economic and social aspects of user fees. In: Economic analysis of multiple-use forestry by H. F. Kaiser and P. Brown (eds.). Proceedings of the XVIII World Congress. International Union of Forest Research Organizations. Ljubljana, Yugoslavia.

Harrington, W. 1987. Measuring recreation supply. Resources for the Future. Washington, DC.

Harris, C. C. and B. L. Driver. 1987. Recreation user fees: Pros and cons. Journal of Forestry 85(5):25-29.

Herfindahl, O. C. and A. V. Kneese. 1974. Economic theory of natural resources. Charles E. Merrill Publishers. Columbus, OH.

Hicks, J. R. 1956. A revision of demand theory. Clarendon Press. Oxford, England.

Hof, J. G. and H. F. Kaiser. 1983. Long-term outdoor recreation participation projections for public land management agencies. Journal of Leisure Research 25(1):1-14.

Hof, J. G., R. D. Lee, A. A. Dyer, and B. M. Kent. 1985. An analysis of joint costs in a managed forest ecosystem. Journal of Environmental Economics and Management 12(4):338-352.

Hof, J. G. and J. B. Loomis. 1983. A recreation optimization model based on the travel cost method. Western Journal of Applied Economics 8(1):76-85.

Hof, J. G. and J. B. Loomis. 1986. A note on marginal valuation of underpriced facilities. Public Finance Quarterly 14(4):489-498.

Hof, J. G. and J. B. Pickens. 1986. A multilevel optimization system for large-scale renewable resource planning. General Technical Report RM-130. Rocky Mountain Forest and Range Experiment Station. USDA Forest Service. Ft. Collins, CO.

Just, R. E., D. L. Hueth and A. Schmitz. 1982. Applied welfare economics and public policy. Prentice-Hall. Englewood Cliffs, NJ.

King, D. A. and J. G. Hof. 1985. Experiential commodity definition in recreation travel cost models. Forest Science 31(2):519-529.

Knetsch, J. L. and J. A. Sinden. 1984. Willingness to pay and compensation demanded: Experimental evidence of an unexpected disparity in measures of value. Quarterly Journal of Economics 99(3):507-521.

Lancaster, K. 1966. A new approach to consumer theory. The Journal of Political Economy 74(2):132-157.

Loomis, J. B. 1980. Monetizing benefits under alternative river recreation use allocation systems. Water Resources Research 16(1):28-32.

Loomis, J. B. 1982. Effect of nonprice rationing on benefit estimates from publicly provided recreation. Journal of Environmental Management 12:283-289.

Loomis, J. B. 1987. Economic efficiency analysis, bureaucrats, and budgets: A test of hypotheses. Western Journal of Agricultural Economics 12(1):27-34.

Losch, A. 1954. The economics of location. Yale University Press. New Haven, CT.

Miller, J. R. and K. C. Higgins. 1981. Comment on "Monetizing Benefits Under Alternative River Recreation Use Allocation Systems" by John B. Loomis. Water Resources Research 17(2):446.

Miller, R. E. and P. H. Blair. 1985. Input-output analysis: Foundations and extensions. Prentice-Hall. Englewood Cliffs, NJ.

Mills, E. S. 1980. Urban economics. Scott, Foresman & Co. Glenview, IL.

Morishima, M. 1959. The problem of intrinsic complementarity and separability of goods. Metroeconomica (11)3:188-202.

Musgrave, R. A. and P. B. Musgrave. 1973. Public finance in theory and practice. McGraw-Hill Publishers. New York, NY.

Peterson, G. L. 1985. Economic growth from forest-based recreation, wildlife, and tourism. In: Proceedings of the 15th forestry forum: Forest resources in regional economic development. Virginia Polytechnic Institute and State University. Blacksburg, VA.

Peterson, G. L. 1988. Estimating recreation values over time. Benchmark 1988 Symposium. Tampa, FL.

Peterson, G. L., B. L. Driver and P. J. Brown. (In Press). The benefits and costs of recreation: Dollars and sense. In: Economic valuation of natural resources: Issues, theory, and applications by R. L. Johnson and G. V. Johnson (eds.). Westview Press. Boulder, CO.

Peterson, G. L., B. L. Driver and R. Gregory. 1988. Valuation of public amenity resources: An integration of economics and psychology. Venture Press. University Park, PA.

Peterson, G. L., J. B. Loomis and C. Sorg. 1985. Trends in the value of outdoor recreation. In: Proceedings of 1985 national outdoor recreation symposium II. Department of Parks, Recreation, and Tourism. Clemson University. Clemson, SC.

Peterson, G. L. and A. Randall. 1984. Valuation of wildland resource benefits. Westview Press. Boulder, CO.

Peterson, G. L. and C. F. Sorg. 1987. Toward the measurement of total economic value. GTR RM-148. Rocky Mountain Forest and Range Experiment Station. USDA Forest Service. Ft. Collins, CO.

Peterson, G. L., D. J. Stynes and J. R. Arnold. 1985. The stability of a recreation demand model over time. Journal of Leisure Research 12(2):121-132.

Pope, C. A. 1985. Agricultural productive and consumptive use components of rural land values in Texas. American Journal of Agricultural Economics 67(1):81-86.

Propst, D. (ed.). 1985. Assessing the economic impacts of recreation and tourism. Southeastern Forest Experiment Station. USDA Forest Service. Asheville, NC.

Randall, A. 1984a. Theoretical bases for nonmarket benefit estimation. In: Valuation of wildland resource benefits by G. L. Peterson and A. Randall. Westview Press. Boulder, CO.

Randall, A. 1984b. Benefit cost analysis as an information system. In: Valuation of wildland resource benefits by G. L. Peterson and A. Randall (eds.). Westview Press. Boulder, CO.

Randall, A. 1987. Total economic value as a basis for policy. Transactions of the American Fisheries Society 116:325-335.

Randall, A. and J. Stoll. 1983. Existence value in a total valuation framework. In: Managing air quality and scenic resources at national parks and wilderness areas by R. E. Rowe and L. G. Chestnut (eds.). Westview Press. Boulder, CO.

Rideout, D. and J. Hof. 1987. Cost sharing in multiple use forestry: A game theoretic approach. Forest Science 33(1):81-88.

Rosenthal, D. H. 1985. Representing substitution effects in models of recreation demand. Ph.D. Dissertation. Colorado State University. Ft. Collins, CO.

Rosenthal, D. H. 1987. The necessity for substitute prices in recreation demand analysis. American Journal of Agricultural Economics 69(4):828-837.

Rosenthal, D. H., J. B. Loomis and G. L. Peterson. 1984. Pricing for efficiency and revenue in public recreation areas. Journal of Leisure Research 16(3):195-208.

Simon, H. A. 1985. Human nature in politics: The dialogue of psychology with political science. The American Political Science Review 79:293-304.

Tversky, A. and D. Kahneman. 1981. The framing of decisions and the psychology of choice. Science 211(30):453-458.

U.S. Department of Agriculture. 1987. Estimating prices for access to opportunities for hunting, fishing, and viewing wildlife on public and private lands. Natural Resources and Environment Steering Committee on Wildlife and Fish Access Prices. Washington, DC.

Vaux, H. J. 1975. The distribution of income among wilderness users. Journal of Leisure Research 7(1):29-37.

von Thunen, J. H. 1826. Der isolierte staat in Beziehung auf Landwirtschaftund National-okonomie. Hamburg, Germany.

Winston, C. 1985. Conceptual developments in the economics of transportation: An interpretive survey. Journal of Economic Literature 23(1):57-94.

14

POLICY DEVELOPMENT AND PROGRAM ADMINISTRATION

Paul V. Ellefson and James R. Lyons

Forest economics has contributed much to the development of policies and programs focused on the nation's forests. The richness of the discipline's contribution is evidenced by the breadth and diversity of literature which addresses the economic and policy concerns of forestry organizations. Equally important has been the application of forest economics to a variety of important state, national, and international forestry issues—applications which have often led to the establishment of some of the nation's most important forestry programs. If the past is any indication, forest economics will continue to be influential in the development of forest policies and the design of programs required to implement them.

The development of policies and programs focused on the nation's forests has grown in complexity in recent years. Likewise, the number of disciplines and the number of organizations expressing an interest in the use and management of forests has also expanded. Ironically, many forest economists lament the fact that their recommendations are often ignored by policy-makers operating in political and administrative environments. In reality, however, such concern implies that the conceptual foundation for policy development has broadened considerably. No longer is economic efficiency a pervasive criterion for judging the merits of forest policies—institutional arrangements, distribution of benefits, and consensually determined standards are often of paramount concern. Furthermore, policy and program development is no longer the task of the few. Interest groups, the media, the general public, and judicial and legislative systems are actively involved in the development of policies. The social science disciplines from which are drawn the conceptual foundations for the development of policy have also expanded—law, sociology, and political science offer significant thoughts to the development of forest policy. Implied by such an environment is that the products created by forest economics researchers have become one of many inputs to the process by which forest policies are developed and implemented. The role and importance of the economics researcher has not become smaller—the tasks to be accomplished have become broader in scope.

228

The research agenda for policy development and program administration must reflect the broadening nature of policy development. To be useful, strategic research directions concerning policy development and program administration would be most appropriate if focused on the following:

- Develop improved understanding of processes by which forest resource policies are developed.
- Acquire a more discerning understanding of participants involved in the development of forest resource policies.
- Acquire improved understanding of budgetary and fiscal investment processes germane to forest resource programs.
- Develop more effective means of planning the use and management of forest resources.
- Develop more effective public program initiatives focused on the use and management of private forest resources.
- Evaluate the efficiency, effectiveness, and distributional consequences of major forest policy issues.

Develop Improved Understanding of Processes by Which Forest Resource Policies Are Developed

The development and application of forest resource policies to issues of technical and social importance are ongoing activities that involve individuals as well as organizations. Such activities can usually be grouped into a consistent pattern of events which in total composes the process by which policies are developed and implemented. Comprehensive research designed to improve the process has been limited within the forestry community. Researchers are prone to focus on the product of the process (i.e., a specific policy); few have researched opportunities for making the process more efficient and, subsequently, the policy product more agreeable. To effectively carryout such research implies an appreciation of policy development as a political activity—a political activity that involves the exercise of power and persuasive tactics, involves intense struggle over values and preferences, surfaces interest in identifying those who will impede (or aid) the adoption of proposed policies, and is concerned over individuals and organizations that are likely to gain or lose if a particular policy is adopted. Such is the stuff of politics.

Research topics concerning the effectiveness and efficiency of policy development and implementation include the following.

Evaluate effectiveness of agenda-setting processes germane to the development of forest policies. Policies and programs are developed in response to problems and issues surfaced by individuals and organizations. Often, issues and problems of immediate concern overshadow less apparent though equally important longer-term concerns. Lacking efficient agenda-setting processes, socially important issues may not be recognized and progressive forest policies are unlikely to surface. How effective are existing agenda-setting procedures (e.g., identification of national issues)? What shapes local, state, and national forestry agendas? What is the nature of persons and organizations involved in agenda-setting processes? How important are political moods of the electorate and political windows of opportunity? Why are certain issues diverted, deferred, or displaced? Under what circumstances are issues ignored and agenda status obstructed? And how can less immediate problems be brought to the attention of policymakers and those who influence them? Except for modest attention to public involvement processes, very little research has been devoted to agenda-setting as a strategic process involving forestry topics (Davies 1974, Ellefson 1990). There exists a rich conceptual base from which to undertake such research (Cobb and Elder 1983, Eyestone 1978, Kingdon 1984).

Assess effectiveness of processes by which forest policies are formulated. Once an issue has been successfully placed on an organization's agenda, the process must turn to identification and evaluation of objectives, policies, and programs that are capable of addressing the issue (i.e., formulation). How effective are formulation processes and how might they be improved? Who is—and is not—involved in the formulation process and from where do ideas for policy options arise? Research focused on the formulation of forest policies and programs is almost nonexistent. The theoretical foundation for formulation is also sparse (Bryner 1984, Polsby 1984). In responding to the Forest and Rangeland Renewable Resources Planning Act of 1974, the USDA–Forest Service's alternative goals and program directions shed some light on this often murky phase of policy development (Forest Service 1981, 1986).

Evaluate alternative means by which forest policies are selected. Displaying the consequences of alternative forest policies is one thing—selecting from amongst the alternatives is another. Forest economists have long been of the opinion that the rules of economic efficiency and rational comprehensive approaches prevail in the selection process. Such is highly unlikely. What is the context within which policies are selected, and how can selection processes—within a political environment—be improved? How appropriate are current selection models to forestry decision environments—rational comprehensive, rational incrementalism, mixed scanning, and organized anarchy? How important is personality and information to forestry decisions, and under what circumstances is group rather than individual decision making more appropriate? What should be the nature and application of strategies and tactics involving bargaining, negotiation, and mediation? And under what circumstances should criteria of various sorts be employed to make judgments about competing policy alternatives (e.g., criteria such as technical-ecological, efficiency-effectiveness, equity-ethical, values-ideologies, procedural). Research on the relative efficiency and effectiveness of policies and programs has a lengthy history within the forest economics community. However, a broader perspective involving non-economic criteria is sorely needed. The conceptual bases for such research is abundant (Bass 1983, Cohen et al. 1972, Lindblom 1959, Reitz 1987).

Evaluate means by which forest policy choices are legitimized. Forest policies may be viewed as desirable, may be agreed to, and may be selected. But selection is no guarantee of implementation and ultimate attainment of laudable goals. Policies must be given some official status—legitimized in a law, a plan, a regulation, a judicial ruling. How are forest policies legitimized and what forms of legitimacy are more effective than others (e.g., law versus agency regulation)? What role does political feasibility play in the legitimizing of forest policies? Specifically, what type of information about political feasibility would be useful under what circumstances (e.g., potential opponents, power base of opponents)? And what is the relative efficiency of various strategies and tactics used to secure legitimacy of a forest policy (e.g., compromise, logrolling, sidepayments)? Formal research on such questions is virtually nonexistent within the forestry community (Ellefson et al. 1981). Unfortunately, the conceptual foundations for such research have yet to be fully developed (Bryner 1984, May 1986, Meltsner 1972).

Assess processes by which forest resource policies are implemented. Implementation of agreed to policies is a most critical step in the policy process. All too often, the forestry community is reminded of the law that failed to meet user expectations or the forest plan that "remained on the shelf." Why do policies and programs fail in the implementation stage, and how might the process be made to function more efficiently? Are there problems in organizing the bureaucracy for implementation purposes (e.g., limited financial resources, limited professional talent)? Are laws and regulations being properly interpreted into feasible directives that will achieve widely agreed-to objectives? And what about follow-through— the routine provision of services, payments, and obligations? Research on the implementation of forest resource policies per se is sparse (Harper n.d.). National Forest System plans, however, are replete with implementation procedures. They represent an especially fertile source of hypotheses for implementation research. Conceptual foundations for implementation research are significant (Bardach 1982, Ripley and Franklin 1982, Scheirer 1981).

Assess means of evaluating forest resource policies and programs. Properly conceived, evaluation is an activity which can make substantial contributions to understanding and improving policy development and implementation. Most evaluations, however, have focused on economic efficiency concerns. Needed is a more broadly based perspective of policy and program worth. (Why, for example, does a program continue to be implemented, even though weighty evidence demonstrates its inefficiencies?) What other mechanisms play a role in evaluation and control of forest resource programs (e.g., legislative and interest group oversight, internal auditing and accounting) and how effective have they been? Through research, might they be designed so as to more effectively facilitate the judging of a policy's pros and cons? For systematic evaluations of program efficiency, what procedures (e.g., project analysis, sector analysis), criteria, and analytical tools are most appropriate? What motivates interest in evaluation (e.g., justification, program change)? Who are (and should be) the clients of evaluation processes and what informational products are desired from an evaluation? Why is there resistance to evaluations? And what role should policy analysts play in the evaluation process (Ellefson 1986a)? Forest economists have a long standing tradition of evaluating the efficiency and effectiveness of forestry programs (e.g., fiscal incentive programs, fire management programs, nondeclining-evenflow timber management programs, forest planning programs, below-cost timber sale issues) (Risbrudt and Ellefson 1983, Gray and Ellefson 1987, Henly and Ellefson 1986). Research of such a nature has been useful to program administrators—but certainly is deserving of more broadly defined concepts of evaluation. The latter are readily available for testing in a forestry context (Datta and Perloff 1979, Abt 1976).

Evaluate processes by which forest resource policies and programs are terminated or succeeded. Termination of policies and programs is a logical—but much neglected—phase in policy development cycles. Interest in the subject has increased in recent years for a number of reasons, including concern over how a program should be ended if the logical outcome of an evaluation suggests termination, or if budgetary retrenchment dictates the cessation of a program (e.g., mandated by a change in political leadership or economic conditions). Seldom are programs completely terminated—some replacement is normally provided. But what should be the nature of efficient processes for termination of programs that are judged to be lacking? What difficulties arise in efforts to terminate programs and how might they be effectively addressed (e.g., legal obstacles, high start-up costs)? How can policies and programs be designed for termination, especially when the policy issue to be addressed has a well defined time dimension, or when termination is expected because of pending changes in economic and political conditions? Research on the termination of forestry programs is almost nonexistent. Such an information void suggests that when termination is immanent, procedures for doing so are likely to be traumatic and less than efficient. Unfortunately, conceptual frameworks for researching program termination are also sparse (Brewer and de Leon 1983, Hogwood and Gunn 1986).

Acquire a More Discerning Understanding of Participants Involved in the Development of Forest Resource Policies

The process by which forest policies and programs are developed and implemented is an important focus for research—it cannot, however, be divorced from the actors or participants who make the process operate. For example, although a policy may be legitimized by a court, administrative response to such an action will be less than effective if legal and judicial processes are not well understood. Similarly, organized interest groups may resist adoption of a policy, but the nature of such resistance will not be appreciated until the objectives, structure, and operation of interest groups are adequately understood. Implied, therefore, is the need for substantial research focused on participants in the policy development process, especially legislatures, courts, interest groups, bureaucracies, and related organizations.

Evaluate the role of legislative systems in the development and implementation of forest resource policies. Legislative systems are a major source of forest resource policies; hence, the manner in which they operate in a forestry context is especially important. Research questions to be posed can be nearly unending. How are legislatures organized to address matters of forest policy and how might they be structured more effectively to do so (e.g., committee structure, staff, specialized offices)? What is the work flow of legislators, and how can forestry interests be facilitated in such an environment? What is the nature of legislative decisions regarding forestry topics, and how might strategies toward consensual decision making be enhanced when forestry issues are addressed? What is the nature of forestry information flows to legislative systems, especially the type of information required, the source of such information, and the means by which it is provided (e.g., legislative hearings, special commissions, bureaucratic or legislative staff)? What is the nature of legislative staff and what role does it play in a forestry context? And what role do legislative systems play in each of the major steps involved in the development and implementation of forest policies (i.e., agenda-setting, formulation, selection, legitimizing, implementation, evaluation, and termination)? Forestry research focused on legislative systems is limited (Council of State Governments 1985, Gray and Ellefson 1987, Ellefson et al. 1981, Ellefson 1985). That which does exist focuses primarily on historical accounts of the political labyrinth through which forest policies have passed on their way to becoming law. Needed is research of a more functional orientation. The conceptual basis for doing so is exceptionally abundant (Bradley 1980, Hammond 1985, Muir 1982).

Assess the role of judicial systems in the development and implementation of forest resource policies. The judiciary has had substantial impact on the course of American forestry in recent years. The laws that have been interpreted and the rulings that have been issued have often been nothing short of revolutionary— witness the land use and state forest practice law rulings in California and the legal interpretation of laws governing the management of the Monongahela National Forest in West Virginia. Ironically, relatively little systematic research has been focused on the judiciary from a forestry perspective. Again, research questions to be raised are nearly endless. What is the fundamental structure of local, state, and federal judicial systems that makes such systems so influential on matters of forestry? Over what subjects does the judiciary have jurisdiction and what is the nature of forestry issues that are likely to appear on judicial agendas? Under what conditions do forestry interests have standing in the eyes of the courts, and what circumstances foster class-action suits? What is the nature of forestry information utilized by the judiciary—what kind, what source, and what quality? To what extent do courts make policy and how do they do so—by reinterpreting laws or constitutions, designing administrative remedies, extending reach of existing law to subjects not previously covered? Should some decisions be excluded from judicial review? How can the forestry community facilitate the actions of judicial systems? And what role does the judiciary play in the various steps leading to the development and implementation of forest resource policies (i.e., agenda-setting, formulation, selection, legitimizing, implementation, evaluation, termination)? Again, forestry research focused on judicial systems has been slim. That carried out has focused on case-by-case reviews of the implication of rulings directed at specific issues. Research focused on judicial processes in a forestry context is very much lacking. As might be expected, the conceptual foundation for such research is especially abundant (Glick and Vines 1973, McLauchlan 1977, Sax 1970).

Evaluate the role of organized special interest groups in the development and implementation of forest resource policies. Organized interest groups active in matters of forest policy have proliferated in recent years. Likewise, their influence on the nature of American forest policies has flourished—to the point of becoming an advocacy explosion. The forestry community has not been totally comfortable with interest groups. Understanding interest group operations and facilitating their involvement in policy development has substantial merit. For example, what is the nature of the interest groups that are relevant to the forestry community (e.g., what groups, how large financially, extent of membership)? What forestry issues are likely to bring forth the involvement of interest groups (e.g., subjects of concern, geographic limits on activities)? In what fashion are interest groups governed and operated? How can they be designed to be administratively more effective? Who governs interest groups and by what means are decisions attained? What is the nature of interest group influence and how can it be organized to the betterment of consensual decisionmaking on matters of forestry? What is the nature of lobbying activities and how can such activities be made more effective within the forestry community? How can the financial and staffing difficulties of interest groups be overcome? And how can organized interest groups become more effective in each of the major steps involved in the development and implementation of forest policies (i.e., agenda-setting, formulation, selection, legitimizing, implementation, evaluation, and termination)? Research concerning organized interest groups of importance to the forestry community has been modest—much remains to be accomplished (Culhane 1981, Ellefson 1986b). In recent years, the conceptual framework for such research has increased substantially (Berry 1984, Ornstein and Elder 1978).

Evaluate the role of bureaucratic systems in the development and implementation of forest resource policies. Bureaucracies are the means by which most forest resource policies are implemented, evaluated—and often proposed. They are the cohesive force which leads to accomplishment of some desired forestry objective. The number of bureaucracies (e.g., state and federal pollution control agencies) and their complexity has grown substantially in recent years. Ironically, very little research has been focused on how forestry and related organizations should be structured and staffed in order to achieve legislative or judicial mandates. How should forestry organizations be structured in terms of labor, authority, and departmentalization? What structures would facilitate communication, decision making, and performance evaluation? How can administrators facilitate intergroup behavior and manage conflict more effectively? What is (and should be) the relationship between politics and administration and between ethics and public service? And what structural and managerial actions will facilitate the often unique relationship that exists between public and private management of forests and forest resources? Research focusing on questions of such a nature has led to a number of classic studies involving the administration of forestry and related organizations (Haskel and Price 1973, Kaufmann 1967, Robinson 1975). The conceptual basis for such research exists throughout extensive literature on organizations and their administration.

Evaluate the role of the media, political parties, and related organizations in the development and implementation of forest resource policies. Journalistic interest in the use and management of forests is keen, yet it is often overwhelming to natural resource professionals. Media interpretation of forestry issues can have a substantial effect on public perceptions of forestry, forest policies, and the forestry profession. Most certainly there is a need and an opportunity for research that would lead to a better understanding of journalistic interest in and effects upon forest policies and programs. Similarly there are other organizations that deserve attention from a forestry perspective (e.g., political parties).

Acquire Improved Understanding of Budgetary and Fiscal Investment Processes Germane to Forestry Programs

The production of forest and related products often requires the investment of land, labor, and capital over unusually long periods of time. Rotation lengths of 100 or more years are not uncommon for the production of timber. Similar time periods are often required to produce the aesthetically pleasing landscape so often demanded by the forest recreationist. The development of wildlife habitat may also require extended periods of time as do many long-term forestry research activities. Because of forestry's often unique requirements for extended periods of time, the need for long-term, sustained levels of investment in forestry and related programs are essential. Financial and professional commitment cannot be erratic if the flow of forest outputs is to have uniformity or sustainability. The problem becomes especially acute during times of austere public budgets. Much neglected by the research community, however, are budgeting and fiscal investment processes. Needed is a concerted effort to understand existing budgetary processes and to develop processes that can accommodate the unique conditions which surface in the management of forests. For example, exactly what difficulties do forestry programs face because of the manner in which budgeting cycles operate? How do budgets and fiscal exigencies influence forest policies and programs? Are their alternative revenue sources more accommodating of the public's interest in forests and how effective have they been (e.g., user fees, bonding programs, private loans)? What problems arise because benefits of the forest are often non-market in nature; how might such problems be addressed in the process of allocating scarce public funds? Are forest planning processes linked to budgeting processes? If not, how can such be effectively accomplished? What impact do managerial sciences (e.g., economics) have on budgetary processes and, if deemed desirable, how can they be made to be more influential? And are there cumbersome administrative reviews of proposed budgets that could be streamlined to encourage greater efficiency? The problems of budgeting and fiscal accounting are significant in a forestry context. They are deserving of research attention (Schultze 1968).

Develop More Effective Means of Planning the Use and Management of Forest Resources

Planning the use and management of forest resources has become an especially important activity of government forestry organizations. Fostered by an abundance of recently established federal and state laws, planning programs have enabled forestry interests to more clearly define agreed to forestry objectives and have enabled forestry professionals to develop forestry programs that are capable of accomplishing such objectives more effectively. The relative newness of comprehensive program and land management planning activities has surfaced a number of planning issues that are deserving of research attention. Such issues do not involve the substance of the plans that are produced; they concern the efficiency and effectiveness of the processes used to produce plans and the administrative uncertainties implied therein. Research on such topics has been very limited within the forestry community (Ellefson 1984, Gray and Ellefson 1987); conceptual foundations are also meager (King 1983).

Evaluate planning objectives and commitment. Planning involves investment of scarce resources. What fundamental purposes are served by planning activities? If benefits are expected, what is their nature and what client groups are likely to receive them? If planning is deemed to be important, what mechanisms can be used to secure agency commitment to planning activities (e.g., legislative directive, agency regulation, professional interest, non-governmental pressure)? And in order to maintain their relevance to the forestry community, how should the effectiveness of planning programs be judged?

Assess alternative planning processes. Planning can employ any one of many procedures. Should the process rely on issue-driven mechanisms, goal-driven mechanisms, or some other approach? How can planning processes be designed so as to accommodate unforeseen events (e.g., change in political leadership)? What is the most effective means of integrating planning processes with other governmental processes (e.g., budgeting, implementation, feedback processes)? And what forest resources (e.g., single- versus multiple-uses), functions (e.g., roads versus fire management), and ownerships (e.g., public versus private) should planning processes be designed to accommodate?

Evaluate resources required to plan. Financial and related resources are consumed by planning activities. Under varying circumstances, what financial commitments are required to carry out planning activities? Are there economies of scale to be captured in the design of planning programs? From what source will financial support originate (e.g., planning unit, fees from serviced unit)? And what type and how much professional talent should be invested in planning activities? And are the investments made in planning worth the benefits obtained?

Assess administrative structure for planning. What unit(s) of government should be responsible for planning (e.g., regional versus national units, planning within functional units, operationally integrated versus independent planning unit)? What agencies should provide input to planning activities (e.g., natural resource, economic development, pollution control, universities)? At what level should such agencies be involved (e.g., governor's staff, department level, division level, field level)? How can coordination of planning within an agency and among agencies be most effectively accomplished? And who at what level is to decide if a plan is acceptable— and by what criteria?

Evaluate public involvement processes. Planning public forestry programs implies accommodation of public desires for the products and services of such programs. What techniques are most effective for securing client and public perspectives on the use and management of forests? What portion of the total planning budget should such activities consume? If the public is to be involved in the planning process, how should such involvement take place (e.g., asked for help, briefed on progress, comment on results)? And what is the most effective means of integrating public input into the preparation of a plan?

Assess information management procedures. Planning activities require the use of much information. What type of information is necessary for the development of plans? How can such information be most effectively generated; who should supply it; and when should it be made available in the planning process? What type of commitment to information management will be required? And should an information management system be unique to a planning effort or should it be designed to serve other purposes (e.g., program budgeting)?

Evaluate implementation and monitoring. What forms of commitment are most effective for securing the implementation of forestry plans (e.g., policy statements)? Who should impose requirements that plans be implemented—legislative directive, administrative rules, interest group pressure? How can plans be effectively integrated with budgeting processes? What flexibility should be built into plans in order to accommodate unforeseen events such as new technologies and new and more reliable information? In what manner are added budget constraints resulting from economic downturns to be addressed? And how is changing program emphasis due to changes in political leadership to be accommodated?

Develop More Effective Public Program Initiatives Focused on the Use and Management of Private Forest Resources

Public forestry programs focusing on private forestry activities are substantial in number and in importance. They are subject of continuing analysis and evaluation—a healthy process that fosters efficiency and effectiveness. Such assessments should occur on a continuing basis and should accommodate technological advances in evaluation procedures.

Evaluate information and service program initiatives. There exists a variety of information and service programs at federal and state levels (e.g., forestry extension, service forestry, research). The effectiveness and efficiency of such programs have been assessed in the past, often resulting in positive administrative responses to improve program design and focus. Evaluations of this sort should continue.

Assess fiscal subsidies and tax program initiatives. Fiscal subsidies designed to encourage the management of private forests have been a significant nationwide activity (e.g., federal cost-share programs). Likewise, tax policies and programs designed to encourage the production of forest products have been met with considerable favor by the forestry community (e.g., state property tax initiatives, capital gain treatment of income from the sale of timber). Except for tax initiatives, such programs have been subject to considerable evaluation (Risbrudt and Ellefson 1983). Tax policies are in need of rigorous evaluation in the spirit of designing tax policies that efficiently achieve public objectives in private timber management.

Evaluate regulatory actions focused on the use and management of forests. Regulatory programs focused on the management of forests and the products and services produced by forests have grown significantly in recent years. For example, the Environmental Protection Agency has wide-ranging power to encourage or discourage the manner in which wood products are manufactured, and it has significant authority to influence state programs which are focused on the manner in which private forestry practices are applied. Similarly, many states have developed programs to legally regulate the forest management activities of private landowners, under the presumption that regulatory programs are more effective than other programmatic alternatives (e.g., extending information, fiscal subsidies, government ownership). Unfortunately, economic evaluations of regulatory programs are few (Henly and Ellefson 1986). They are certainly deserving of research attention— again in the spirit of designing more efficient programs to meet a variety of public interests.

Evaluate the Efficiency, Effectiveness, and Distributional Consequences of Major Forest Policy Issues

Forest policy issues of national and regional importance arise on a continuing basis. Research forest economists often have much to contribute to the resolution of such issues—indeed, they have a responsibility to do so. Not only can economists contribute to the substance of debates over forest policy but their involvement also often entails opportunity to try untested analytical procedures and processes—all to the betterment of future analyses. Economic and policy researchers should continue their role in assessing the implications of a wide array of issues—ranging from atmospheric deposition to international trade and from below-cost timber sales to public wilderness land management.

Summary and Conclusions

The development of forest resource policies and the administration of programs designed to implement such policies has increasingly become acknowledged as an important area of research. In recent years, however, the scope of the policy development arena has been significantly broadened, requiring the talents of a variety of social sciences—including forest economics. To be most effective in the years ahead, policy-type research should strategically focus on improving the forestry community's understanding of the processes by which policies and programs are developed, the participants involved in the development of forest resource policies, the means by which the use and management of forests are planned, the budgeting and fiscal processes important to forestry, and the innumerable public programs which are focused on the use and management of private forests. In addition, researchers should be actively involved in the evaluation and analysis of appropriate forestry issues. All such activities should be undertaken in the spirit of developing more effective policies and programs focused on the nation's public and private forests.

Literature Cited

Abt, C. C. 1976. The Evaluation of Social Programs. Sage Publications. Beverly Hills, CA.

Bardach, E. 1982. The implementation game: What happens after a bill becomes a law? MIT Press. Cambridge, MA.

Bass, B. M. 1983. Organizational decision making. Richard D. Irwin Publishers. Homewood, IL.

Berry, J. M. 1984. The interest group society. Little, Brown and Company. Boston, MA.

Bradley, R. B. 1980. Motivations in legislative information use. Legislative Studies Quarterly 5(3):393-406.

Brewer, G. D. and P. de Leon. 1983. The foundations of policy analysis. The Dorsey Press. Homewood, IL.

Bryner, G. C. 1984. Regulatory rule making and the process of policy formation. In: Public policy formation by R. Eyestone (ed.). JAI Press. Greenwich, CT.

Cobb, R. W. and C. D. Elder. 1983. Participation in American politics: The dynamics of agenda-building. Johns Hopkins University Press. Baltimore, MD.

Cohen, M., J. March and J. Olsen. 1972. A garbage can model of organizational choice. Administrative Science Quarterly. 17(March): 1-25.

Council of State Governments. 1985. Forestry in the South: A survey of state legislative and executive roles and information needs. Lexington, KY.

Culhane, P. J. 1981. Public lands politics: interest group influence on the Forest Service and the Bureau of Land Management. Johns Hopkins University Press. Baltimore, MD.

Datta, L. E. and R. Perloff. 1979. Improving evaluations. Sage Publications. Beverly Hills, CA.

Davies, J. C. 1974. How does the agenda get set? In: The governance of common property resources by E. T. Haefele (ed.). Johns Hopkins University Press. Baltimore, MD.

Ellefson, P. V., J. O'Laughlin and R. A. Skok. 1981. Minnesota timber development options: A classification of alternatives contained in the Legislative Commission on Minnesota Resources timber development study. Station Bulletin 543 (Forestry Series 38). Agricultural Experiment Station. University of Minnesota. St. Paul, MN.

Ellefson, P. V. 1984. Organizational patterns and administrative requirements for development of forest plans: The case of state governments in the United States. Proceedings of IUFRO symposium on forest management planning and managerial economics. Tokyo, Japan.

Ellefson, P. V. 1985. Congressional testimony: Who presents wood-based industrial interests? Journal of Forestry 83(5):300

Ellefson, P. V. 1986a. Decisions about public forestry programs: The role of policy analysts and analytical tools. Journal of Resource Management and Optimization 4(1):65-77.

Ellefson, P. V. 1986b. Forestry and related political action committees. Journal of Forestry 84(5):20-26.

Ellefson, P. V. 1990. Forest resource policy: Process, participants and programs. McGraw-Hill Publishers. New York, NY (forthcoming).

Eyestone, R. 1978. From social issues to public policy. John Wiley and Sons. New York, NY.

Forest Service. 1981. Alternative Goals. 1985 Resources planning Act Program. Program Aid No. 1307. U.S. Department of Agriculture. Washington, DC.

Forest Service. 1986. Resources planning act program 1985-2030: Final environmental impact statement. FS 403. U.S. Department of Agriculture. Washington, DC.

Glick, H. R. and K. N. Vines. 1973. State court systems. Prentice-Hall. Englewood Cliffs, NJ.

Gray, G. J. and P. V. Ellefson. 1987. Statewide forest resource planning programs: An evaluation of program administration and effectiveness. Bulletin 582. Agricultural Experiment Station. University of Minnesota. St. Paul, MN.

Hammond, S. W. 1985, Legislative staffs. In: Handbook of legislative research by G. Loewnberg, S.C. Patterson and M.E. Jewell (eds.). Harvard University Press. Cambridge, MA.

Harper, J. (no date). A model for implementation of a statewide forest resource plan. Division of Forestry. Minnesota Department of Natural Resources. St. Paul, MN.

Haskell, E. H. and V. S. Price. 1973. State environmental management: Case studies of nine states. Praeger Publishers. New York, NY.

Henly, R. K. and P. V. Ellefson. 1986. State forest practice regulation in the U.S.: Administration, cost and accomplishments. Bulletin AD-SB-3011. Agricultural Experiment Station. University of Minnesota. St. Paul, MN.

Hogwood, B. W. and L. A. Gunn. 1986. Policy analysis for the real world. Oxford University Press. Oxford, England.

Kaufman, H. 1967. The forest ranger: A study in administrative behavior. Johns Hopkins University Press. Baltimore, MD.

King, W.R. 1983. Evaluating strategic planning systems. Strategic Planning Journal 4:263-277.

Kingdon, J. W. 1984. Agendas, alternatives and public policy. Little, Brown and Company. Boston, MA.

Lindblom, C. E. 1959. The science of muddling through. Public Administration Review. 19(2):79-88.

May, P. J. 1986. Politics and policy analysis. Political Science Quarterly 101(1):109-125.

McLauchlan. W. P. 1977. American legal processes. John Wiley and Sons. New York, NY.

Meltsner, A. J. 1972. Political feasibility and policy analysis. Public Administration Review 32 (November/December).

Muir, W. K. 1982. Legislature: California's school for politics. The University of Chicago Press. Chicago, IL.

Ornstein, N. J. and S. Elder. 1978. Interest groups, lobbying and policymaking. Congressional Quarterly Press. Washington, DC.

Polsby, N. W. 1984. Political innovation in America: The politics of policy initiation. Yale University Press. New Haven, CT.

Reitz, H. J. 1987. Behavior in organizations. Irwin Publishers. Homewood, IL.

Ripley, R. B. and G. A. Franklin. 1982. Bureaucracy and policy implementation. The Dorsey Press. Homewood, IL.

Risbrudt, C. D. and P. V. Ellefson. 1983. An economic evaluation of the 1979 forestry incentives program. Bulletin 550. Agricultural Experiment Station. University of Minnesota. St. Paul, MN.

Robinson, G. O. 1975. The Forest Service: A study in public land management. Johns Hopkins University Press. Baltimore, MD.

Sax, J. L. 1970. Defending the environment: A strategy for citizen action. Alfred A. Knopf Publishers. New York, NY.

Scheirer, M. A. 1981. Program implementation: The organizational context. Sage Publications. Beverly Hills, CA.

Schultze, Charles L. 1968. The politics and economics of public spending. Brookings Institution. Washington, DC.

15

RESOURCE ASSESSMENT, INFORMATION MANAGEMENT, AND COMMUNICATIONS TECHNOLOGY

Thomas E. Hamilton

Forest resource economics has much to say about how resources are assessed, how information is managed, and how processed information is communicated. In the past, however, resource economics research has been heavily oriented to timber-related issues and the information requirements thereof. Historically, timber was the major renewable resource produced for the open market; thus it was very suitable to the application of economic methods. Timber was also the only renewable resource physically removed from the land and subsequently delivered to consumers—a process logically raising questions about the economics of doing so. Timber, along with water in some regions, was also one of the first resources to be identified as in short supply relative to expected demands. And timber was (and continues to be) a resource that is easily measured; thus it was readily approachable with techniques unique to economic analysis.

Inventory and analysis of forest resources was included as a provision of the Organic Act of 1897, which established the national forest system. Via the McSweeney-McNary Act of 1928, such efforts were expanded to include all the nation's forested land. While both laws spoke to natural resources in general, their implementation concentrated on the nation's timber resource. Over the past 20 years, however, forest and rangeland management has become more multiple-use oriented, and economics research has struggled to incorporate these multiple uses into analyses and assessments. The process of doing so has created a number of challenging problems for the analyst and the economics researcher. Applying economic concepts to multiresource questions has demonstrated that

- Not only are basic data inadequate, but units of measure are not available for consistent analyses within or among resources.
- Cost allocation is a major problem when evaluations of individual resources are needed.

- Benefits are difficult to quantify, particularly when many diverse uses need a common denominator.
- Assignment of benefits to specific resource functions is difficult when such benefits are derived from several activities.
- Future supply-demand interaction is difficult to specify because many diverse values are associated with amenity resources.

Challenges of the above nature have brought about significant reorientation of research related to the collection, management, and use of information for purposes of analysis. Much additional work is needed. Specifically, research is needed in the following strategic directions:

- Develop more effective means of structuring and classifying renewable resource information.
- Develop more effective techniques for maintaining data and information consistency over time.
- Develop more effective means of organizing and administering information management.
- Develop more effective means of anticipating future renewable resource information needs.
- Develop more effective means of evaluating multiresource interactions and effects.

Develop More Effective Means of Structuring and Classifying Renewable Resource Information

Organization of renewable resource information, commonly viewed as land and resource classification, has grown in complexity proportional to the complexities involved in resource analysis, management, and planning. Increased data collection costs combined with declining budgets have heightened concern over the issue. Some organizations have addressed the problem by using information from a variety of sources rather than relying on their own collection apparatus as had often been done in the past.

Decision-makers and analysts largely agree that classification of land and resource characteristics is necessary. There is less agreement, however, on the type of classification system needed, especially the level of detail required. The problem was highlighted by a series of land and resource classification articles which appeared in the *Journal of Forestry* (1978). An overriding message throughout the series was that although "a single, multipurpose system has considerable intellectual appeal, the user community is not [so] structured" (Hirsch, Cushwa, Flach, and Frayer 1978). Posed was a very real dilemma for the design of classification systems. On one hand, purpose should drive the design of a classification system; systems poorly designed for a specific purpose often meet with considerable user objection. In contrast, the practicality (e.g., limited financial resources) of collecting information for a variety of purposes implies that some user objectives be compromised. Within such extremes, renewable resource information must be organized.

Research can help solve the classification problem by providing information that improves the forestry community's understanding of why and how renewable resource information relating to forest and rangelands should be structured and classified. Research on topics of the following nature is needed.

Evaluate hierarchical approaches to classification. Hierarchical approaches to classification should be designed and evaluated. Such systems should enable the linking of information from the broadest levels of generalization to the narrowest levels of specificity. Such would improve opportunities for users of classification systems to find a point of commonality to which diverse sets of information can be related.

Develop efficient, easy-to-use systems for processing large amounts of data from various sources. Technology is available to process large amounts of complex information and to easily transfer information from one system to another. Special problems relating to resource information need to be merged with this capability.

Develop land and resource measurement units for all renewable resources. Measurement units for some resources (e.g., timber, range forage) are well developed. Others (e.g., recreation) are less uniform and therefore cannot always be accommodated by current classification systems. The problem is definitely in need of research attention.

Assemble and evaluate existing classification systems. Until all major classification systems are identified and evaluated, past classification difficulties are likely to be compounded as new resource problems make broader demands on information sources. A recent comparison of five national land classification maps is an excellent starting point (Forest Service 1988).

Develop classification systems that minimize uncertainty in estimating responses to management. Renewable natural resources are subject of an extremely wide range of management prescriptions. Needed is a classification system that is capable of recognizing such management activities and that can provide clearly defined resource responses to management activities.

Develop classification systems that are capable of demonstrating multiresource interactions. Renewable natural resources are multi-faceted, that is management and use of one resource often has decisive implications for other resources. Needed is a classification system that can accurately portray such interactions.

Develop More Effective Techniques for Maintaining Data and Information Consistency over Time

Renewable resources information is collected for one of two basic purposes: resource analysis and resource management. Over time, the relative importance of information for purposes of analysis has increased. In an absolute sense, however, management needs for inventory information have also grown. Such increases can be attributed to recent increases in analytical capabilities and to increases in the informational needs of complex management models.

Growth in the amount of data required is not the only challenge—data collection has become increasingly complex. Multiresource considerations, for example, are becoming an increasingly important part of resource decisions. Only in recent years has information from inventory processes been specifically designed to depict an entire array of renewable resources. In the past, existing information (primarily designed for timber decisions) has been force-fit for other renewable resource considerations. In addition to a base level of information useful for a variety of resources, recent analyses have been designed to demonstrate joint products and the effects of differing management strategies. In many cases, the design and collection of data has not kept pace with analytical needs.

Multiuser needs for information is also a factor that has increased the complexity of data collection. For a variety of reasons (e.g., cost, consistency within government, magnitude of data collection task), public agencies have increasingly relied on agency counterparts for some information needs. Differences in information definitions, dissimilar timing of inventories, variation in geographic specificity, and organizational rigidity are but some of the difficulties associated with joint data collection and information sharing.

The complexity of data collection has also been affected by changes in management sophistication. As research and experience refine ability to anticipate the results of management actions, the need for more specific resource and area information has grown. The need to direct more attention to environmental effects has also increased information needs. New technology, however, has the potential to significantly alter traditional data collection processes. Remote sensing, for example, has gradually been replacing on-the-ground measurements as a way of collecting data. Such technology offers significant potential for advancing information detail and the frequency of information updates.

Changes in the management of renewable resources are occurring at an ever increasing rate. Research can contribute to an orderly transition in data collection by providing techniques which respond to emerging needs while maintaining consistency with data and information that were collected in the past. Some specific research needs include the following.

Evaluate the applicability of remote sensing technologies to renewable resource problems. Remote sensing techniques may have considerably greater application to renewable resources assessments than previously thought. The technology may be particularly useful since it focuses the attention of renewable resource researchers on their areas of expertise while relying on the actions of experts in remote sensing technology.

Develop standard parameters for describing supplies of each renewable resource. Over time, the key measures for resources such as timber have been defined and broadly accepted. Measures for other resources are much less uniform and not always agreed upon. Efficiency and consistency in data collection depends on securing such agreement.

Evaluate sampling methods to improve efficiency of multiresource data collection. Improving the efficiency of multiresource data collection should provide for retaining relationships with past inventories (trend analyses) while enabling incorporation of new information characterizing other resource supply situations.

Develop geographically hierarchical systems of data collection. Geographically hierarchical systems of data collection should be designed to enable display and analysis of data at various regional levels, and to provide for supplemental data collection where analysis for smaller geographic areas is desired. Such systems would allow additions to ongoing inventory activities rather than initiation of completely new inventories for analyses at local levels.

Develop More Effective Means of Organizing and Administering Information Management

Classification systems and uniformity of collected information promote exchange of renewable resources information. Neither, however, assure that information sharing will occur. Past attempts to exchange information have frequently been unsuccessful for reasons other than agreement on common terminology or techniques. Equally important problems lie in the areas of inventory timing, organizational design, and established public expectations. In response to mutual interests in encouraging the exchange of information, five federal agencies with major renewable resource inventory responsibilities (USDA–Forest Service, USDA–Soil Conservation Service, USDI–Bureau of Land Management, USDI–Fish and Wildlife Service, and USDI–Geological Survey) agreed in 1978 "to minimize duplication and overlapping efforts and to enhance and encourage overall data collection, data sharing, appraisal efficiency, program compatibility, and expedite technology transfer." A five agency policy group has periodically met to further such objectives.

Problems associated with information exchange sometimes are administrative and, occasionally, bureaucratic. Others, however, lend themselves to research. In the latter case, research can examine alternative strategies for organizing and administering information management. Consider the following.

Evaluate time schedules for collection of renewable resources information. Time schedules for collection of renewable resources information should be assessed with an objective of considering costs, problems of combining data from various inventories, updating information to a common point in time, and a variety of organization structures. The results should help determine where cost effective changes can be made while accommodating specific agency objectives.

Assess opportunity for combining interagency inventory activities. In need of assessment are opportunities to combine interagency inventories (e.g., contractual arrangements, joint data collection efforts) in a manner that will accomplish multiple agency objectives. The assessment should focus on sample design, existing inventory expertise, budget availability, potential shifts in agency workforce, and control of the process.

Evaluate opportunity for development of centrally managed data bases. Centrally managed data bases for joint maintenance and response to individual needs should be evaluated. At present, users are often left to use whatever information is available—or information they currently have knowledge of. The five agencies previously mentioned have prepared a summary of agency inventory activities. Unfortunately, the data often resides in a variety of places and is not always readily usable in a joint fashion. A common manager could have as an objective the development of commonality among many and varied bases of information.

Develop compatible systems for managing and analyzing renewable resources information. The technology for handling information is evolving rapidly; a key advancement in recent years has been ease of interchange among various types of equipment. What is needed to complete the interchange of information is commonality of systems. Research could help by identifying opportunities for gaining this commonality.

Develop More Effective Means of Anticipating Future Renewable Resource Information Needs

Rapid evolution of demand for renewable resources has been a significant factor in generating needs for new research. Anticipation of such demands is critically important to research planning, especially definition of problems, conduct of research, and implementation of research results. This problem is accentuated in the area of data collection and analysis because of the often lengthy time periods which occur between resource inventories. For example, an average target for USDA–Forest Service inventory updates is 10 years. In many areas of the nation, fluctuations in funding have resulted in considerably longer inventory cycles.

Research is needed to anticipate future renewable resource information needs and to update the total base of resource data to commonly agreed upon points in time. In addition, there is a need to bring information and analytical techniques to a common level of sophistication. Data for some resources has been developed and refined over many years; generally this has occurred for tangible resources such as timber, minerals, and water. However, much less is known about other forest resource outputs which are less easily measured and are often not physically removed from the land. Some specific research needs are as follows.

Develop meaningful measures of forest resource supply. Required are meaningful measures of supply that lend themselves to measurement of management effectiveness, economic and environmental analyses, and comparison with other resource uses.

Develop estimates of demand for nonmarket forest resource outputs. Not all forest outputs are readily valued by market transactions. Needed are estimates of demand for resources where markets are not established, where supplies are provided as free goods, or where substitutes are not clearly defined.

Develop systems for evaluating and prioritizing information and data requirements. Information needs for analysis and management can be nearly infinite. Required are more effective systems for identifying and evaluating information requirements. A logical step toward development of such a system is identification of data and information (resource-by-resource basis) currently used by managers and analysts. Also in order would be evaluation of manager and analyst information requirements (resource-by-resource). Where, for example, are additional needs the greatest and where will improvements in data availability have the greatest payoff? Information evaluation systems should also be capable of determining where additional data will have negligible benefits. Some decisions may not be changed, even if more information were available. Additional information may also simply add to imbalances in analytical capability among resources.

Develop information accuracy standards for various information uses. Sampling systems should be designed that will readily lend themselves to supplementation as new management and policy questions arise.

Develop More Effective Means of Evaluating Multiresource Interactions and Effects

Renewable resource management requires a clear understanding of a variety of resources, including recreation, water, forage, timber, and fish and wildlife. Multiresource evaluation greatly expands the complexity of information collection, information management, and information analysis. In part, this increased complexity arises because of the joint product nature of many renewable resources, that is, the process by which combinations of inputs work to produce combinations of outputs. Certain combinations of production factors (labor, capital, and technology) may be needed to produce a desired result on a given area of land. If that desired result is a primary objective (e.g., timber), benefits or losses in other resource areas may occur as a bonus or a constraint. Joint output problems are intensified when multiple resource objectives are sought. In such situations, multiple resource management prescriptions must be provided. Unfortunately, the data necessary to develop such prescriptions is generally not available.

Multiresource evaluation problems can also arise from more broadly defined multiple effects of resource management and use. Concern goes well beyond management effects on the availability of major renewable resources. Included are a host of direct physical, biological, social, and economic effects of a particular action. These broadly defined effects can be direct—occurring on the land base or involving agents directly associated with a management action—or indirect—effects that are once removed from the primary action (e.g., community stability, regional employment).

Multiresource evaluation has been addressed in a piecemeal fashion for many years (emphasized by the interest in the 1960s in multiple use of forest resources; sharply focused by the environmental movement of the 1970s). Inability to properly evaluate multiresource interactions has been a direct reason for many, if not most, disagreements, legal actions, and added management requirements facing owners and managers of forest and range land. Although not capable of solving the value-laden portion of management actions, research can contribute by clarifying many of the fact-related questions which surround debates over multiresource interactions. Some topics in need of immediate research include the following.

Develop a rationale for assigning joint costs of management to individual resource outputs. Despite the movement toward multiple output management prescriptions, there is still substantial decision-making interest in information about the cost of producing specific resource outputs. Such information is especially difficult to develop for supporting activities (e.g., forest roads).

Develop a rationale for assigning secondary resource outputs or benefits to actions associated with production of a primary resource. Accounting for benefits (or losses) from one resource as a result of actions taken to manage or use another resource is an especially perplexing problem. For example, what timber or water effects can be expected as a result of vegetative manipulation undertaken for wildlife habitat purposes? How should such effects be assigned in investment accounting systems?

Develop clearly defined data requirements for multiresource management prescriptions. Available data and information for single resource management actions may not be appropriate to multiresource management prescriptions. For example, adjustment of categories depicting tree distribution by diameter class might make such data more useful for nontimber evaluations—without detracting from usefulness for timber evaluations.

Develop multiresource effects of management and use for all forest and range lands. In response to the National Forest Management Act of 1976, national forest system planning activities have enabled significant gains in identifying multiresource effects. Evaluation of the latter, however, indicates that a good deal more work is needed to gain an acceptable level of consistency and accuracy. In addition, there is no similar, widespread information counterpart for lands other than national forest system lands. Until such information is developed, assessments will be unable to present a complete analysis of multiresource effects.

Develop techniques for analysis and display of multiresource assessments. Conceptually, a multiresource assessment of the future supply-demand situation for renewable resources would be a substantial step forward in setting the stage for management planning. To date, however, efforts to develop such a capability have been inconclusive: Data are lacking, magnitude and complexity of data are great, and organization and display problems are acute. The latter area would be a good starting point for research.

Identify criteria enabling use of interaction information in decisionmaking. Multiresource interactions and effects include a broad array of information measured in a wide variety of ways. Decisionmakers need guidance on how complex sets of information can be effectively used in the decisionmaking process.

Summary and Conclusions

Renewable resource assessment and information management are activities of critical importance to the efficient use and management of forest resources. Much of the information required to undertake economic evaluations is predicated on the existence of accurate and readily available data and information. Unfortunately, appropriate data is not always available, the cost of obtaining it may be very high, and once available it may not accurately reflect the management conditions in need of evaluation. To overcome informational problems facing managers of renewable resources, research should be strategically focused on developing more effective means of structuring and classifying information, maintaining data and information consistency over time, anticipating future resource information needs, evaluating multiresource interactions and effects, and organizing and administering information management. Research activities focused on such subjects would go far toward the establishment of information bases necessary for the development of effective programs and policies.

Literature Cited

Hirsch, A. C., T. Cushwa, K. W. Flach and W. E. Frayer. 1978. Land classification: Where do we go from here? Journal of Forestry 76(10): 672-673.
Journal of Forestry, 1978. Land Classification Series. Journal of Forestry 76(10):644-673.

Forest Service. 1988. A comparison of five national land classification maps. Agricultural Handbook 672. U.S. Department of Agriculture. Washington, DC.

Bibliography

Alston, R. 1983. Conceptual and technical problems in the analysis and definition of forest balance. In: Forest inventory for improved management. Helsinki, Finland.

Bettwy, L. A. 1978. Resource inventories and agency decisions. In: Proceedings of integrated inventories of renewable natural resources: A national workshop. General Technical Report RM-55. Rocky Mountain Forest and Range Experiment Station. USDA Forest Service. Ft. Collins, CO.

Bones, J. T. 1986. Current development in the federal forest inventory program in the United States. In: Proceedings of Forest Resource Monitoring Working Party. XVIII World Congress. International Union of Forest Research Organizations. Ljubljana, Yugoslavia.

Bostrom, K. 1987. Land requirements for industrial forestry. In: Land and resource evaluation for national planning in the tropics. GTR WO-39. USDA Forest Service. Washington, DC.

Buckman, R. E. 1974. Resolving conflicts. In: Inventory design and analysis. Rocky Mountain Forest and Range Experiment Station. USDA Forest Service. Fort Collins, CO.

Caballero, M. D. 1980. Benefits and costs of natural resource inventories of arid and semiarid lands. In: Arid land resource inventories: developing cost-efficient methods. General Technical Report WO-28. USDA Forest Service. Washington, DC.

Carey, H. H. and D. Nickolas. 1984. Analysis and display of multiresource interactions in the RPA assessment. Unpublished report prepared for the Pacific Southwest Forest and Range Experiment Station. USDA Forest Service. Berkeley, CA.

Driscoll, R. S. 1978. Forest inventories for national assessment in the United States. In: National Forest Inventory. Bucuresti, Romania.

Ffolliott, P. F. 1978. A multifunctional inventory approach to multiple-use analysis. In: Integrated inventories of renewable natural resources: A national workshop. General Technical Report RM-55. Rocky Mountain Forest and Range Experiment Station. USDA Forest Service. Ft. Collins, CO.

Glascock, H. R., Jr. 1978. Need for integrating inventories: Moderator's comments. In: Integrated inventories of renewable natural resources: A national workshop. General Technical Report RM-55. Rocky Mountain Forest and Range Experiment Station. USDA Forest Service. Ft. Collins, CO.

Hahn, J. T. 1987. The role of a data base management system in regional and national planning. In: Land and resource evaluation for national planning in the tropics. GTR WO-39. USDA Forest Service. Washington, DC.

Hamilton, T. E. 1978. National integrated inventories: Is what you need what you do? In: Integrated inventories of renewable natural resources: A national workshop. General Technical Report RM-55. Rocky Mountain Forest and Range Experiment Station. USDA Forest Service. Ft. Collins, CO.

Hoekstra, T. W., L. A. Joyce, and T. E. Hamilton. 1987. Estimation of regional multiresource interactions. In: Land and resource evaluation for national planning in the tropics. GTR WO-39. USDA Forest Service. Washington, DC.

Hof, J. and T. Baltic. 1988. Forest and rangeland resource interactions: A supporting technical document for the 1989 RPA assessment. General Technical Report RM-156. Rocky Mountain Forest and Range Experiment Station. USDA–Forest Service. Ft. Collins, CO.

Husch, B. 1978. Why carry out National Forest inventories? In: National Forest Inventory. Bucuresti, Romania.

Kaiser, F. H. 1987. The importance of multiple use values in designing inventories. In: Land and resource evaluation for national planning in the tropics. GTR WO-39. USDA Forest Service. Washington, DC.

LaBau, V. J. 1987. Interagency cooperation in state and subregional inventories. In: Land and resource evaluation for national planning in the tropics. GTR WO-39. USDA Forest Service. Washington, DC.

Lea, G. D. 1980. Policy and program development. In: Arid land resource inventories: Developing cost-efficient methods. General Technical Report WO-28. USDA Forest Service. Washington, DC.

Lewis, B. J. and P. V. Ellefson. 1983. Information for timber management planning: An assessment of availability and adequacy in Minnesota. Station Bulletin No. 547. Agricultural Experiment Station. University of Minnesota. St. Paul, MN.

Lund, H. G. 1986. A primer on integrating resource inventories. General Technical Report WO-49. USDA Forest Service. Washington, DC.

McCurdy, D. R. and C. C. Myers. 1978. Methodologies for designing resource inventories to support management information systems. In: Integrated inventories of renewable natural resources: A national workshop. General Technical Report RM-55. Rocky Mountain Forest and Range Experiment Station. USDA Forest Service. Ft. Collins, CO.

Pulford, W. J. 1978. Regional integrated resource inventories: a place for integration. In: Integrated inventories of renewable natural resources: A national workshop. General Technical Bulletin RM-55. Rocky Mountain Forest and Range Experiment Station. USDA Forest Service. Ft. Collins, CO.

Singh, K. D. 1987. Land evaluation for forest resources development planning at national level. In: Land and resource evaluation for national planning in the tropics. GTR WO-39. USDA Forest Service. Washington, DC.

Forest Service. 1981. An assessment of the forest and rangeland situation in the United States. Forest Resource Report No. 22. U.S. Department of Agriculture. Washington, DC.

Forest Service. 1987. Forest Service Resource Inventory: An overview. Forest Inventory and Economics Research. U.S. Department of Agriculture. Washington, DC.

Wikstrom, J. H. 1978. Developing inventories to support resources assessments. In: National Forest Inventory. Bucuresti, Romania.

Wikstrom, J. H. 1980. The analytical basis for integrated forest and rangeland ecosystem inventory. In: Arid land resource inventories: Developing cost-efficient methods. General Technical Report WO-28. USDA Forest Service. Washington, DC.

Wikstrom, J. H. 1983. Our changing concepts of sustained-yield and forest balance. In: Forest inventory for improved management. Helsinki, Finland.

Wunderlich, R. E. 1978. Integration of data information between agencies. In: Integrated inventories of renewable natural resources: A national workshop. General Technical Report RM-55. Rocky Mountain Forest and Range Experiment Station. USDA Forest Service. Ft. Collins, CO.

16

FORESTRY SECTOR
ENVIRONMENTAL EFFECTS

J. E. de Steiguer

The quality of natural environments has become an especially important social issue in various settings throughout the world. Although not new to the agenda of governments (public distress over environmental degradation was recorded in European medieval times), concern over the quality of the environment carries a sense of urgency heretofore unseen (Freeman, Haveman, and Kneese 1973; Seneca and Taussig 1974). Perhaps the single most significant agenda-setting event was the 1962 publication of Rachel Carson's *Silent Spring*. This book, which discussed the detrimental effects of pesticides in the environment, popularized the notion that something was seriously and insidiously wrong with the way modern industrialized societies regarded natural surroundings. Earth Day 1970 further underscored public concern over environmental conditions (Hay 1987). From such beginnings, the "ecology movement" arose as a cause to be quickly embraced by a wide range of individuals and organizations, including scholars who began generating a vast body of learned works on the subject.

The environmental literature of the 1960s and early 1970s is replete with suggested causes of environmental problems, including population pressures, technological changes, economic growth, and market failures (Rees 1985). Economists, however, quickly pointed out that population pressure, technological change and economic growth per se were not the causes of declines in environmental quality. They argued that when growth and change are accompanied by market failure, environmental problems of various sorts begin to arise (Freeman, Haveman, and Kneese 1973).

Identifying market failures as the cause of environmental problems implies an understanding of well-functioning market systems. In order to function efficiently, markets must have certain characteristics, namely, atomistic competition; homogeneous products; perfect knowledge of market conditions and prices; and freely mobile resources with assignable property rights (Ferguson and Maurice 1974). If all such conditions exist, a market is said to be perfectly competitive and will result in an efficient, socially optimal allocation of resources. If one or more of the assumptions of perfect competition is violated, market failure occurs and the allocation of resources becomes less than socially optimal.

Crucial to efficient environmental management is the existence of well-defined and assignable property rights. In order for markets to function properly, the ownership of all resources must be clearly defined and enforceable; individuals wishing to use a resource cannot do so without compensation to the owner. Such is clearly not the case with many environmental resources—most of which have traditionally been regarded as common property. Air, for example, has value not only for its ability to sustain life but also as a receptor for and assimilator of waste materials. Because of the common property nature of air (belonging to everyone, hence no one), polluters are able to use air as a waste receptor without incurring costs. External costs are passed on to other parties in the form of impaired health and degraded forests. The presence of such externalities constitutes a form of market failure which results in over-production of the pollutant and under-production of a quality environment. Social costs exceed private costs by the amount of the external cost, thus causing a less than socially optimal allocation of resources. Market failure (externalities) as the economic cause of environmental problems stems in large measure from research undertaken by A. C. Pigou in 1932 (Fisher 1981). Unfortunately, Pigou's contribution passed unnoticed for several years until revived by Scitovsky (1954) and Bator (1958). By the 1960s, the concept of externalities was widely accepted as an explanation for environmental degradation and associated misallocation of resources.

The imposition of charges on producers of pollutants is one means of addressing environmental concerns (charge equals the difference between private and social costs) (Seneca and Taussig 1974). Things are not, however, quite so simple. Determination of appropriate pollution control levels requires comparison of pollution control costs with the value of pollution caused damages. Assessment of the latter is often the most troublesome phase of such analyses. Implied are difficulties in estimating dose-response functions and in valuing unpriced goods and services (Freeman, Haveman, and Kneese 1973). The optimal level of pollution control is that which makes the marginal costs of control equal to the marginal value of damages (Siebert 1987). As the word "optimal" suggests, eliminating all pollution damages is typically not economically feasible—a fact often ignored by strident environmentalists (Just, Hueth, and Schmitz 1982).

Economists have also examined the attributes of a variety of policy instruments useful for correcting market failures which cause environmental problems. Seneca and Taussig (1974) categorized such instruments as regulation and prohibition; taxes, subsidies, and effluent charges; and government production of environmental services. Of these policy instruments, taxes, subsidies, and effluent charges have received the strongest support from both environmentalists and professional economists. Such are argued as providing the greatest degree of economic efficiency and equity while, at the same time, generating the fewest administrative problems. Also, their actual implementation has, in some instances, been met with remarkable success.

Environmental Economics of Forestry

Environmental economics is the subject of a significant body of literature—some of which addresses forestry and forestry related activities. Authors addressing environmental economics in general have written dozens of textbooks on the subject, and organizations (e.g., Resources for the Future, Inc.) have published hundreds of monographs and research papers addressing a multitude of economic concerns important to environmental interests. Each year, some two dozen scholarly journals publish over 250 environmental economics articles. Although environmental economics in a forestry context is of major concern here, the general body of environmental economics literature is rich in ideas and approaches useful to forestry. Research by economists such as the following are especially noteworthy: Arrow and Fisher (1974), Baumol (1972), Bohm (1970), Crocker (1971), d'Arge and Kogiku (1973), Fisher and Krutilla (1975), Fisher and Peterson (1976), Freeman (1974), Kneese (1976), Krutilla and Cichetti (1972), Mishan (1974), and Smith (1979). Such research provides important insight into topics such as externalities, property rights, environmental preservation, and natural resource policy. Textbooks can also be an excellent source of information about environmental economics (Howe 1979, Fisher 1981, and Seneca and Taussig 1974).

What of our understanding of environmental economics as related to forestry issues? A decade ago, the effect of air pollutants on forests was scarcely mentioned in the literature; recently, however, the subject has become of great concern. Under the National Acid Precipitation Assessment Program (NAPAP), forest scientists have begun to explore complex relationships between air quality and forest health. The most up-to-date summary of current research on the subject is contained in the NAPAP Terrestrial Effects Peer Review (1987) document. The fact remains, however, that even though interest in the forestry effects of air pollutants is high, the economic assessment of such effects is very limited. Assessments which have been carried out are based on very limited information. Adams (1986) reviewed three unpublished studies on the subject. One, examining the effects of acid deposition on forest ecosystems of the eastern U.S., concluded (with sparse dose-response information) that damages from current levels of acid deposition were $1.75 billion. The second study estimated air pollutant damages to the forests of eastern Canada to be $1.5 billion. The third used TAMM and arbitrary growth reductions to estimate possible impacts of all air pollutants on forest product markets. De Steiguer (1987) highlighted the extreme difficulty of conducting economic research on the forestry consequences of air pollutants. Of special concern were problems involving estimation of dose-response functions.

On a global scale, changes in the levels of atmospheric carbon dioxide and depletion of the ozone layer are environmental issues which have also become of interest to forest scientists. Solomon and West (1985) have discussed the potential effects of carbon dioxide-induced climate changes on world forests. Their review, while not truly an economic analysis, does present information on the economic importance of the resource at risk. Sedjo and Clawson (1983) discussed tropical deforestation as a possible contributor to the carbon dioxide problem and, furthermore, cited common property as the principal cause of the deforestation dilemma. Forest successional patterns of northern hardwoods as affected by global climate warming have been modeled by Binkley and Larson (1987). The latter provide information which could be helpful to an economic assessment of changing climates.

The effects of various silvicultural practices on water quality have been examined by a number of forest economists. Hickman and Jackson (1979) developed a linear programming model to assess the economic consequences of soil loss from silvicultural activities in eastern Texas. Kirby and Rupe (1987) investigated the cost of avoiding sedimentation on forested watersheds in Idaho. In an agriculture example relevant to the forestry sector, Taylor (1975) examined the use of marketable rights for fertilizers as a means of achieving water quality standards. Sutherland (1982) and Russell and Vaughan (1982) estimated the regional and national recreational fishing benefits due to improved water quality. Grobey (1985) examined several past studies of the economic impacts of logging on anadromous fish stocks. And Ellefson and Miles (1985) evaluated the economic consequences of undertaking forestry practices thought conducive to the production of quality water on nine national forests in the Midwest.

Literature concerning forest protection economics provides some especially valuable information about procedures for valuing forest resources and determining optimal levels of damage control. A model for determining the optimal level of expenditure needed to control timber damage caused by the southern pine beetle was developed by de Steiguer, Hedden and Pye (1987). Leuschner and Young (1978) estimated the economic impact of southern pine beetle on reservoir campsites. Thompson and others (1979) conducted an evaluation of alternative forest management systems as a means of controlling the spruce budworm. Methods for estimating the economic impacts of wildfire on timber stands in the Rocky Mountains were developed by Mills and Flowers (1986). Parlar and Vickson (1982) investigated procedures for determining the optimal level of forest fire control.

Several researchers have also explored legislative attempts to regulate environmental degradation by means of public policy (Cubbage and Siegel 1985, Cubbage and Ellefson 1980, Ellefson 1988, Siegel 1974). Henly and Ellefson (1986) specifically addressed the public and private sector costs of such legislation. Regulatory laws are often motivated by a desire to protect forest environments from harmful externalities and to maintain the productivity of forests for future generations. State forest practice laws usually represent some compromise between environmental and timber interests. However, environmental protection has increasingly become the dominant concern.

In sum, economists have generated a large body of information on the subject of environmental economics. Several authors have made important contributions toward a better understanding of the economic implications of environmental concerns pertinent to forestry. Yet, such advancements notwithstanding, considerable economics research remains to be accomplished. To be most effective, such research should focus on the following strategic areas.

- Develop better understanding of nonforestry externalities imposed on the forestry sector.

- Develop improved understanding of forestry externalities imposed on the forestry sector.
- Acquire better understanding of forestry externalities imposed on nonforestry sectors.
- Develop improved understanding of public policy instruments available for managing environmental externalities.
- Acquire better understanding of population and growth as a source of forestry-related environmental problems.

Develop Better Understanding of Nonforestry Externalities Imposed on the Forestry Sector

Production and consumption activities which occur outside the forestry sector can generate external costs which affect forestry. An example is the adverse forest productivity effects of pollutants generated by nonforestry industrial activities (e.g., energy production, chemical production). Nonforestry externalities (e.g., atmospheric deposition) recently have received significant news media attention; consequently, they are often foremost in the mind of the public. Topical areas in need of research attention include the following.

Evaluate the economic effects of air pollutants on forest productivity. Research into the economic impacts of air pollution on forests is a multidimensional problem of considerable complexity which will require a series of related studies. Research should be undertaken to determine the effects of various pollutants (e.g., gaseous, acid deposition, heavy metals) on the productivity of both market and nonmarket goods and services produced by several major forest ecosystems. One of the greatest difficulties to be faced by investigators involved in such research will be development of reliable dose-response functions for various pollutants and forest types. This challenge will require interdisciplinary investigations involving biologists and biometricians.

Evaluate the impacts of global climate changes on forest succession and productivity. Depletion of atmospheric ozone and the resultant warming of global climate is a topic of great concern to scientists. Research is needed to assess the effects of warming trends on the succession of forest types and to evaluate the economic impacts of such trends. As in the case of air pollution, such research will require collaborative efforts among several disciplines, including ecology, biometrics, and economics.

Evaluate the economic impacts of a variety of natural hazards on forestry. The extensive forest damage caused by the recent eruption of the Mount St. Helens volcano serves as a dramatic reminder of the manner in which natural hazards can affect forestry. Examples of other natural agents which damage forests include landslides, avalanches, and lightning strikes. Research is needed to assess the potential risk of natural hazards and to develop methods for assessing their economic impacts. Such research could, in the case of landslides and avalanches, consider the cost and benefits of preventing damage.

Evaluate the economic impacts of exotic biological agents on forest ecosystems. Many foresters are well acquainted with the disastrous consequences of chestnut blight on American forestry. Kudzu, first introduced as an erosion control measure, is also an example of an exotic biological agent gone awry. Research is needed to investigate the possible economic impacts on forest ecosystems of introduced plants, animals, insects, and diseases.

Evaluate the economic impacts of altered ground and surface water regimes on forestry. Large-scale hydrologic projects undertaken for purposes of flood control or power generation often have severe impacts on forest ecosystems. Forests are altered or, in some instances, entirely eliminated due to flooding, draining, or changing water tables. Research is needed to determine the economic effects of such hydrologic engineering projects.

Develop Improved Understanding of Forestry Externalities Imposed on the Forestry Sector

Economics research should also be concerned with forest sector production and consumption activities that negatively affect forest resources. An example is timber harvesting activities which degrade the long-term productivity of forest sites. Issues of this sort have a long history of concern among forest managers. Surprisingly, however, research on the economic impacts of such externalities is limited. Possible research topics which could enhance understanding of the problem are as follows.

Evaluate the economic effects of silvicultural activities on insect and disease populations. Since economic efficiency is increased via the lowering of regeneration and harvesting costs, the often preferred silvicultural system for many commercial timber species is an even-aged monoculture. Yet biologists recognize that such management methods may increase the risk of disease and insect epidemics. Research is needed to investigate the relationship between single species management and the incidence of epidemics and to examine the economic benefits of greater species diversity. Research into this important topical area could have an especially significant impact on current forest management practices.

Evaluate the effects of timber harvesting practices on forest stand and site conditions. Timber harvesting involves a number of activities (e.g., skidding and road building) which can be detrimental to harvested sites. Such adverse impacts include soil compaction, nutrient loss, damage to the residual stands and alteration of the water table. Research is needed to assess the economic effects of harvesting activities on long-term forest productivity. Such research is a broad topic that will require a series of conceptually related studies.

Evaluate the effects of forest regeneration practices on site productivity. Forest regeneration practices used in modern operations (e.g., bulldozing, windrowing, and burning) can severely impact forested sites. Economic studies are needed to ascertain the potential effects of such regeneration methods.

Evaluate the impacts of recreation activities on forest stand and site conditions. Like timber harvesting, forest recreation activities can result in adverse site impacts. Visitor use can cause soil compaction, improperly constructed and maintained trails can lead to soil erosion problems, and auto emissions can damage vegetation. Economic investigations are needed to examine and suggest economically efficient means of controlling recreational damages to forests.

Evaluate the effects of recreation overuse on the quality of recreation experiences. Many forestry professionals are aware of the extreme environmental consequences associated with congestion of recreation areas in such places as Yosemite Valley and the Great Smoky Mountains. Crowding and overuse of forested areas is a problem which diminishes the value of recreational experiences. Economics research is needed to assess the consequences of relieving congestion problems via users fees and better designed recreational facilities.

Evaluate the effects of timber harvesting on forest recreation activities. Site disturbances associated with timber harvesting can have adverse impacts on forest recreation experiences (e.g., decreased visual aesthetic enjoyment, foul bodies of water, increased noise pollution). Potential conflicts in resource use have grown as forests are expected to be providers of both timber and recreation resources. Research is needed to determine the economic impacts of timber production activities on various forest recreation uses.

Acquire Better Understanding of Forestry Externalities Imposed on Nonforestry Sectors

Economics research is needed to assess production and consumption activities occurring within the forestry sector that have adverse effects on nonforestry environments. Such activities include pesticide and herbicide use and prescribed burning, both of which can have detrimental impacts on air and water quality outside forested environments. Interest in forest-nonforestry externalities probably began, in earnest, during the early 1970s when public concern heightened over dioxin-contaminated herbicides. Economics research is needed on topics of the following nature.

Evaluate the effects of pesticide and herbicide use on water and air quality. Both pesticides and herbicides play increasingly important roles in forest management. For example, managers of pine seed orchards would find their jobs almost impossible without access to pesticides for control of cone and seed insects. Similarly, managers of commercial pine plantations regard herbicides as an essential tool for controlling undesirable hardwood competition. Despite the benefits provided by pesticides and herbicides, there are some potentially serious side-effects on human and animal populations. Economic research is needed to measure the social costs of such chemicals as they enter ground and surface water and the air. Such research could be one of the most important topics addressed by forest economists.

Evaluate the effects of prescribed burning on air quality. Like pesticides and herbicides, fire is an important tool used in the preparation of sites for planting and in the control of vegetative competition in forest stands. However, the release of gases and particulate matter into the surrounding air can have adverse impacts on property, aesthetics and human health. For example, not uncommon are auto accidents occurring on roads obscured by smoke from forest fires. Research is needed to assess the economic effects of forest burning on nonforested environments.

Evaluate the impacts of timber harvesting activities on water quality. Road construction and skidding activities associated with timber harvesting can have negative impacts on water quality via sedimentation and nutrient loading (both in and out of the forest environment). The problem has been extensively researched as a result of legislation addressing nonpoint sources of water pollution. Nevertheless, it remains as an area of public concern and, hence, deserves additional attention by forest economists. Research will be needed to examine the effects of decreased water quality on game and fish population as well as human health. Ideally, such research will suggest economically efficient means of controlling forestry-related water pollution.

Evaluate the economic effects of desertification and deforestation on global climate change. A great deal of concern has recently been focused on increased levels of carbon dioxide in the earth's atmosphere. One hypothesis is that the extensive elimination of forests in the Third World, particularly in West Africa, is a major contributor to the problem. Involving some formidable global modeling problems, this is an especially complex research topic dealing with common property issues in subsistence economies. However, it looms large as an important economic research topic to be addressed by forest economists.

Evaluate the impacts of pulp and paper production on water and air quality. Pulp and paper production continues to be an important source of local and regional employment in many portions of the world. However, such production facilities can be a source of very noxious environmental insults. Economic research is needed to assess the benefits of reductions in air and water pollution resulting from paper production and to weigh these benefits against the costs of pollution control.

Develop Improved Understanding of Public Policy Instruments Available for Managing Environmental Externalities

Although research is needed to measure the external costs of environmental pollution, attention must also be directed to research that will suggest means of limiting such externalities via efficient policy instruments. Such research has largely been the domain of economists outside the forestry community; little in the way of rigorous, theoretical attention has been directed at the problem by forest economists. Following are some research topics related to policy means for controlling environmental degradation.

Evaluate government regulation and prohibition as a means of reducing environmental pollution. Regulation and prohibition can be policy instruments used by governments to correct market failures which lead to environmental pollution. Regulation involves the establishment of public agencies and commissions to control production and consumption processes, while prohibition means the outright banning of certain pollution-causing activities. Research is needed to investigate the economic efficiency and equity of regulation and prohibition as means of limiting forestry-related externalities.

Evaluate government fiscal incentives and disincentives as a means of reducing environmental pollution. The imposition of taxes and the granting of subsidies are also means by which externalities can be limited. Such methods, while not as direct as regulation or prohibition, can be incentives to reduce pollutants via their ability to alter benefits and costs associated with certain economic activities. Economic research is needed to examine the role of fiscal and tax policies as means of managing environmental pollution.

Evaluate government production of environmental services as a means of reducing environmental pollution. Through outright purchase of resources or the exercise of eminent domain, government can provide certain environmental services which alleviate environmental quality problems. Sewage and water treatment are two prime examples of government provision of environmental services. Research is required to assess the role that government-provided services can play in improving environmental quality related to forestry activities.

Acquire Better Understanding of Population and Growth as a Source of Forestry-Related Environmental Problems

Most economists tend to view environmental degradation as a problem of market failure rather than a problem of increased population or of industrial development. Nevertheless, a few notable economists (e.g., E. J. Mishan) have suggested that population growth and economic expansion are important sources of environmental quality problems. As such, they deserve the serious attention of economic researchers. Possible study topics related to population and economic growth issues are as follows.

Evaluate the effects of population growth and movement on forest resources. Research is needed to examine the relationship between population trends and the demand for environmental services from forests. Investigation of this topic would certainly need to distinguish between developing and developed nations, since differing population problems occur in each nation and different sorts of demands are placed on each nation's forest resources. Economic models of demographic change as it influences resource usage and environmental degradation would be important contributions.

Assessment of the environmental consequences of economic growth in the forestry-based sector. Research is needed to evaluate the effects on environmental quality of expanded sawmilling, paper production, and other forest-based industries. Expansion in such industries is important for the economic well-being of local and regional economies, but there are attendant problems related to environmental degradation and the quality of life of an area's residents. Economic research is needed to assess the tradeoffs and to suggest optimal industrial development patterns.

Summary and Conclusions

Environmental economics has made some important advances in the past two decades, a significant portion of which has relevance to forestry and related activities. The time has come, however, for the forest economics community to vigorously apply principles of environmental economics (especially empirical investigations) to a variety of environmental problems of importance to the nation's forestry sector.

Strategic research directions of considerable merit include determination of the economic consequences of externalities imposed on, by, and within the forestry sector. Also important is furthering the forestry community's understanding of policy instruments appropriate to the management of environmental externalities and the consequences of population and economic growth as a source of forestry-related environmental concerns. Focusing economics research on strategic directions of such a nature could lead to an information base that will further public interest in environmentally sound management of the nation's forests.

Literature Cited

Adams, R. M. 1986. Agriculture, forestry and related benefits of air pollution control: A review and some observations. American Journal of Agricultural Economics 68(2):464-472.

Arrow, K. J. and A. C. Fisher. 1974. Environmental preservation, uncertainty and irreversibility. Quarterly Journal of Economics 88:312-319.

Bator, F. 1958. The anatomy of market failure. Quarterly Journal of Economics 72:351-379.

Baumol, W. J. 1972. On taxation and the control of externalities. American Economic Review 62:307-322.

Binkley, C. S. and B. C. Larson. 1987. Simulated effects of climate warming on the productivity of managed northern hardwood forests. In: Forest Decline and Reproduction: Regional and Global Consequences. International Institute for Applied System Analysis. Laxenburg, Austria.

Bohm, P. 1970. Pollution, purification and the theory of external effects. Swedish Journal of Economics 72:153-166.

Carson, R. 1962. Silent Spring. Houghton Mifflin Company. Boston, MA.

Crocker, T. D. 1971. Externalities, property rights and transaction costs: An empirical study. Journal of Law and Economics 14:451-464.

Cubbage, F. W. and P. V. Ellefson. 1980. State forest practice laws: A major policy force unique to the natural resources community. Natural Resources Lawyer 13(2):421-468.

Cubbage, F. W. and W. C. Siegel. 1985. The law regulating private forest practices. Journal of Forestry 83(9):538-545.

d'Arge, R. C. and K. C. Kogiku. 1973. Economic growth and the environment. Review of Economic Studies 40:61-78.

de Steiguer, J. E. 1987. Methods for economic assessment of atmospheric pollution impacts on forests of the eastern United States. In: NAPAP Terrestrial Effects Task Group Peer Review Summaries. National Acid Precipitation Assessment Program. Washington, DC.

de Steiguer, J. E., R. L. Hedden and J. M. Pye. 1987. Optimal level of expenditure to control the southern pine beetle. Research Paper SE-263. Southeastern Forest Experiment Station. USDA Forest Service. Asheville, NC.

Ellefson, P. V. 1988. Private forest, Public interest. Habitat: Journal of Maine Audubon Society 5(4):28-30.

Ellefson, P. V. and P. K. Miles. 1985. Protecting water quality in the Midwest: Impact on timber harvesting costs. Northern Journal of Applied Forestry 2(2):57-61.

Ferguson, C. E. and S. C. Maurice. 1974. Economic analysis. Richard D. Irwin Publishers. Homewood, IL.

Fisher, A. C. 1981. Resource and environmental economics. Cambridge University Press. Cambridge, MA.

Fisher, A. C. and J. V. Krutilla. 1975. Resource conservation, environmental preservation and the rate of discount. Quarterly Journal of Economics 89:358-370.

Fisher, A. C. and F. M. Peterson. 1976. The environment in economics. Journal of Economic Literature 14:1-33.

Freeman, A. M. 1974. On estimating air pollution control benefits from land value studies. Journal of Environmental Economics and Management 1:74-83.

Freeman, A. M. III, R. H. Haveman and A. V. Kneese. 1973. The economics of environmental policy. John Wiley and Sons Publisher. New York, NY.

Grobey, J. H. 1985. Politics versus bioeconomics: Salmon fishery and forestry values in conflict. In: Forest land public and private by R. T. Deacon and M. B. Johnson (eds.). Pacific Institute for Public Policy Research. San Francisco, CA.

Hay, S. P. 1987. Beauty, health and permanence: Environmental politics in the United States, 1955-1985. Cambridge University Press. Cambridge.

Henly, R. A. and P. V. Ellefson. 1986. State forest practice regulation in the U.S.: Administration, cost and accomplishments. Station Bulletin AD-SB-3011. Agricultural Experiment Station. University of Minnesota. St. Paul, MN.

Hickman, C. A. and B. D. Jackson. 1979. Economic impacts of controlling soil loss from silvicultural activities in east Texas. Forest Science 25:627-640.

Howe, C. W. 1979. Natural Resource Economics. John Wiley and Sons Publishers. New York, NY.

Just, R. E., D. L. Hueth and A. Schmitz. 1982. Applied welfare economics and public policy. Prentice-Hall Publishers. Englewood Cliffs, NJ.

Kirby, M. W. and J. B. Rupe. 1987. The cost of avoiding sedimentation. Journal of Forestry 85(4):39-40.

Kneese, A. V. 1976. Natural resource policy 1975-1985. Journal of Environmental Economics and Management 3:253-288.

Krutilla, J. V. and C. J. Cichetti. 1972. Evaluating benefits of environmental resources with special application to the Hell's Canyon. Natural Resources Journal 12:1-29.

Leuschner, W. A. and R. L. Young. 1978. Estimating the southern pine beetle's impact on reservoir campsites. Forest Service 24(4):527-542.

Mills, T. J. and P. J. Flowers. 1986. Wildfire impacts on the present net value of timber stands: Illustrations in the northern Rocky Mountains. Forest Science 32(3):707-724.

Mishan, E. J. 1974. What is the optimal level of pollution? Journal of Political Economy 82:1287-1299.

National Acid Precipitation Assessment Program. 1987. Terrestrial Effects Task Group Peer Review Summaries. Washington, DC.

Parlar, M. and R. G. Vickson. 1982. Optimal forest fire control: An extension of Park's model. Forest Science 28(2):345-355.

Pigou, A. C. 1932. The economics of welfare. Macmillan and Company. London, UK.

Rees, J. 1985. Natural resources: Allocation, economics and policy. Methuen. London, UK.

Russell, C. S. and W. J. Vaughan. 1982. The national recreational fishing benefits of water pollution control. Journal of Environmental Economics and Management 9(4):328-354.

Scitovsky, T. 1954. Two concepts of external economics. Journal of Political Economy 62:143-151.

Sedjo, R. A. and M. Clawson. 1983. How serious is tropical deforestation? Journal of Forestry 81(12):792-794.

Seneca, J. J. and M. K. Taussig. 1974. Environmental economics. Prentice-Hall Publisher. Englewood Cliffs, NJ.

Siebert, H. 1987. Economics of the environment: Theory and policy. Springer-Verlag. Berlin, FRG.

Siegel, W. C. 1974. State forest practice law today. Journal of Forestry 72:208-211.

Smith, V. K. 1979. Natural resource scarcity: A statistical analysis. Review of Economics and Statistics 61:423-427.

Solomon, A. M. and D. C. West. 1985. Potential responses of forests to CO_2-induced climate change. In: Characterization of information requirements for CO_2 effects. DOE/ER 0236. U.S. Department of Energy. Washington, DC.

Sutherland, R. J. 1982. A regional approach to estimating recreation benefits of improved water quality. Journal of Environmental Economics and Management 9(3):229-247.

Taylor, C. R. 1975. A regional market for rights to use fertilizer as a means of achieving water quality standards. Journal of Environmental Economic and Management 2(1):75-91.

Thompson, W. A., C. S. Holling, D. Kira, C. C. Huang and I. Vertinsky. 1979. Evaluation of alternative forest system management policies: The case of spruce budworm in New Brunswick. Journal of Environmental Economics and Management 6(1):51-68.

17

COMMUNITY AND REGIONAL ECONOMIC GROWTH AND DEVELOPMENT

Henry H. Webster and Daniel E. Chappelle

Forestry has much to offer community and regional interests in economic growth and development. Its ability to contribute, however, depends in large measure on the availability of information provided by well designed research programs. Forest economics research can further community-wide understanding of forestry's contribution to economic growth and can provide a foundation for judging the effectiveness of various approaches suggested as means of maximizing such growth. The products of research can be especially useful as means of encouraging awareness of the processes by which communities and regions grow and develop; of promoting clear visions of the several kinds of contributions forestry can make to such processes; and of enhancing understanding of how existing economic analyses can contribute to economic development.

Forestry Contributions to Development

Forestry can contribute to community and regional economic growth and development in a number of ways. Consider three primary methods—all of which contribute to employment and income.

Direct source of employment. Forestry can be a direct source of employment, wherein people are directly involved in the management and protection of forest resources. Such can involve professional managers (source of technical advice and general program direction) as well as laborers (carry out specific management and protection practices). In times of enduring and widespread economic distress within a community or the national economy, public-works type programs have been especially important to forestry. Such programs have provided a meaningful source of opportunities for purposeful work (e.g., reforestation, fire protection). In the history of the United States, there have been a number of times when persons employed by public-works programs have contributed substantially to forestry via the forest management practices accomplished and the recreational facilities constructed.

Source of raw material. Forestry can also be a source of raw material for industrial development, which in turn becomes an important source of income and employment. Historically, a number of important links have been made between raw material, development, and employment. For example, Civilian Conservation Corps participants planted trees more than 50 years ago—these trees now provide the raw material base for many forestry enterprises. Likewise, trees planted by authority of the Agricultural Soil Bank Program (1950s) have become the foundation of many industrial forestry activities.

Support for tourism. Forest resources can also provide support for tourism in forested regions. Part (admittedly modest) of the employment and income generated by such tourism can be legitimately attributed to forests. In some cases, forest resources are the essential ingredient to wildland recreational experiences.

Forestry as Part of Economic Systems

Forestry and forestry enterprises are part of community and regional economic systems. Although often the primary sector of a community or region's economy, forestry enterprises have important links to other sectors of the economy. When a change (e.g., raw material shortages) occurs within the forestry sector, the impacts of such a change are transmitted forward to secondary and tertiary sectors. Similarly, when a change occurs in secondary or tertiary sectors (e.g., increase or decrease in final product demand), the impacts may well be transmitted back to the forestry sector. Also of special note is the spatial nature of markets for primary forest enterprises (they occur relatively close to raw material sources). In contrast, products linked to secondary and tertiary sectors (including forest products sectors) are traded region-wide or may be traded at national or even international levels. Also noteworthy is the fact that most forest products can be produced in final form by any area possessing appropriate natural resources, including foreign nations. Hence, opportunity exists for strong substitution effects, even though significant immobility exists in industrial capacity (high asset values and low salvage values). In addition, many substitute inputs exist—inputs which are correlated to some extent within geographic areas (e.g., substitute tree species in the manufacture of wood products).

Output demand patterns and strong substitution effects at the resource level pose special circumstances for forestry enterprises. Because of such patterns and effects, the use of raw materials supplied by community and regional economies is extremely sensitive to final demands—but not to resource supplies. Also, demand patterns and substitution effects place forestry enterprises in the position of being price-takers (rather than price-makers) for products manufactured and sold. On the input side, however, forestry firms tend to be spatial monopolists (price-makers), since they require more rural land area per unit of product than average enterprises. For various reasons, the size and shape of supply areas vary a great deal (e.g., differences in types of inputs, the density of establishments, and economic power of firms).

The regional patterns described above have special significance for analysis of forestry in a community and regional context. They make it imperative that the structure of a region's entire economy be understood (including leakages) and that attention be focused on both supply and demand factors. Analysis of community and regional systems requires careful determination of whether scarce input supplies are most important or whether scarcity of demand is most significant. From an analysis standpoint, the scarce factor is usually considered exogenous; hence it has the effect of driving models of forestry systems (as can be demonstrated by input-output models). Of course, reliability of forecasts from such models depends on the accuracy of the predictions for exogenous variables.

Analyses of communities or relatively small geographic areas are more difficult to carry out than analyses of higher levels in the spatial hierarchy. Although a small area may be more homogeneous and compact, analysis is often arduous because economic activity within the area may be dominated by imports and exports. Further complicating small-area analysis are problems with secondary data. Generally such data are not sufficiently detailed for use in analyses, and the collection of primary data is impeded by the need to insure confidentiality of respondents to economic surveys.

Review of Previous Analytical Research

A substantial amount of research concerning community and regional economic development has been accomplished over the years. Much is directly relevant to forestry; some is only partially applicable to forestry interests. Because the breadth of information and its complexity is substantial, an ordered simplification is needed. Ten interrelated topical areas have been identified. Research within each has been assessed to determine relevance to economic growth and development, type of analytical methods used, principal contributions of research and examples of research.

Economic Structure of Forestry and Associated Sectors

Relevance to Economic Growth and Development: Description of community or regional economic structure serves as a basis for analysis of impacts resulting from changes in an economy. By linking forestry sectors with other regional sectors, the contributions of such sectors to the region's economy and to economies outside the region can be visualized and more clearly understood. *Analytical Methods*: Economic base models, location quotients, coefficients of localization, and regional input-output accounts. *Research Contributions*: Understanding specific linkages between forestry sectors and other sectors of an economy. Identify communities and regions that depend heavily on forestry enterprises. *Example Research*: Maki, Schallau, and Beuter (1968), Kaiser (1968 and 1969), Kaiser and Dutrow (1971), Maki and Schweitzer (1973), Youmans et al. (1974), Darr and Fight (1974), Brodie et al. (1978), Flick et al. (1980), Diamond and Chappelle (1981), and Keegan and Polzin (1987).

Industrial Concentration and Diversification

Relevance to Economic Growth and Development: Concentration and diffusion of economic activity, and measures of regional growth shares (distinct from national share and industrial mix share) are important to understanding regional economic systems. Given such information, comparisons can be made of growth in economic activities in different regions and in growth of economic activities between different sectors (e.g., forestry versus nonforestry). *Analytical Methods*: Shift-share analysis, measures of diversification, and measures of concentration. *Research Contributions*: Increased understanding of forestry's contribution to regional economic growth. Estimation of trends in such contributions. Increased understanding and estimation of interregional flows and competition between regions for a specific sector of the economy. *Example Research*: Dutrow (1972), Beuter and Olson (1980), Bilek and Ellefson (1984), and Schallau and Maki (1986).

Regional and Community Stability

Relevance to Economic Growth and Development: Knowledge about the stability of economic systems is critical to judgments about suggested changes in such systems. Analyses of regional and community stability also provide a framework for assessing the causes of economic instability. *Analytical Methods*: Economic base models, measures of diversification, measures of concentration, and forecasting methods. *Research Contributions*: Identification of stable and unstable regional structures and the causes of such instability. *Example Research*: Schallau, Maki, and Beuter (1969), Waggener (1977), Schallau (1980), Haynes (1983), and Schallau and Polzin (1983).

Stability of Natural Resource Sectors

Relevance to Economic Growth and Development: Knowledge of economic sector stability assists in the selection of specific sectors to be targeted for economic assistance. Also facilitates determination of how stability might be increased. Required is the forecasting of both supply and demand factors. *Analytical Methods*: Economic base models, measures of diversification, measures of concentration, and forecasting methods. *Research Contributions*: Identification of methods for stabilizing economic sectors (e.g., sustained yield units). *Example Research*: Schallau, Maki, and Beuter (1969), Waggener (1977), Schallau (1980), Beuter and Olson (1980), Haynes (1983), and Schallau and Polzin (1983).

Economic Impacts

Relevance to Economic Growth and Development: Provides a measure of forestry's economic contributions in specific circumstances (e.g., output, income, employment). Also provides some indication of the distribution of impacts. *Analytical Methods*: Enumeration of firms and their levels of employment and income generation. Interindustry input-output techniques to determine multipliers (which determine indirect and induced impacts). Economic base models. Econometric models. *Research Contributions*: Substantial clarification of forestry's economic contributions. Consequence has been a reduction in "competitive exaggeration" (e.g., established fact that multipliers are in the range of two to three). Research has been conducted at various spatial levels and in various locations throughout the nation. Now possible to generalize to other geographic locations for many secondary forest product sectors. *Example Research*: Bromley et al. (1968), Kalter and Lord (1968), Hughes (1970), Laurent and Hite (1972), Darr and Fight (1974), Connaughton and McKillop (1979), Flick et al. (1980), Schooley and Jones (1980), and Chappelle et al. (1986).

Forecasts of Future Economic Activity

Relevance to Economic Growth and Development: Levels of overall economic activity strongly affect contributions of all economic sectors, including forestry. Such effects must be known before forecasts of future forestry contributions can be made. *Analytical Methods*: Examination of trends, simple regression analysis, and econometric analyses. *Research Contributions*: Numerous specific forecasts of forestry's contribution to particular geographic areas have been established. Forecasts of future economic activity have been used as input to other types of economic analyses (e.g., economic impact analysis). *Example Research*: Beuter et al. (1976) and Bell (1977).

Interregional Flows of Inputs and Outputs

Relevance to Economic Growth and Development: Interregional flows provide a measure of external trade that may impact or influence a regional economy. Such measurements identify the effect of exports and imports and provide a basis for judging the economic performance of a community or region. *Analytical Methods*: Interregional economic accounts, interregional input-output models, transportation models, and mathematical programming models. *Research Contributions*: Information on the magnitude of external flows to and from regions has been developed. Also, decisions as to which specific sectors might realistically be expanded in reaction to external flows have been guided. Understanding of where a region fits within a system of regions has been furthered. *Example Studies*: Holland and Judge (1963), Kaiser (1972), Schallau (1980), and Kallio et al. (1987).

Location of Economic Activities

Relevance to Economic Growth and Development: Regional analysis enables identification of specific sectors in which to expand or contract economic activity. Location analysis, however, provides insight as to which specific locations within a region should be the recipient of such activities. *Analytical Methods*: Various location models oriented to agricultural crops, industrial activities, and settlements (central places). Simulation and related mathematical programming approaches. *Research Contributions*: Provided guidance to location of specific economic activities and has minimized high cost of misallocations, including subsequent failure of economic activities. Provided link to industrial targeting programs. *Example Research*: Hagenstein (1964), Holly (1970), and Chappelle et al. (1982).

Valuation of Non-Priced Goods and Services

Relevance to Economic Growth and Development: Forestry systems involve outputs priced by organized markets as well as outputs for which market prices are not typically determined. For many types of economic analyses, values for both types of outputs must be known. To guide decisions, there is need to have readily available measures of value for non-priced goods and services. Such values must be compatible with market prices if the two are used in combination. *Analytical Methods*: Various types of shadow prices derived by various methods, including mathematical programming analysis. *Research Contributions*: Shadow prices have been developed by various means. Although considerable shadow-price variation has been found for specific outputs, the range of values has proven useful to analyses at community and regional levels and for targeting economic activities. *Example Research*: Kalter and Lord (1968), Brown and McMillan (1977), and Sutherland (1981).

Estimation of Population and Employment

Relevance to Economic Growth and Development: Estimates of population and employment are essential ingredients to measurement of forestry's economic contributions. In order to quantify demand for goods and services provided by forestry systems, population estimates must be available. Also, to properly quantify regional economic structures, employment composition must be accessible. *Analytical Methods*: Demographic and employment accounting procedures. *Research Contributions*: Data base for numerous analyses (e.g., economic impact analysis) has been developed. *Example Research*: Marcin (1972), and Wall and Oswald (1975).

As the above review demonstrates, forest economics research focused on community and regional economic growth and development has progressed significantly in recent years. However, much additional research remains to be accomplished. To be effective, such research should concentrate on the following strategic directions:

- Develop improved understanding of forestry's contribution to economic growth and development.
- Develop effective means of maximizing forestry's contribution to economic growth and development.
- Develop improved understanding of the process by which economic development occurs within the forestry sector.

Develop Improved Understanding of Forestry's Contribution to Economic Growth and Development

The contribution of forestry to economic growth and development is often unknown or is highly uncertain. Such is certainly not conducive to the development of effective programs considered necessary to achieve the public's interest in high levels of employment and long-run economic stability. Needed is research that will lead to improved understanding of actual contributions made by forestry in a regional and community setting.

Define and develop structure of regional economies with an emphasis on forestry sectors. Required is development of a generalized framework that can be used to describe the structure and operation of economies that have a natural resources component. Do common structural factors exist within and between regional economies? Can such factors be described and subsequently applied to other regions and communities? For example, if a pulp mill is proposed to be part of a region's economic base, can structural generalities be used to imply something about its wood procurement area and the type and magnitude of economic contributions that will be made to surrounding communities? Could similar structural-type information be applied to a pulp mill's labor procurement area? Furthermore, will a pulp mill's regional contributions be similar to that of a steel mill, an auto plant, a large sawmill, or a fast-food restaurant? Research interest should attempt to discover fundamental structural relationships that would be applicable to pulp mills whether located in California or in West Virginia. Intersector differences must, however, be recognized. The 100 additional jobs created by increased production at a pulp mill will most certainly result in community and regional economic effects that are very different from those resulting from 100 additional jobs created by a fast food establishment in the same regional economy.

Research should emphasize determination of how best to portray natural resource sectors in a structural framework relevant to a region's economy. Furthermore, research should work toward delineation of natural resource areas on the basis of economic and ecological relationships; the intent being to delineate ecological-economic regions. The latter would serve as the basis for the development of integrated production plans. Such would permit implementation of sector management plans that are consistent with forest product marketing realities and with the ecological capabilities of a region. Examples of ecological-economic regions are the Chesapeake Bay area, the Great Lakes region, the Pacific Northwest, and the Canadian prairies. A strong regional tie between ecological and economic considerations could have a very positive influence on judgments about the appropriateness of various types of economic forestry enterprises. An example of actions in this direction are the efforts of three Midwest states (Michigan, Minnesota, and Wisconsin—which make up the Upper Great Lakes Region) to develop a regional forestry alliance (i.e., Lake States Forestry Alliance). The states involved are similar ecologically and economically, especially in their northern portions (e.g., similar forest types). The states could eventually move toward joint production plans such as collective business attraction plans and coordinated timber management plans near state borders. The important research question is how such multi-state areas should be characterized.

Determine forestry's position in regional structures and its contribution to growth and development within such structures. Focusing on forestry sectors per se (not generalized regional economies as discussed above), what is the fundamental nature of regional structures involving forestry? For example, do notions of regional ecological-economic structure have relevance? And if so, what growth and development contributions can be expected from forestry within such a context? Of interest is comprehensive modeling of actual economies with the intent of identifying structural components common to forestry (e.g., comprehensively determining the economic and community effects of a pulp mill being placed at a specific geographic location). Of concern to research should be forestry's diverse contributions in various regions and a comparison of forestry's contributions between different types of regions.

Determine demand areas and input supply areas for natural resource sectors of regional economies. Research on regional economies and how forestry fits into and contributes to such economies is certainly warranted. However, also of importance is the internal workings of forestry sectors per se. What, for example, can be said of input supply areas for forestry enterprises in general and for specific communities and regions in particular? Supply areas broader than wood fiber should be assessed (e.g., labor, transportation). Similarly, evaluations should be directed at demand areas for enterprises in general and for specific communities and regions.

Measure net regional economic growth and development. Research should be undertaken to further the quantification of net economic growth. Needed, for example, is quantification of linkages, tradeoffs, and countervailing forces which subtract from gross regional economic growth. Estimates are needed of *net* economic growth—not growth per se. For example, construction of a sawmill may not result in "real" economic development if the new mill adversely impacts the operations of three adjacent mills. Research is also needed to improve understanding of structural diversification in natural resource sectors and communities. By what processes do natural resource sectors become diversified, and how might such diversification be encouraged—if deemed appropriate? Analyses should be undertaken to determine the extent to which diversification of a region's economy insulates the region from shocks induced by actions in national and international economies. How stable or noncyclical are resource industries? Comparisons with other industries in particular regions are vital to answering such a question. For example, timber industries in the Lake States may be relatively stable in comparison to heavy industries, while in the Pacific Northwest they may be relatively unstable relative to newer technology-based industries (e.g., electronics, weapons production, aircraft construction).

Develop Effective Means of Maximizing Forestry's Contribution to Economic Growth and Development

Forestry offers significant potential as a contributor to the economic growth and development of many regions and communities. Of concern, however, is access to effective means of harnessing such potential and subsequently maximizing forestry's contribution to regional interests. Needed is research that will enable effective targeting of industrial development efforts and the encouragement of development that leads to substantial gains in employment with only modest disruption of natural and man-made environments.

Evaluate means of more effectively targeting industrial development programs. What information currently exists that would enable more effective targeting of industrial development programs? Is the information sufficiently detailed so that it can be applied with precision? How can and should information be used to guide industrial targeting? What additional information would make targeting of development programs more effective?

Assess the impact of social variables on regional growth and development. How do social attributes and variables affect regional growth and development? How should they be accounted for? The difficulty (or ease) of establishing wood-fired generators in different communities has pointed to the importance of understanding the impact of social variables on growth and development.

Evaluate the impact of physical and biological variables on regional economic growth and development. The relevance of various technologies to growth and development varies through time. How might communities be guided to recognize technologies that have potential to enhance their economic well being? Needed is information to help communities recognize and capitalize on available assets (e.g., timber, labor) in light of newly available technologies.

Evaluate the impact of forest sector economic growth and development on environmental quality. In what manner and to what extent do forestry enterprises affect the quality of regional and local environments? How do such impacts compare with the environmental impacts of alternative forms of development (e.g., pulp mill versus steel mill)? How can communities equate environmental harm resulting from development activities to benefits realized from such activities? By what means might communities plan for remedial actions?

Determine means of integrating economic impact models with resource management and development models. Research should be undertaken to devise means of integrating models of forest management activities with models depicting the economic impacts of such activities. This integration would enable establishment of public-sector resource management strategies on the basis of expected economic and related outcomes. Ultimately, such integration could be significant to the allocation of public budgets. Regional impact models integrated with resource management models would be especially beneficial to public land management and investments therein. It could eliminate or reduce situations where a region's demand for public timber is low yet large amounts of timber are purchased, while an adjacent region's demands for timber are high yet virtually no timber is sold.

Develop Improved Understanding of the Process by Which Economic Development Occurs Within the Forestry Sector

The process by which economic development occurs within a forestry sector is not fully understood. For example, how are opportunities for economic development originally recognized? By what process is an initial (and subsequent) processing facility located? What are the employment and community effects of development at various steps in the process? Research focused on such questions could prove especially important to further understanding of industrial development processes. Although such research may be descriptive in nature, the application of analytical techniques will be required. Case studies may be an especially useful approach. Nationwide, there exist a number of areas where major forest-based economic growth has occurred (is occurring) in recent years. Locations that have several new plants could yield especially fruitful case studies that examine interactions between plants (e.g., complementary relationships concerning the supply of labor and raw material). Without question, case studies can be an important avenue for increasing knowledge about processes by which economic development occurs within forestry sectors.

Summary and Conclusions

Forest resource economists can make a substantial contribution to enhanced understanding of forestry's role in regional economic development (Webster and Chappelle 1987). They can do so by undertaking rigorous economic analyses and by communicating the results of their research. Of special concern to economics research should be improvements in the understanding of forestry's contributions to economic development, the means by which forestry can maximize economic development contributions, and the process by which economic development occurs within forestry sectors of various types.

Accurate measurement of forestry's several economic contributions is extremely important to realistic efforts concerning community and regional development. Forest economics research can provide accurate, well-founded estimates of economic contribution. By so doing, departures can be made from all-too-common patterns of "competitive exaggeration," patterns in which competing sectors attempt to prove that each is bigger and better than the other in the world of economic development. Competitive exaggeration obscures reasonable judgments about useful directions for development efforts. Accurate measurement with comparable methods from one sector to another can notably improve the basis on which such judgments are made.

Research enabling more effective efforts to maximize forestry's economic contributions can also contribute to a sense of direction in the world of economic development. The products of research can help identify specific sectors that offer realistic potential for growth. Such can be especially useful in a practical sense. For example, local community development organizations often exhibit excessive enthusiasm for development—enthusiasm which can dissipate development efforts on sectors that have little realistic hope for development. Such enthusiasm also encourages treatment of each sector and each development effort as wholly separate from others, thus undercutting possible synergistic effects. Research enabling the design of effective development programs can lead to achievement of a better and a more integrated focus for development.

Effective communication of the products of economic development research is needed if they are to contribute to practical affairs. Such communication can take many forms. The essence of them all involves analysts communicating directly and regularly with people in development organizations; communicating in simplified terms about major implications of well-executed analyses. Such involves working with an interpretive reducing-glass rather than an analytic microscope. Only then will the products of research further social interest in economic development via the forestry sector.

Literature Cited

Bell, E. F. 1977. Estimating effect of timber harvesting levels on employment in the western United States. Research Note INT-237. Intermountain Forest and Range Experiment Station. USDA Forest Service. Ogden, UT.

Beuter, J. H., K. N. Johnson and H. Lynn Sheurman. 1976. Timber for Oregon's tomorrow: An analysis of reasonably possible occurrences. Research Bulletin 19. Forest Research Laboratory. Oregon State University. Corvallis, OR.

Beuter, J. H. and Douglas C. Olson. 1980. Lakeview federal sustained yield unit—Fremont National Forest: A review 1974-1979. Department of Forest Management. Oregon State University. Corvallis, OR.

Bilek, Edward M. and Paul V. Ellefson. 1984. Employment in Minnesota's wood-based industry: A shift-share analysis. Staff Paper Series Number 46. Department of Forest Resources. University of Minnesota. St. Paul, MN.

Brodie, D., R. McManon and W. Gavelis. 1978. Oregon's forest resources: Their contribution in the state's economy. Resource Bulletin 23. School of Forestry. Oregon State University. Corvallis, OR.

Bromley, Daniel W., G. E. Blanch, and H. H. Stoevener. 1968. Effects of selected changes in federal land use on a rural economy. Bulletin 604. Agricultural Experiment Station. Oregon State University. Corvallis, OR.

Brown, N. and J. A. McMillan. 1977. Recreation program development impacts: A dynamic regional analysis. American Journal of Agricultural Economics 54(4):750-754.

Chappelle, D. E., R. C. Miley and R. Gustafson. 1982. Land use models. In: Guiding land use decisions: Planning and management for forests and recreation by D. W. Countryman and D. M. Sofranko (eds.). Johns Hopkins University Press. Baltimore, MD.

Chappelle, D. E., S. E. Heinen, L. M. James, K. M. Kittleson, and D. D. Olson. 1986. Economic impacts of Michigan forest industries: A partially survey-based input-output study. Research Report 472. Agricultural Experiment Station. Michigan State University. E. Lansing, MI.

Connaughton, K. P. and W. McKillop. 1979. Estimation of "small area" multipliers for the wood processing sector: An econometric approach. Forest Science 25(1):7-20.

Darr, D. R. and R. D. Fight. 1974. Douglas County, Oregon: Potential economic impacts of a changing timber resource base. Research Paper PNW-179. Pacific Northwest Forest and Range Experiment Station. USDA Forest Service. Portland, OR.

Diamond, J. D. and D. E. Chappelle. 1981. Application of an input-output model based on secondary data in local planning: The case of Manistee County. Research Report 409. Agricultural Experiment Station. Michigan State University. E. Lansing, MI.

Dutrow, G. F. 1972. Shift-share analysis of southern forest industry, 1958-1967. Forest Products Journal 22(12):10-14.

Flick, W. A., P. Trench, III and J. R. Bowers. 1980. Regional analysis of forest industries: Input-output methods. Forest Science 26(4): 548-560.

Hagenstein, P. R. 1964. The location decision for wood-using industries in the northern Appalachians. Research Paper NE-16. Northeastern Forest Experiment Station. USDA Forest Service. Broomall, PA.

Haynes, R. W. (ed.). 1983. Competition for National Forest timber: Effects on timber-dependent communities. General Technical Report PNW-148. Pacific Northwest Forest and Range Experiment Station. USDA Forest Service. Portland, OR.

Holland, I. I. and G. G. Judge. 1963. Estimated interregional flows of hardwood and softwood lumber. Journal of Forestry 61(7):488-497.

Holley, L. D. 1970. Location of the softwood plywood and lumber industries: A regional programming analysis. Land Economics 46(2): 127-137.

Hughes, J. 1970. Forestry in Itasca County's economy: An input-output analysis. Miscellaneous Report 95. Agricultural Experiment Station. University of Minnesota. St. Paul, MN.

Kaiser, H. F., Jr. 1968. Interindustry model of the U.S. forest products economy. Forest Products Journal 18(11):15-18.

Kaiser, H. F., Jr. 1969. Input-output analysis of the southern forest economy, 1963. Research Paper SO-43. Southern Forest Experiment Station. USDA Forest Service. New Orleans, LA.

Kaiser, H. F., Jr. and George F. Dutrow. 1971. Structure and changes in the southern forest economy, 1958-1967. Research Paper SO-71. Southern Forest Experiment Station. USDA Forest Service. New Orleans, LA.

Kaiser, H. F., Jr. 1972. Multiregional input-output model for forest resource analysis. Forest Science 18(1):46-53.

Kallio, M., D. P. Dykstra and C. S. Binkley (eds.). 1987. The global forest sector: An analytical perspective. John Wiley & Sons Publishers. London, UK.

Kalter, R. J. and W. B. Lord. 1968. Measurement of the impact of recreation investments on a local economy. American Journal of Agricultural Economics 50(2):243-256.

Keegan, C. E., III and P. E. Polzin. 1987. Trends in the wood and paper products industry: Their impact on the Pacific Northwest economy. Journal of Forestry 85(11):31-36.

Laurent, E. A. and J. C. Hite. 1972. Economic-ecologic linkages and regional growth: A case study. Land Economics 47(1):70-72 (Feb.).

Maki, W., C. H. Schallau and J. H. Beuter. 1968. Importance of timber-based employment to the economic base of the Douglas-fir region of Oregon, Washington, and northern California. Research Note PNW-76. Pacific Northwest Forest and Range Experiment Station. USDA Forest Service. Portland, OR.

Maki, W. and D. L. Schweitzer. 1973. Importance of timber-based employment to the Douglas-fir region, 1959 to 1971. Research Note PNW-196. Pacific Northwest Forest and Range Experiment Station. USDA Forest Service. Portland, OR.

Marcin, T. C. 1972. Projections of demand for housing by type of unit and region. Agriculture Handbook No. 428. U.S. Department of Agriculture. Washington, DC.

Schallau, C. H. 1980. Stages of growth theory and money flows from commercial banks in timber-dependent communities. Research Paper PNW-279. Pacific Northwest Forest and Range Experiment Station. USDA Forest Service. Portland, OR.

Schallau, C. H. and P. E. Polzin. 1983. Considering departures from current timber harvesting policies: Case studies of four communities in the Pacific Northwest. Research Paper PNW-306. Pacific Northwest Forest and Range Experiment Station. USDA Forest Service. Portland, OR.

Schallau, C. H., W. Maki and J. Beuter. 1969. Economic impact projections for alternative levels of timber production in the Douglas-fir region. Annals of Regional Science 3(1):96-106.

Schallau, C. H. and W. R. Maki. 1986. Economic impacts of interregional competition in the forest products industry during the 1970's: The South and the Pacific Northwest. Research Paper PNW-350. Pacific Northwest Research Station. USDA Forest Service. Portland, OR.

Schooley, D. C. and J. G. Jones. 1983. An input-output model for measuring impacts of the Oklahoma forest products industry on the state's economy. Res. Rep. P-838. Agricultural Experiment Station. Oklahoma State University. Stillwater, OK.

Sutherland, R. J. 1981. A regional approach to estimating recreation benefits of improved water quality. Journal of Environmental Economics and Management 9(3):229-247.

Waggener, T. R. 1977. Community stability as a forest management objective. Journal of Forestry 75(11):710-714.

Wall, B. and D. Oswald. 1975. A technique and relationships for projections of employment in the Pacific coast forest products industries. Research Paper PNW-189. Pacific Northwest Forest and Range Experiment Station. USDA Forest Service. Portland, OR.

Webster, H. H. and D. E. Chappelle. 1987. Can forest resource economists help solve societal problems? Renewable Resources Journal 5(4):13-18.

Youmans, R. C., D. R. Darr, R. Fight, et al. 1974. Douglas County, Oregon: Structure of a timber county economy. Circular 645. Agricultural Experiment Station. Oregon State University. Corvallis, OR.

18

TAXATION OF FOREST PRODUCTS AND FOREST RESOURCES

W. David Klemperer

Forest tax research has focused on four major categories of taxes applied to the use and management of forest land, namely local government taxes (property taxes and taxes in lieu thereof), federal and state income taxes, death taxes, and special severance taxes. Since colonial times, property taxes have been an important source of revenue for local governments. As such, the earliest writings about forest taxation dealt with the impacts of the property tax on forests (Fairchild 1908). Fairchild's research (1908, 1935) maintained that the annual property tax on full market value (the unmodified property tax) was inherently biased against investments with long payoff periods (e.g., forestry). His work provided the theoretical foundation for modified property taxes on forests, wherein assessments or tax rates are reduced or another form of tax imposed. Trestrail (1969), Lindholm (1973), and Pasour and Holley (1976) found fault with Fairchild's analyses and maintained that the property tax was not biased against forestry. Their findings were based on the assumption that property taxes are fully shifted into higher stumpage prices. However, Klemperer (1977) maintained that given the degree of competition in wood production nationally and worldwide, property taxes are more likely to be shifted back into lower land values. Refining Fairchild's analysis, Klemperer (1974) found that property taxes are not always biased against forestry, but are likely to be—the bias would be against land uses with high establishment costs and long payoff periods. Boyd (1986) has summarized these arguments. Other forest property tax studies have examined the land use implications of different assumptions about tax shifting (Stier and Chang 1983) and the theoretical impact of various taxes upon optimal rotation age (Chang 1982).

Several states have repealed the unmodified property tax on forests and have applied local property tax rates to forest productivity values based on capitalized income from a hypothetical sustained yield forest. For any given land quality, such a tax (productivity tax) is levied on the same taxable value regardless of timber stocking. Williams and Canham (1972) point out the problems in determining taxable productivity values, and Klemperer (1983) notes that the typical productivity valuation formula—if correctly applied—could result in a tax more burdensome to forestry than the unmodified ad valorem tax. Other states have attempted to solve forest property tax problems by levying yield taxes as a percent of harvest value and exempting timber from all other taxes (Clements et al. 1986). Marquis (1952) provided one of the first summaries of yield tax theory and practice. In a forestry region with many timber ages and roughly equal harvests, Klemperer (1976) noted that for any given amount of regional tax revenue raised, the yield tax—among all tax types—is likely to be the least discouraging to forest land use. Rideout and Hof (1986) refined this analysis to include tax impacts upon optimal rotation ages, and Gamponia and Mendelsohn (1987) provided an alternate present value view of yield tax equivalency which was no less burdensome than an ad valorem tax but that raised more tax revenue.

State income taxes are usually levied on timber income which is the same or similar to that defined under federal income tax law (McGee et al. 1984). Before 1987, capital gains income from sale of assets (including timber) held longer than 6 months to a year received a lower federal income tax rate than other income. Some states also gave capital gains preferential treatment. Sunley's (1972) analyses criticized this preferential treatment of timber income. The U.S. General Accounting Office (1981) decried the scarcity of analytical work providing theoretical justification for preferential treatment of timber income and found no firm evidence of resulting conservation or reforestation benefits. Similarly, Chang (1983) found little evidence that lower capital gains tax rates favored reforestation among nonindustrial private forest owners. Using a general equilibrium approach, Boyd and Daniels (1985) estimated that in 1979 the national welfare loss due to preferential capital gains tax treatment of timber income was in the range of about $240 to $360 million.

A possible justification for reduced capital gains tax rates is that inflation makes any given tax rate relatively more burdensome on capital gains than on ordinary income. However, a fixed tax rate reduction is a highly inaccurate way to correct for this inflationary bias. For a wide range of conditions, Klemperer and O'Neil (1987) showed how indexing the tax-deductible basis could exactly compensate for the negative effect of inflation on income taxes. Effective January 1987, the Tax Reform Act of 1986 eliminated preferential tax treatment of all capital gains and continued to disallow indexing the basis for inflation. While capital gains and ordinary income now receive the same tax treatment, the arguments for preferential rates on capital gains, especially timber, continue to be heard. On the other hand, efficiency rationales for uniform taxation of all income, timber included, can be made—especially if indexing for inflation were allowed (Klemperer 1987a). Recently, several analyses have examined impacts of the 1986 Tax Reform Act and similar proposals upon selected forestry investments (Leuschner and Haney 1985, Siegel 1987, Dangerfield and Gunter 1986, Guertin and Rideout 1987).

Since the 1950s, rising timberland values caused concern that state and federal death taxes could have negative impacts on continuity of management on non-corporate forests. Most research in this area has focused on impacts of death taxes on selected forest ownerships and methods by which owners can minimize such taxes (Sutherland and Tedder 1979, Olson et al. 1981, Howard 1985).

In light of current information and recent research activities, several strategic directions seem promising for future research on the economics of forest taxation:

- Develop improved focus and understanding of forest taxation goals.
- Develop more effective local government tax programs focused on forestry.
- Develop more effective state and federal income tax programs focused on forestry.
- Develop improved understanding of tax shifting as germane to forestry.
- Develop improved understanding of death taxes as applied to forestry.
- Develop effective methods of appraising forest components for tax purposes.
- Determine forestry implications of newly proposed taxes and risk considerations.

Develop Improved Focus and Understanding of Forest Taxation Goals

Forest tax policy is typically altered with the intent of achieving a new goal or of achieving an existing goal more efficiently. Unfortunately, most forest tax goals are poorly defined or are conflicting in purpose (Klemperer 1980). Additional research is needed to define consistent and rational forest tax goals—goals beyond that of raising government revenue.

Assess appropriateness of tax policies aimed at increasing forest outputs. A frequently voiced forest taxation goal is to encourage production of forest outputs. But such a goal is difficult to implement. The most encouraging tax for forestry would be no tax at all—generally a politically unpalatable approach. We are left with the question: How much should existing tax structures be altered (e.g., property taxes or income taxes) in order to favor the production of additional forest outputs? A policy of reducing taxes to encourage the production of timber implies that the market produces suboptimal quantities of wood. Here, suboptimal means that more investment in wood production and less in some other productive activity would increase net benefits to society. Research should more critically examine this premise which underlies a number of tax policies. Does the free market indeed produce suboptimal quantities of wood in the U.S.? And if so, is tax policy the most effective means to increase wood output? Are tax subsidies effective at all in boosting wood supplies? If tax subsidies are aimed at increasing nonwood outputs (e.g., scenic beauty, soil and water protection, recreation opportunities), research should critically evaluate the effectiveness of such subsidies. If low investment forestry (e.g., natural seeding and revegetation to hardwoods or brush) brings the same nonmarket outputs as intensive timber management investment, then why offer the subsidy? More broadly, is the market producing suboptimal quantities of nonwood outputs from private forest lands, and if so, what policies (not limited to tax incentives) could most effectively remedy this market failure?

Assess the neutrality of taxes focused on forestry. If the market system provides optimal quantities of forest outputs, many economists suggest imposing neutral taxes that neither encourage nor discourage forestry investment relative to other ventures. This is an economic efficiency argument: If the market without taxes tends to allocate land, labor, and capital to their highest returns, do not impose a nonneutral tax which would distort this efficient allocation. Such an approach suggests research to more clearly identify nonneutralities in tax structures and to recommend changes to improve neutrality. Moreover, where markets are not maximizing social welfare, could selectively nonneutral taxes correct the problem or would direct regulation be more efficient? How important are such problems in forestry? Much research on neutrality has been accomplished in the forestry area and elsewhere; the analyses could, however, be refined.

Develop More Effective Local Government Tax Programs Focused on Forestry

Evaluate equity guides for taxation of forest property. In rejecting the unmodified property tax on forests, taxing authorities have abandoned the traditional equity guide of taxing the same percent of market value for all classes of property. However, there is little agreement on what equity guide should replace the traditional guide when property taxes are modified or replaced by yield taxes or productivity taxes. More research is needed to assist policy makers in deciding what is a fair level of any given type of timber tax relative to ad valorem taxes borne by other properties.

Determine effective means of transition from one tax system to another. When changing from one forest tax system to another, tax-shifts between taxing districts and between forest owners are likely to occur. In addition, uneven revenue flows from yield taxes are often a problem for governments. What are the best means of gradually phasing from an existing tax to a new tax and of smoothing revenue-flows from a yield tax? (An approach is suggested by Klemperer and Clements 1988.) To what extent are forest tax revenue-smoothing problems likely to subside as a result of greater public reliance on financing education with statewide levies that are distributed to counties according to need? The 1970s saw several court cases involving such topics. In California, for example (Serrano versus Priest, 1971. Sup., 96 Cal. Rptr. 601), plaintiffs charged that using the property tax to fund education creates a situation where the quality of a child's education depends upon the wealth of the district wherein the child resides. The argument was that such a condition discriminated against low-income districts and violated the Fourteenth Amendment's equal protection clause. While the plaintiffs failed to prevail, the controversy remains alive; its ultimate resolution could have serious implications for forest property taxes. The issue and its implications warrant further research.

Assess the effectiveness of site value taxes. Touted by Henry George (1879) as the ideal single tax, the site value tax is levied annually as a percent of market value of all land, excluding the value of improvements. It is a property tax on unimproved land value, exempting all improvements from taxation. In theory, the site value tax is neutral with respect to land use decisions and, in a well-functioning market, will foster efficient resource allocation. In its pure form, the tax has not been widely used in the U.S. It has, however, been successfully implemented in portions of Australia and New Zealand (Heilburn 1966). Although the forest productivity tax may appear to be a site value tax, it is not, since taxable productivity value formulas typically include some timber value with the land. Some states specifically exempt timber from property taxation, making their forest property tax a site value tax. However, a true uniform site value tax exempts all improvements such as buildings, roads, and fences for all land uses. Exempting timber but not other improvements, as suggested by Cartwright and Dowdle (1979), might be seen as an inefficient bias by some. If the general property tax were repealed, and if site value taxation were instituted nationwide or regionwide, what would be the land use implications? Would a more or less efficient investment pattern result? Would the tax base be adequate to meet government budgets? In the absence of comparable bare land sales, how could unimproved land beneath improvements be accurately valued? Theoretical arguments in favor of site value taxation and against the general property tax suggest the need for additional research.

Assess opportunity for a property tax on land plus establishment cost. In place of a property tax on land and timber, Klemperer (1982) has suggested a tax on land plus establishment cost (including the value of improvements such as roads, drainage, site preparation, and afforestation). The tax on establishment value could be either an annual property tax or a yield tax with the same rotation-start present value. Although some theoretical evaluations of such a tax have been made, further analyses using empirical data would be useful.

Assess use valuation under the property tax. In order to prevent property taxation on unwarranted development values, or to help maintain forest use where development values do exist, many states have allowed current use valuation of forest lands (Hickman 1983). More research is needed to determine how effective such laws are in maintaining forest land use (if indeed such is beneficial), and what alternative courses of action might be more desirable. Also, alternative tax rollback provisions implemented when land use is changed should be more fully evaluated.

Identify and evaluate the cost of administering local government taxes on forests. Few studies show comparative costs of administering different types of local government taxes on forests—costs to government as well as the landowner. Such information from numerous states would be useful to governments interested in judging the administrative efficiency of various tax programs. Research could also be undertaken to compare tax receipts net of costs, given either government valuations or owner-reported taxable values with government spot checks. Such research would be especially applicable to yield taxes.

Develop More Effective State and Federal Income Tax Programs Focused on Forestry

Now that both capital gains and ordinary income are taxed at the same rate, a stronger case can be made for basis-indexing to remove the negative impact of inflation on income taxes. Research could examine the potential impacts of basis-indexing on a wide range of forestry investments, given different inflation rates. The tax revenue losses as a result of indexing should also be examined under different scenarios. In addition, what income tax rate increases would be required to offset these losses to governments? Although of questionable validity, the notion of indexing the gain rather than the basis should also be evaluated. Such has been suggested by some timber interests.

When mature forests are offered for sale, the asking price for a forest is likely to exceed bid price, after consideration of income taxes (shown mathematically by

Kovenock and Rothschild [1983]). This lock-in effect of an income tax has been theoretically verified for loblolly pine and Douglas fir under several scenarios (Klemperer 1988). However, more research is needed to empirically document the extent to which the income tax discourages sale of immature timber for postponed harvest and causes market inefficiencies. If income taxes were levied annually on increased forest value, regardless of sale, no difference would exist between bid and ask because both buyer and seller would face identical future taxes. Moreover, some suggest that such a tax on accrued income, although difficult to administer, would be more neutral than current approaches involving taxation of realized income (Chisholm 1975, Diamond 1975, and Kovenock and Rothschild 1983). More research is needed to evaluate accrued income taxes and their potential impacts upon forestry.

Appropriate treatment of costs under income taxation continues to be an area of disagreement. Should costs such as reforestation, site preparation, annual costs, fertilization, and timber stand improvement be expensed or capitalized? Additional theoretical work is needed to evaluate these options, both from an efficiency perspective and from an economy-wide neutrality perspective. Also a subject for continuing disagreement is the return to preferential income tax rates for long-term capital gains. Additional research documenting the advantages or disadvantages of such a policy from an economy-wide, general equilibrium view would be especially useful to policy makers. Empirical analyses could examine impacts on hypothetical forest investments.

Develop Improved Understanding of Tax Shifting as Germane to Forestry

Analyses of the theoretical impacts of any proposed or existing forest tax are highly sensitive to assumptions about tax shifting (i.e., does a tax increase result in increased timber prices, decreased forest values, and decreased interest rates used by forest buyers to discount future net revenues, or does a combination of shifting occur?). For most sales taxes or value-added taxes, laws permit sellers to increase product prices by the amount of the tax—forward shifting is clearly the case. For property taxes in the agricultural sector, the opposite has been found to be true. Pasour (1973) found that such taxes in agriculture are shifted largely into lower farmland values. For rental dwellings, property taxes appear to be substantially shifted forward to tenants (Hyman and Pasour 1973).

Very little research to evaluate the impact of selected forest taxes on timber prices and land values has been undertaken—even though statistical techniques for doing so are readily available. This lack of research attention may be due to inadequate data. Nevertheless, research is needed, since sharply different conclusions are often reached under alternative shifting assumptions. Pasour and Holley (1976) found that an annual property tax would pose no problem to forestry when the tax is shifted forward into higher stumpage prices. However, when the property tax is shifted into lower land values, it appears to be biased against capital intensive land uses such as forestry (Klemperer 1974).

Opportunities for empirically testing the impacts of selected taxes on forest land values may exist when similar forest regions bear different tax burdens.

Statistical analyses might reveal whether differences in forest land values or stumpage prices in such regions could be attributed to tax differences. Either time series or cross-sectional data, or a combination, could be used for such analyses.

Develop Improved Understanding of Death Taxes as Applied to Forestry

Whenever death tax laws are changed, numerous reports are prepared to advise on planning strategies, ownership forms, and transfer of property rights to legally minimize state and federal death taxes. Such topics are certainly of significant concern to private noncorporate owners. However, the following questions are rarely addressed in a forestry context: If one of the objectives of death taxes is to reduce concentrations of family wealth and to redistribute income, how successful have such taxes been in doing so? To what degree have individual forest estates been fragmented, and, when this occurs, has the result of fragmentation been a corresponding increase in the concentration of wealth elsewhere? Do forest land sales stimulated by death taxes lead to greater corporate ownership of forests? Would such a transfer tend to bring more or less efficient use of forest resources? Broadly viewed, do death taxes lead to smaller or larger returns—both private and public—from the nation's forests, regardless of the distribution of such returns? Such questions have efficiency as well as equity implications. Most certainly the issues raised by such questions are deserving of additional research.

Research should also be undertaken to identify and evaluate the effectiveness of strategies whereby landowners can borrow money to pay death taxes. Forests earning competitive rates of return should be in a position to pay interest on substantial loans. If forest income is competitive but deferred, loans with delayed payments could be explored (some already exist through deferred tax payment plans at the federal level). If current or projected forest income is inadequate to retire major loans to pay death taxes, researchers should determine whether governments should shelter such enterprises by softening the death tax burden. Again the efficiency question: Do death taxes stimulate more or less efficient use of forest capital?

The federal government and some states allow forest land valuation based on current use rather than market value for death tax purposes; thus substantially reducing taxes where market values exceed typical forest land values. More research should be undertaken to determine the long run implications of such policies (such are similar to current use valuation under the property tax). If indeed the potential use for the property could bring more income than forestry, why should government policies foster the less efficient use? On the other hand, if the government's estimate of true market value exceeds the actual potential, suggested is the need for research on methods to accurately estimate market values when current use is below the best market potential. However, where a current forest use brings nondollar benefits or a higher use yields nondollar costs such as pollution, justification exists to foster forest use. Under what conditions and to what extent are such arguments valid, and, when they are, would broad tax policies be the most effective means to correct the market failure? Would such failures be more efficiently corrected by zoning, land use regulations, or public purchase of selected sites? Issues raised by such questions are certainly deserving of research.

Develop Effective Methods of Appraising
Forest Components for Tax Purposes

Market values of standing timber and land are required for death tax and property tax purposes. However, sufficient market transactions seldom exist to establish such values separately for land, for individual age classes of timber, or in aggregate for a single property. Methods for determining values of bare land have been suggested. Some of these methods include abstraction method and regression analysis (Bare and McKetta 1977), the relative market value method (Rickard 1976), and a simulation approach (Klemperer 1979, 1981). However, such methods have not been applied and subsequently compared in the same forested region. Particularly useful would be comparisons in a region where bare land market values were known. An assessment could then be made of which valuation technique most closely predicts actual market values.

Valuation of standing timber also presents problems for forest taxation. When lacking evidence of standing timber sales (not scheduled for immediate cutting), it is not appropriate to value large tracts of standing timber by multiplying volume times current stumpage prices. Depending upon the timber's growth rate and projected harvest dates, market value may be less than, equal to, or greater than stumpage values of currently harvested timber (Klemperer 1985, 1987b). To value timber not scheduled for immediate cutting, a capitalized income approach is usually suggested. However, arguments abound on the appropriate capitalization rate and the assumed rate of harvest; the computed value is highly sensitive to such inputs. To alleviate these problems, research should explore opportunities to derive capitalization rates from actual timberland sales and, subsequently, apply such rates to discount projected harvest incomes based on harvest patterns typical for a subject property. Analogous approaches have been successfully used to value commercial real estate for which comparable sales are unavailable. The market-derived capitalization rate is the interest rate which discounts projected net income to exactly equal the purchase price for sold properties. A key point is that the consistent projected price and cost assumptions used in deriving the capitalization rate should also be applied to the property to be valued. This approach, which has not been developed for timber, is mentioned briefly in Klemperer (1985). As yet unsolved are special problems which exist in separately valuing land and timber when considering even-aged versus uneven-aged forests.

Determine Forestry Implications of Newly
Proposed Taxes and Risk Considerations

Assess potential impacts of value-added tax applied to forestry. Over the past 20 years, Congressional interest has been repeatedly focused on a national value-added tax (VAT). Such a tax has never been enacted into law. Recently, however, the value-added tax has been suggested as a partial solution to federal deficit problems, a stimulus to savings (the VAT is consumption-based), and to improve balance of trade (the VAT can be excused on exports). Although levied as a percent of the value added to commodities at each stage in production, the value-added tax is much like a national sales tax which increases product prices by the VAT percent and is borne by consumers.

Given renewed interest in a value-added tax (popular in Europe and several other foreign countries), research on the potential forestry effects of such a tax would be timely. One study involving value-added tax is underway at Virginia Polytechnic Institute (Klemperer and O'Neil 1982). However, numerous questions remain unanswered. Under various assumptions, what would be the likely effects of different national value-added tax rates on forest products consumption, stumpage markets, international trade in forest products, land use decisions, and efficiency of resource allocation throughout the economy? During value-added tax deliberations, such prior knowledge would be crucial to informed decisionmaking.

In addition to a value-added tax approach, there are opportunities for developing countless combinations of existing and new forms of forest taxes. For each alternative, questions need to be answered about effects on forest investment decisions, direction of tax shifting, neutrality and national welfare, administrative feasibility, and equity relative to other property classes. Given repeated criticisms of existing forest taxes, more research on alternative taxes is certainly warranted.

Assess risk as a basis for preferential tax treatment. Boyd and Daniels (1986) suggest evaluating the notion that forestry investments involve unusually high risks (e.g., fire, insect, and disease) and therefore deserve preferential tax treatment. Does forestry tend to be more risky than average industrial investments, and what effect have past preferential tax policies in general had upon risk taking? Additional research on risk and its implications for forest taxation is warranted.

Summary and Conclusions

Research addressing forest tax policies and their implementation deserves more attention in the years ahead. Important focal points for such research include enhanced understanding of forest taxation goals, tax shifting phenomena, value appraisal methods, death taxes, and the variety of forest tax programs implemented by governments at local, state, and federal levels.

In all aspects of forest tax research, far more work is needed in estimating the long-range incidence and welfare implications of tax alternatives. Less emphasis should be placed on descriptive studies and on analyses of selected tax impacts imposed on only the forestry sector. Such approaches have dominated forest tax research too long. More emphasis should be placed on quantitative general equilibrium approaches (Boyd and Daniels 1985) which do not examine forestry in a vacuum but consider interaction between sectors of the economy.

Forest tax policy makers need to know not only the effects of tax alternatives upon forestry or a particular forest sector but also the aggregate economy-wide effects. In many cases, the negative (or positive) effects of a tax-change in one sector are accompanied by positive (or negative) effects elsewhere. Researchers need to examine and quantify *net* effects in order to know whether society as a whole gains or loses from a given tax policy. Research that considers forestry investments alone or that pursues taxes which promote forestry will tell only half the story.

Literature Cited

Bare, B. B. and C. W. McKetta. 1977. Forest land valuation in Washington State: Comparison of abstraction and regression. The Real Estate Appraiser 43(1):25-31.

Boyd, R. 1986. Forest taxation: Current issues and future research. Research Note SE-33. Southeastern Forest Experiment Station. USDA Forest Service. Asheville, NC.

Boyd, R. and B. J. Daniels. 1985. Capital gains treatment of timber income: Incidence and welfare implications. Land Economics 61(4):354-362.

Cartwright, P. W. and B. Dowdle. 1979. An economics analysis of forest taxation in Washington State. College of Forest Resources. University of Washington. Seattle, WA.

Chang, S. J. 1982. An economic analysis of forest taxation's impact on optimal rotation age. Land Economics 58:310-324.

Chang, S. J. 1983. Reforestation by nonindustrial private landowners: Does the capital gains tax matter? In: Nonindustrial private forests: A review of economic and policy studies by J. R. Royer and C. D. Risbrudt (eds.). School of Forestry and Environmental Studies. Duke University. Durham, NC.

Chang, A. H. 1975. Income taxes and investment decisions: The long-life appreciating asset case. Economic Inquiry 13:565-578.

Clements, S. E., W. D. Klemperer, H. L. Haney, Jr. and W. C. Siegel. 1986. Current status of timber yield and severance taxes in the United States. Forest Products Journal 36(6):31-35.

Dangerfield, C. W. and J. E. Gunter. 1986. How the tax reform act will affect tree farmers. Forest Farmer. November-December:14-15.

Diamond, P. A. 1975. Inflation and the comprehensive tax base. Journal of Public Economics 4:227-244.

Fairchild, F. R. 1908. The taxation of timberlands in the United States. In: Proceedings of second conference of National Tax Association. National Tax Association. Washington, DC.

Fairchild, F. R. 1935. Forest taxation in the United States. Misc. Publication 218. U. S. Department of Agriculture. Washington, DC.

Gamponia, V. and R. Mendelsohn. 1987. The economic efficiency of forest taxes. Forest Science 33(2):367-377.

General Accounting Office. 1981. New means of analysis required for policy decisions affecting private forestry sector. EMD-81-18. U. S. Congress. Washington, DC.

George, H. 1879. Progress and poverty. Robert Schalkenback Foundation (1954). New York, NY.

Guertin, D. S. and D. B. Rideout. 1987. 1986 Tax Reform Act and forest investments. Journal of Forestry 85(9):29-32.

Heilbrun, J. 1966. Real estate taxes and urban housing. Columbia University Press. New York, NY.

Hickman, C. A. 1983. Use value assessment of forest lands in the United States. International Real Estate Journal 1(4):62-70.

Howard, T. E. 1985. Estate planning for nonindustrial forest owners. Land Economics 61(4):363-371.

Hyman, D. N. and E. C. Pasour, Jr. 1973. Property tax differentials and residential rents in North Carolina. National Tax Journal 26(2):303-307.

Klemperer, W. D. 1974. Forests and the property tax: A reexamination. National Tax Journal 27:645-651.

Klemperer, W. D. 1976. Impacts of tax alternatives on forest values and investment. Land Economics 52:135-157.

Klemperer, W. D. 1977. An economic analysis of the case against ad valorem property taxation in forestry: A comment. National Tax Journal 30(4):469.

Klemperer, W. D. 1979. Segregating land and timber values from sales of uneven-aged forests. The Appraisal Journal 47(1):16-25.

Klemperer, W. D. 1980. Equity considerations in designing substitutes for unmodified forest property taxes. Proceedings of symposium on state taxation of forest and land resources. Lincoln Institute of Land Policy. Cambridge, MA.

Klemperer, W. D. 1981. Segregating land values from sales of forested properties under even-aged management. Forest Science 27(2):305-315.

Klemperer, W. D. 1982. An analysis of selected property tax exemptions for timber. Land Economics 58(3):293-309.

Klemperer, W. D. and C. J. O'Neil. 1982. Potential implications of a value added tax for forestry. Bulletin FWS-4-82. Proceedings of forest taxation symposium. Department of Forestry. Virginia Polytechnic Institute and State University. Blacksburg, VA.

Klemperer, W. D. 1983. Ambiguities and pitfalls in forest productivity taxation. Journal of Forestry 81(1):16-19.

Klemperer, W. D. 1985. Effects of inflation and debt on old growth timber appraisal. The Appraisal Journal 53(1):90-114.

Klemperer, W. D. 1987a. Income tax reform and the forest economy. Tax Notes. July 6.

Klemperer, W. D. 1987b. Deviations between stumpage value and the value of young timber scheduled for future harvest. The Appraisal Journal 55(4):535-547.

Klemperer, W. D. and C. J. O'Neil. 1987. Effects of an inflation adjusted basis on asset-values after capital gains taxes. Land Economics 63(4):386-395.

Klemperer, W. D. 1988. The theoretical excess of ask over bid prices for immature timber after income taxes. Unpublished manuscript. Department of Forestry. Virginia Polytechnic Institute and State University. Blacksburg, VA.

Klemperer, W. D. and S. E. Clements. 1988. Smoothing revenue distributions from a forest yield tax. Journal of State Taxation. January (in press).

Kovenock, D. J. and M. Rothschild. 1983. Capital gains taxation in an economy with an "Austrian sector." Journal of Public Economics 22(2):215-256.

Leuschner, W. A. and H. L. Haney, Jr. 1985. Southern forestry and a modified flat tax. Southern Journal of Applied Forestry 9(10):201-205.

Lindholm, R. W. 1973. Taxation of timber resources to maximize equity and wood fiber production: An Oregon case study. Publication No. 5. Bureau of Business and Economic Research. University of Oregon. Eugene, OR.

Marquis, R. W. 1952. Forest yield taxes. Circular No. 899. U.S. Department of Agriculture. Washington, DC.

McGee, G. T., H. L. Haney, Jr., and W. C Siegel. 1984. State income tax implications for nonindustrial private forestry in the United States. Forest Products Journal 34(3):27-36.

Olson, S. C., H. L. Haney, Jr. and W. C. Siegel. 1981. State death tax implications for private nonindustrial forestry. Forest Products Journal 31(7):28-38.

Pasour, E. C., Jr. 1973. Real property taxes and farm real estate values in North Carolina. Economic Research Report No. 25. North Carolina State University. Raleigh, NC.

Pasour, E. C., Jr. and D. L. Holley. 1976. An economic analysis of the case against ad valorem property taxation in forestry. National Taxation Journal 29:155-164.

Rickard, W. 1976. A consideration of forest land valuation alternatives. Report to the Washington Forest Protection Association. Tacoma, WA.

Rideout, D. and J. G. Hof. 1986. A reevaluation of the site burden concept when forest land value is maximized. Forest Science 32(2):511-516.

Siegel, W. C. 1987. Implications of the 1986 Federal Tax Reform Act for forestry. National Woodlands. March-April:10-19.

Stier, J. C. and S. J. Chang. 1983. Land use implications of the ad valorem property tax: The role of tax incidence. Forest Science 26(Dec.):702-12.

Sunley, E. M. 1972. The federal tax subsidy of the timber industry. In: The Economics of Federal Subsidy Programs. Joint Economic Committee. U.S. Congress. Washington, DC.

Sutherland, C. F. and P. L. Tedder. 1979. Impacts of federal estate taxation on investments in forestry. Land Economics 55(4):510-20.

Trestrail, R. W. 1969. Forests and the property tax: Unsound accepted theory. National Tax Journal 22:347-356.

Williams, E. T. and H. O. Canham. 1972. The productivity concept in forest taxation. Assessor's Journal 7:29-51.

19

DISTRIBUTION AND MARKETING OF FOREST RESOURCE PRODUCTS

William G. Luppold and Gilbert P. Dempsey

Forests are one of the nation's most valuable natural resources. They are extensive, renewable, and produce high-quality, high-valued products that are increasingly demanded both at home and abroad. As such, U.S. forest products are traded in a competitive global market economy. The magnitude of domestic and foreign trade in forest products, whether these products are traded in the form of logs and lumber, intermediate products, or consumer goods, is partially determined by the creativity and efficiency with which they are conceived, produced, and marketed. The form in which U.S. forest products are traded also reflects their domestic value added (i.e., the goods and services added by U.S. labor and capital to the value of the product prior to its exportation) and, thus, the level of product returns for reinvestment in the economy. And since the markets for wood products are ever-changing in a competitive environment, the functions of time, place, and price are also critical on a current basis.

Analyses of forest products markets, both domestic and foreign, are key ingredients to increasing the efficiency and competitiveness of the U.S. forest products. The latter is composed of a complex set of interrelated markets that resist easy analytical access by economists, policymakers, and the industry. Still, notable advances have been made by several researchers who have developed successful analyses of segments of the forest products market. For example, early work by McKillop (1967) successfully applied econometric techniques to the wood products market to estimate the structural parameters of aggregate demand and supply relationships, demand and supply elasticities, influences determining consumption and price levels, and to forecast consumption and price levels in the United States. Subsequent models permitted periodic assessments of long-term supply and demand for forest resources (Adams and Haynes 1980), including not only timber consumption and production but also equilibrium prices and policy simulations. Most recently, Buongiorno (1981) and Gilless and Buongiorno (1987) succeeded in the development of a spatial equilibrium model of the North American pulp and paper industry that provides long-range projections of production, consumption, imports, exports, equilibrium prices, and fiber inputs. Results of the latter's application provides a solid indication of the role that an industry will play in the future of the U.S. forest economy.

A major challenge to meeting national interests in industrial forestry goals is the organized, collaborative, and structured development of technical and socio-economic information pursuant to the markets that affect the present and future production, distribution, and consumption of forest products. Meeting this challenge will require national commitment, realization of the positive contribution of research, additional resources, and expediting the necessary actions to compete intellectually, technologically, and economically in the world market. The alternative is to be out-maneuvered by other economies that place heavy emphasis on the development of production technology and market information, strategies, and tactics to increase their share of current markets for wood products and to develop new markets.

The aforementioned authors and others have mainly used econometric, simulation, and mathematical programming to answer policy questions. Many other scientists have combined approaches to also make valuable contributions to the understanding of forest-based markets and their interactions. However, due to the scarcity of research investments in the past, there are several important questions that have not been adequately explored. Future research concerning the marketing of forest-based products should address the following strategic concerns:

- Develop improved understanding of consumer and industrial demand for forest products.
- Acquire enhanced understanding of production costs and competitiveness of forest products industries.
- Develop improved understanding of market operations and trade in forest products.
- Develop effective means of producing and marketing nonwood-based forest products.

Develop Improved Understanding of Consumer and Industrial Demand for Forest Products

Consumer demand drives a capitalistic economy. Consumers' actions ultimately determine what will be produced and how much will be produced. Therefore, any estimation of future use of forestry resources and research into new technology to better utilize such resources should first consider the ultimate user. Among the major considerations are examining the impact of changes in family composition, income, and lifestyle on demand for forest products; assessing the impact of consumer sovereignty on forest products demand; and developing an understanding of worldwide consumer demand for wood products.

Examine the impact of changes in family composition, income, and lifestyle on the demand for forest products. When forecasting future demands for forest products and resources, forest economists may dwell on certain forestry aspects (e.g., growth and yield) and make broad-based assumptions about the impact of changing population and income on demand. Such assumptions may be totally valid when looking at forest products in the aggregate. However, because social and cultural characteristics of society are constantly changing, broad-based assumptions based on past behavior may be inappropriate for predicting future demand for specific types of forest products. For example, between 1970 and 1985, family size decreased by 10 percent; the proportion of households composed of two adults and children decreased from 40 percent to 28 percent of all households. While real income for married couples has on the average increased, income of families where the wife does not work has actually decreased. Such changes probably have had a large impact on price and income elasticity of forest products and products made from forest products. As a result, use factors that relate wood material use per unit of output and material use consumed per unit of population are changing.

There are actually two forces acting simultaneously that could be shifting aggregate demand curves for forest products: (1) the changing proportions of subgroups within the population, and (2) changes in the utility functions within subgroups. Such changes will affect own price, cross price, and income elasticity of forest products. The impact of social and economic changes on forest products demand may be implicit by changing the use factor and wood use to population multipliers. Critical to the development of national policy as well as to the forest products industry's future marketing strategies is a comprehensive understanding of

- The effect of population characteristics and attitudes on the demand for particular forest products and for forest products in general.
- The effect of industrial and organizational buyer behavior on the demand for various types of forest products.

Assess the impact of consumer sovereignty on forest products demand. Another factor that will impact future demand for wood products is the way consumers perceive wood products. Such perceptions can be the result of primal factors, cultural or traditional habits, and manipulation through advertising. Given limited exposure, the perceptions caused by cultural and traditional factors tend to remain constant for a given subgroup of a population. However, perceptions manipulated through advertising are related to the degree of consumer sovereignty for a group of products and the size of the added budget. The impact of advertising on demand for wood products has never been fully analyzed. Three questions germane to this issue are

- Does advertising affect the demand for structural products (e.g., oriented strand board), generic consumer products (e.g., writing paper), and big ticket consumer goods (e.g., furniture)?
- Does advertisement expand markets for a particular set of wood products or does it just cause change in the market shares for individual products?
- What are the effects of advertising and promotional efforts on demand elasticities as well as increases in demand?

Furthermore, consumers' perceptions toward a product can be changed by either direct advertising to consumers or by indirect advertising to individuals who advise consumers. In the forest products industries, these advisors include interior designers, architects, contractors, engineers, magazine editors, and others. Three important questions in regard to consumer advisors are

- To what degree do the consumer advisors influence consumers' decisions?
- To what degree are consumer advisors influenced by promotional efforts and product propaganda?
- What measure (including definition) should be used to determine the effectiveness of investments in various promotional efforts focused on individual markets?

Develop an understanding of worldwide consumer demand for wood products. While the United States is and will probably remain the major market for domestically produced wood products, the international markets for U.S. wood products have grown steadily over the last 20 years. Between 1970 and 1985, softwood products exports increased by 27 percent while hardwood exports increased by 76 percent. Softwood exports are primarily construction products such as plywood and framing lumber. Hardwood exports tend to be higher grade furniture lumber, face veneer, and veneer logs. Although there are distinct differences in softwood products and hardwood products exports, both are basically dependent on family formation, construction activity, and income levels in developed and developing economies. However, the need for dwellings relative to the need for furniture is much greater in economically developing countries than in developed countries. Therefore, critical to long- and short-term demand projections are

- Comparative studies of the type of materials demanded by developed versus developing countries.
- Studies examining the relative income elasticities of demand for economically developed versus developing countries.

Although developing countries offer great potential for future softwood and hardwood export demand, the developed countries of Asia, North America, and Europe are currently the most profitable markets for domestic producers. However, cultural practices, level of population growth, access to forest products from domestic or international sources, and distance makes each of these markets different. In order to anticipate short-term and long-term changes in exports to these markets, there is a need for

- Comparative analyses of differences and similarities between markets in Asia, Europe, and North America.
- Development of short-term and long-term demand projections for selected products within Asian, European, and North American markets.

Acquire Enhanced Understanding of Production Costs and Competitiveness of Forest Products Industries

Production costs have a significant impact on the ability of forest products industries to remain competitive. Important to the latter are multiproduct production processes, legally imposed costs of production, and a number of hidden production costs. Such conditions are deserving of research attention.

Identify and evaluate the allocation of costs in multiproduct processing. Forest products markets contain numerous examples of firms with multiproduct production processes. For example, hardwood forests are composed of numerous tree species normally of various ages and quality. The roundwood products that can be derived from these forests include veneer logs, sawlogs, and small diameter roundwood. Hardwood stumpage is often purchased for sawlogs of a specific group of species. The volume of small diameter roundwood and sawlogs of other species will affect the stand's price, but its effect could be positive or negative depending on availability of markets, harvesting methods, or restrictions by the landowner. The cost of an individual product is unclear because of the multiproduct nature of the sale.

The multiproduct nature of stumpage sale is not unique to hardwoods. Generalizations can be applied throughout the forest products industry. Multiproduct considerations make it difficult to assign costs associated with harvesting a specific species or product. Any number of accounting procedures could be employed to assign a value and cost to an individual product or species resulting in an apparent loss associated with one product or species and a huge profit associated with another product or species. Three researchable problems resulting from this multiproduct situation in stumpage markets are

- How do stumpage purchasers accurately assess the value and cost associated with an individual species or product?
- How can the supply and demand of products in multiproduct situations be accurately estimated?
- What degree of theoretical accuracy is necessary in the estimations of supply and demand?

Multiproduct production processes are not unique to the roundwood market. For instance, a producer of southern pine lumber also produces pulp chips. Normally chips are considered a byproduct. If chips are indeed byproducts, then their cost of production is the marginal cost associated with chip production plus any opportunity cost associated with diverting materials used in chip production to alternative uses. However, if the value of chips produced from the log is internalized in the price of the log, then chip production cannot be considered a byproduct but is a co-product. Two questions to be asked in this regard are

- To what degree are chip prices internalized into sawlog prices?
- What effect does this price internalization have on the true supply of lumber and chips, especially southern pine?

Another variation of multiproduct production concept occurs in many medium size and larger hardwood sawmills that sell graded lumber. These mills have historically been in the business of producing high grade lumber. However, a certain proportion of lower grade lumber is always produced in this process. A common statement heard in the hardwood sawmilling industry is that profits are made on production of the higher grade lumber, while lower grade lumber is produced at a loss. Such statements are made because the roundwood cost assigned to the production of low grade lumber is a proportional cost of the log on a volume basis.

Multiproduct production is of two basic forms: technically complementary products and technically competing products (Carlson 1956). A simple definition of complementary products is the case where co-products are developed simultaneously. A simple definition of competing products is where there exists a choice of varying the production of one product at the expense of another. In the case of competing products, opportunity costs must be considered. Two questions meaningful to the complementary multiproduct situations versus competing multiproduct situations are

- How do producers of forest products maximize profits in a multiproduct situation?
- What is the ultimate effect of multiproduct production on the demand for inputs and the supply of output?

Furthermore, questions regarding the benefits of new technology that changes yields of one product cannot be accurately answered unless formats are developed to accurately divide the costs incurred in a multiproduct situation.

Examine the impact of environmental and safety related costs on forest products industries. Issues dealing with environmental impact and worker safety are old topics that have taken on new meaning for the American economy over the last 20 years. The need for legislation addressing any such issues is value dependent—with the general population being the ultimate decision maker. The cost of addressing environmental and safety issues in terms of higher price, reduced demand for domestically produced products, and loss of jobs to competing producer countries is to some extent measurable.

Environmental legislation may have reduced the supply of timber available to forest industries in the short run; conversely, some environmental laws have increased the supply of recreational areas to the general public and have set aside public wilderness areas for sake of generations yet to come. The economic trade-off that has taken place in such situations is the potential loss of industrial forestry jobs in a specific locale versus a potential increase in service sector jobs in the same locale. Two areas of future research are

- To analyze the cost or benefits of changing local employment patterns resulting from the implementation of environmental programs that restrict timber producing or harvesting activities.

- To assess the change environmental legislation has on the socio-economic characteristics of the locale, given shifts in forest land use.

When a particular region reduces production of forest products, other regions will benefit through higher profits and increased production. While there could be a net loss associated with the relocation of productive capacity from one region to another, relocation may provide the impetus necessary to install new, more efficient equipment and other production resources. Or, the region benefiting from relocation may face a greater demand for timber and increased profits to timber landowners. A question to be addressed is

- What is the extent and net cost of industry relocation due to environmental restrictions?

Environmental and safety legislation increases the cost of logging, primary processing, transportation, and secondary processing. Certain forestry laws have apparently raised both the fixed and the variable cost of manufacturing forest products. However, the impact of these two issues probably has had unequal effects on the various production levels as well as the various production processes within each production level. For instance:

- What has been the impact of environmental legislation on logging versus sawmilling?
- What has been the impact of safety legislation on hardwood sawmilling versus softwood sawmilling?

If equipment and actions required by safety legislation simply add to cost without adding to productive efficiency, then profits, employment, and supply of forest products will drop. National legislation of such a nature tends to impact the industry nationwide, thus, the relocation of productive capacity could be to countries that are lacking such constraints. Such relocation will be more apparent in secondary processing because of the relative cost of transporting logs and the increased tendency of other timber-producing countries to reduce exports of both primary and secondary forest products. Indeed a major shift in wood household furniture production has occurred since the late 1970s with imports increasing from 5 percent of the market to 20 percent of the market. The main beneficiary of this change has been Taiwan. Research should be conducted on

- The impact of environmental and safety policies on the competitiveness of domestic versus foreign forest product producers, especially furniture imports.
- The impact of environmental and safety policies on domestic forest products industry employment.

The term "externality" is used widely in environmental economics when discussing an unintended byproduct of a production process. There have also been externalities associated with environmental and safety legislation. Some positive externalities could be increased efficiency due to introduction of new technology and reduced worker absence resulting from increased safety. Some negative externalities include unemployment of workers lacking technical or social mobility and the loss of productive capacities to other countries. The need for environmental and safety legislation is an inarguable question—society puts a great deal of value on both human life and natural beauty. Major research questions are

- To what degree and to what cost should environmental and safety policies be pursued?
- Has past environmental and safety legislation specific to the forest industry accomplished its objectives?

Assess hidden costs affecting the competitiveness of forest products among themselves and alternative products. The term "competitiveness" has been coined in recent years to describe the relative cost advantage or disadvantage of one product, industry, or country over another. Although most economists are well versed in the multiple meanings of the word "cost," many technical-oriented, forestry-related professionals may have more limited definitions. Because of this, cost concepts may not always be correctly applied in the development and promotion of new or existing forest-based products. This can reduce the competitiveness of new and existing forest products. Research should be undertaken to

- Examine prospective nonwood-based products and determine implications for the production and marketing of wood-based products.

The cost associated with the use of a product is not only the cost of that product, per se, but also the cost of other products that must be used in conjunction with the product. This interrelationship is explicit in a derived demand function that contains prices of technical complements as independent variables. Because of the technical nature of this type of cost, it is perhaps the easiest to explain to the noneconomist. Still there are some costs associated with risk and opportunity costs that are missing from most feasibility analyses. Therefore, research should be completed that will

- Examine ways in which risk and opportunity costs can be included directly in feasibility analyses.

The cost of time is an important concept in the construction industry. Not only do construction firms have deadlines to consider but also must face the cost for construction float loans. Any technique that uses less labor or less costly material may still increase overall cost if it delays the time to build a structure. Therefore, there should be developed

- Procedures that can measure the cost of time when developing and evaluating new forest construction products.

Technologies that are developed for timber-based industries are often evaluated on the basis of internal rates of return. Although internal rates of returns do take into account multiperiod cash flow problems, interest rates, and the time value of money, they do not usually consider variations in sales or business cycles. The risk of a business cycle can be internalized in terms of a higher interest rate, but a drop-off in sales at the beginning of the payback period is much more expensive than a drop-off in sales at the end of a payback period. Therefore, there should be developed

- Procedures to define the effects of variable cash flows on the internal rate of return of forestry investments.

Although risk is not a cost, the risk of borrowing money for equipment in a cyclical market can be seen in terms of average costs. A firm that purchases equipment with borrowed funds in order to reduce average variable costs could be at a disadvantage during periods of lower demand. The disadvantage results from the fact that other, less-leveraged firms in the industry can sell closer to their average variable cost while leveraged firms have to cover the proportion of the fixed cost that is owed to the lender plus variable costs. Furthermore, unleveraged firms can shut down at virtually no cost while the leveraged firm must still pay a monthly note. Accordingly, research is needed to

- Analyze the significance of firm leverage in market efficiency.

Opportunity costs, or the amount a factor of production could earn in some alternative use, should be examined in feasibility analyses. For instance, given current prices for No. 2 Common (2C) red oak lumber, current costs of kiln drying and processing, and current prices of red oak dimension, it can be demonstrated that turning green 2C red oak into dimension is profitable. However, it can also be shown that selling the 2C red oak after kiln drying and foregoing the dimension process altogether may be even more profitable. Development of production technology that excludes the profit potential of intermediate products may show profitability but stands a chance of being ignored by the intended users (industry). Therefore

- Methods should be developed that allow for the inclusion of opportunity costs in feasibility analyses.

Another example that combines the value of time, opportunity costs, and return to capital is the age old question of using lower grade lumber in production of furniture parts. An examination of current prices of 2C and No. 1 Common red oak indicates a $380 per thousand board feet price difference. Yet the difference in yield and the additional cost of processing 2C seems to be much less than this figure. However, because the total output of furniture parts per 8-hour shift is decreased using 2C lumber, there is an opportunity cost associated with lost sales. This problem could be examined by

- Researching the return to existing or proposed capital investment using alternative lumber inputs.

Procedures developed during the course of such an examination would prove useful to utilization scientists in all forest products industries.

The cost structure and competitiveness of the forest products industry is affected by a number of general economic conditions and by technical innovations. Research should be addressed at

- Assessing general economic conditions affecting the production of forest products (e.g., inflation, balance of payments, energy supplies, productivity, access to capital).
- Evaluating technological innovations affecting markets and marketing of wood-based products (e.g., computers, product design, information systems, transportation).

Develop Improved Understanding of Market Operations and Trade in Forest Products

Marketing and trade of forest products entail a number of critical activities important to efficient product distribution. Important is the free flow of market information, a well-organized market structure, an efficiency-producing competitive arrangement, and a spatial condition that encourages efficiency of production and resource use. Topics of such nature are deserving of research.

Develop, analyze, and provide market information. Market information and market certainty are two necessary attributes for an efficiently operating market. Information and certainty can be examined as two separate concepts, namely, perfect information (i.e., knowledge of current and past market prices and quantities) and certainty (i.e., knowledge of future market prices and quantities). Without information and certainty, an element of risk is introduced, causing firms to diverge from profit-maximizing positions (Anderson, Dillon, and Hardaker 1980). The impact of imperfect information in forest products markets may differ because of the degree of concentration. A researchable question is

- Are competitive firms, which rely heavily on the market signals, more adversely affected by poor market information than concentrated firms that have more direct control over their markets?

The end product of poor market information is risk. However, the effect of risk may differ depending on the objective of the firm. A firm that is service-oriented may have a tendency to overproduce or carry large inventories so as not to lose customers. An industry that is in a very cost competitive environment may have the tendency to underproduce and carry no inventory. Better information is needed on

- What are the effects and costs of risk under alternative firm objectives?

- To what extent does risk influence the objectives of firms in wood products markets?

One of the most noticeable aspects of the hardwood lumber industry versus forest industries in general is the lack of publicly owned firms. Although economies of scale in hardwood lumber production allow for small producers, the lack of publicly owned hardwood sawmills may result from an informational barrier to entry. Such a barrier would exist if lack of information reduces the availability of capital or the obtainable market. Research dealing with the extent and cost of information barriers to entry would allow better calculation of the costs and benefits of better information gathering systems. Questions that should be analyzed are

- To what extent is the lack of information a barrier to entry?
- To what degree has imperfect information caused inefficient production, price variation, and lack of growth?
- What is the impact of insufficient market information on the supply, demand, and competitiveness of domestically produced forest products?
- If increased market information is desirable, who should pay the cost of information collection?

Identify and evaluate the relative efficiency of alternative market environments. In the forest products industry, elements of monopoly, monopsony, oligopoly, and oligopsony can be found. Forest products firms may also be integrated into two or more market levels. The extent to which such structures influence interactions between firms and overall efficiency of forest products industries is largely unknown. Macroeconomic theory indicates that the prices paid and received for an input or output will vary depending on the degree of concentration of the selling or purchasing firm (Scherer 1970). However, these differences are indeterminant on a theoretical basis when neither seller nor buyer are perfectly competitive. The market solution in such cases is negotiated between the parties involved. Given the current nature of the forest products market, researchable topics are

- Relative to competitive market prices, to what degree are forest products prices different under varying degrees of market concentration?
- To what degree do negotiated settlements benefit the more concentrated firm?

Competitive conditions between buyers and sellers do not have to exist for the most efficient market solution. Most market concentration results from spatial factors such as in the case of a spatial monopoly or economies of size. Any attempt to increase the number of firms in these situations will most likely result in a socially less desirable solution than in the case of concentrated firms. A researchable question, therefore, is

- Do the production efficiencies that allow concentrated firms to exist offset any increases in price resulting from market concentration?

One situation that is prevalent in forest products markets, but has not been fully analyzed, is the existence and impact of spacial monopsonies. A spacial monopsony could exist when a sawmill, pulpmill, or other primary forest products firm may be the only major buyer in a given area. If these mills act like monopsonists, they will pay less for roundwood inputs than a competitive mill. However, since most forest products firms are in business for the long term, there is a disincentive to take a short-term monopsonistic profit. Therefore, researchable questions are

- To what extent do spacial monopsonies exist in primary forest products markets?
- What is the impact of spacial monopsonies on long-term timber supplies?

Many firms within the forest products industry are vertically integrated. Reasons for vertically integrating productive processes include the need to ensure steadiness in the supply and quality of raw material and to add value to the product. However, there are hidden costs associated with the management of vertically integrated firms. Are forest products firms actually gaining what they want through vertical integration? And what is vertical integration costing them in terms of direct and opportunity costs? An analysis of the costs and benefits associated with vertical integration and the factors that facilitate or discourage vertical integration would help economists better understand the nature of the forest industry and allow the industry to make better organizational decisions. Needed are analyses of

- The costs and benefits associated with vertical integration.
- The factors that facilitate or discourage vertical integration.

An issue related to vertical integration concerns market channels. Whereas vertically integrated firms have internal market channels, other firms buy and sell through brokers, wholesalers, and distributors—middlemen often accused of doing nothing but making money. The need for such firms, however, is demonstrated by their continued existence in a competitive economy. Whether current market channels are as efficient as possible is another question. An analysis of the efficiencies and inefficiencies of forest products market channels that utilize middlemen would allow industry to develop more cost efficient distribution channels and thus increase U.S. competitiveness in world markets. Required is

- Comparative analysis of the efficiencies and inefficiencies of forest products marketing channels and the institutions involved therein (e.g., cooperatives, concentration yards, wholesalers, retailers).
- Evaluations of alternative marketing structures to be used by forest product enterprises (e.g., territorial marketing, functional marketing, product marketing, customer marketing).
- Assessment of legal implications of various types of marketing channels and associated activities (e.g., exclusive dealing, tying contracts).

Evaluate the spacial, temporal, and form issues affecting efficiency and competitiveness. Since forest products are normally bulky and are consumed some distance from the point of production, spacial issues involving transportation of forest products have been an area of general concern by forest economists. However, the related issues of space, time, and form have received less attention. During the period between harvesting timber and the installation or use of the final wood product, there are several separate steps in which products are transported, stored, and manufactured into a new form. At any point in this process inefficiencies could occur, including the wrong sequencing of production, storage, or transportation activity. Furthermore, the most efficient sequencing of activities may be dependent on exogenous forces such as transportation costs, interest rates, and labor costs. A classic space, time, and form issue is plant location. The most efficient site or sites to place one plant or several plants can be determined through the minimization of the cost of collecting input, producing and distributing the output (Bressler and King 1970).

The minimization of the previously mentioned costs allows for trade-offs between economies of scale and transportation costs. The multicost minimization format has been used in the location of primary forest processing facilities. Still, there may be several other applications of this procedure to forest-based industries. One area where minimization of multiple costs may prove enlightening is in dry kiln location for the hardwood industry. A researchable topic in this area is

- When does it make sense to locate a dry kiln adjacent to a sawmill and when does it make sense to develop a separate concentration yard with dry kilns and collect green lumber from several mills?

Furniture production is a multistep procedure involving drying, dimensioning, assembling, finishing, and distributing. Currently there are furniture firms that accomplish all such steps at one site while other firms use a multiplant (multisite) approach. Several factors can influence the decision to use a single plant or several plants, including size of the operation, transportation and location costs, number and nature of product lines being distributed, the opportunity cost of selling or purchasing intermediate products (dimension, composite panels, etc.), or management objectives. A researchable topic is, subject to management objectives, to

- Develop procedures to determine the number of plant sites required to minimize direct and opportunity costs.

Develop Effective Means of Producing and Marketing Nonwood-Based Forest Products

Forests are important sources of a variety of nonwood forest products. For example, forest resources are frequently an important contributor to the recreational camping experiences of a significant segment of the nation's public. Important industries supporting such interests have been developed (e.g., private campgrounds). Similarly, a significant portion of the economic vitality of the tourism industry is dependent on the existence of forest and related resources. Even industrial owners of timberland express interest in making their forest land available for purposes of recreation and hunting. The production and distribution of nonwood-based forest products and services often entails problems, many of which could benefit from economic and marketing research. Examples of research are

- Determine type and extent of potential markets for forest-based recreation (e.g., foreign tourists, hunting lodges, recreational camping) as might be provided by owners of forest land, including private industrial and private nonindustrial owners.
- Assess the effectiveness by which current marketing channels enable distribution of forest-based recreation.
- Evaluate advertising, sales promotion and mass publicity as means of increasing the efficiency with which nonwood-based forest products and services are brought to the attention of consumers.
- Develop more effective means of pricing forest-based recreation.

Summary and Conclusions

Marketing of forest-based products is a multifaceted activity of special importance to the forestry community and the nation. Entailed are a number of complex activities involving the transformation of raw materials into useful products and their subsequent distribution to industrial and final consumers. Such activities vary in terms of efficiency and effectiveness; they are deserving of research attention. To be most effective, research activities should be designed to improve understanding of (1) consumer and industrial demand for forest products, (2) production costs and competitiveness for forest products industries, (3) market operations and trade in forest products, and (4) marketing of nonwood-based forest products and services. Research investments in topics of such a nature could go far toward improving the efficiency with which forest-based products are produced and distributed.

Literature Cited

Adams, D. M. and R. W. Haynes. 1980. The 1980 softwood timber assessment market model: Structure, projections, and policy simulations. Monograph 22. Forest Science. Society of American Foresters. Bethesda, MD.

Anderson, J. R., J. L. Dillon and B. Hardaker. 1980. Agricultural decision analysis (2nd ed.). The Iowa State University Press. Ames, IA.

Bressler, R. G., Jr. and R. A. King. 1970. Markets, prices and interregional trade. John Wiley & Sons Publishers. New York, NY.

Buongiorno, J. 1981. Outline of a world model of the pulp and paper industry. In: Analysis of world trade in forest products. International Institute for Applied Systems Analysis. Laxenburg, Austria.

Carlson, S. 1956. A study on the pure theory of production. Kelley & Millman Publishers. New York, NY.

Gilless, J. K. and J. Buongiorno. 1987. Papyrus: A model of the North American pulp and paper industry. Monograph 28. Forest Science. Society of American Foresters. Bethesda, MD.

McKillop, W. L. M. 1967. Supply and demand for forest products: An econometric study. Hilgardia 38(1):1-132.
Scherer, F. M. 1970. Industrial market structure and economic performance. Rand McNally College Publishing Co. Chicago, IL.

Bibliography

Bethel, J. S. (ed.) 1983. World trade in forest products. University of Washington Press. Seattle, WA.
Bliss, C. J. 1975. Capital theory and the distribution of income. North-Holland Publishers. Amsterdam, Netherlands.
Bressler, R. G. and R. A. King. 1970. Markets, prices and interregional trade. John Wiley and Sons Publishers. New York, NY.
Ellefson, P. V. and R. N. Stone. 1984. U.S. wood-based industry: Industrial organization and performance. Praeger Publishers. New York, NY.
Ferguson, C. E. 1971. The neoclassical theory of production and distribution. Bently House, Syndic of Cambridge University Press. New York, NY.
Gregory, G. R. 1987. Resource economics for foresters. John Wiley and Sons Publishers. New York, NY.
Hicks, J. R. 1965. Capital and growth. Oxford University Press. New York, NY.
Houthakker, H.S. and L. D. Taylor. 1970. Consumer demand in the United States: Analyses and projections. Harvard University Press. Cambridge, MA.
Lancaster, K. J. 1968. Mathematical economics. Macmillan Publishers. New York, NY.
Liu, T.C. and G.H. Hildenbrand. 1965. Manufacturing production functions in the United States, 1957. Cornell University Press. Ithaca, NY.
Mead, W. J. 1966. Competition and oligopsony in the Douglas-fir lumber industry. University of California Press. Berkeley, CA.
Panshin, A. J., E.S. Harrer, J. S. Bethel and W.J. Baker. 1962. Forest products: Their sources, production and utilization. McGraw-Hill Publishers. New York, NY.
Philps, L. 1974. Applied consumption analysis. North-Holland Publishing Company. Amsterdam, Netherlands.
Robinson, J. 1964. The economics of imperfect competition. St. Martin's Press. New York, NY.
Scherer, F. M. 1980. Industrial market structure and economic performance. Houghton Mifflin Publishers. Boston, MA.
Shepard, W. G. 1979. The economics of industrial organization. Prentice-Hall Publishers. Englewood Cliffs, NJ.
Wold, H. O. A. and L. Jureen. 1952. Demand analysis. John Wiley and Sons Publishers. New York, NY.

20

FOREST RESOURCES LAW
AND LEGAL PROCESSES

Benjamin V. Dall

Review of the forestry-law literature reveals a number of patterns which have relevance to the development of a research agenda for forest law. The clearest pattern emerges from the historical division between public and private law. Public law is defined in terms of rules that are intended to establish and define communal or social interests. Private law generally refers to rules which structure the decisions of individuals or firms in actions with others. An example of public law would be workman's compensations statutes, while private law can be illustrated by rules which govern landlord-tenant relations. In respect to forestry, the record is replete with examples of legislation expressing the public interest. Indeed the history of American public forestry can be outlined by a chronological listing of statutory law. But it is interesting to note that the few books addressing timber law deal exclusively with private law subjects such as contracts, land title, and survey law. This development is a reflection of fact that forest lands provide a broad range of public goods and services, and at the same time serve as the essential medium for forest commodities production.

Since colonial days there have been continuous legislative efforts to give expression to the public interest in forests. During important conservation periods, both state and federal governments exercised proprietary powers to become owners of forest land, or they adopted regulatory controls over privately owned forest land. Law was the instrument used to create the National Forest System (also to specify its administration), and law was the instrument used to regulate the use of pesticides on both public and private forest lands. At the same time, private sector investment in the production of forest products required the development of a rich body of land and commercial law. From the nation's earliest history, governments used their authority to construct a dependable institutional framework for private enterprise. During our agrarian period, a refinement of land law was vital to land-intensive enterprises such as forestry. Considerations such as these help in understanding the long history of forestry-law connections, the particular forms of forestry-law literature, and the basis for development of a forestry-law research agenda. While not complete, the accompanying bibliography illustrates the type of forestry publications which report research on the legal aspects of problems important to forestry.

Legal research, in the ordinary sense, is quite different from scientific research and, moreover, legal studies may take different forms. Practical legal research consists of a bibliographic search for rule statements to answer questions raised by a particular set of facts. The process is guided by the need to solve a problem which is specified by the facts of a case. It progresses by closure to find the most precise rule statements for addressing the issue in question. In this respect, legal research can be described as an exercise in deductive logic or a search for deductive validity.

Legal research typically faces two major challenges. First is that of overcoming information retrieval problems (e.g., imprecise language, multiple system data storage, historical tracing of legal rules). Legal rules have various forms; they can be located by reference to several bibliographic sources according to the type of law in question and the institutions from which the rules emanate. The most commonly searched sources are statutory codes, court reporters, and codes of regulations of administrative agencies. Statutory codes contain legislative law organized and indexed in a topical framework; codes of regulations provide a similar arrangement for rules promulgated by administrative agencies. Court reporter systems consist of a chronological publication of appellate court opinions without subject matter organization. Law libraries typically have several law finders or search books that can be used by the researcher to locate an appropriate statement of law. The legal profession has developed very elaborate systems to assist in the task of locating laws and regulations. Detailed indices, tables, digests, books of words and phrases (replete with cross references) have been designed to guide the search. After finding an appropriate statement of law, the search must examine the record to determine if the rule has been modified by subsequent decisions.

The second challenge facing legal research involves creativity on the part of the researcher. Legal rules are typically very general in nature, while legal questions raised by a particular case are very specific. Rarely does a rule statement give an absolutely unambiguous answer. The facts and circumstances of the case must be developed by analysis and construction to achieve a reasoned argument for or against control. The training, skill, and experience of the researcher determines the merits of the result. The process can be labeled "legal analysis" for purposes of examining the record of forestry-law studies and prospects for future research.

Forestry-law studies take many forms, most notably scholastic legal research, descriptive studies, interdisciplinary studies, and case studies. "Scholastic legal research" is a label assigned to the bulk of studies published in legal journals. Every law school has a law journal or law review which serves as a vehicle for the publication of professional legal studies. Articles in such periodicals address general subjects via the presentation and analysis of all relevant case law, statutory law or regulations according to topics of interest. An article, for example, about the National Forest Management Act of 1976 may appear in such periodicals. In all likelihood, it will have been written by a lawyer (who acquired a working knowledge of forestry) for the legal profession. For a variety of reasons, the article may not be entirely responsive to the needs of professional foresters (e.g., it may focus excessively on legal processes, assume too much for a nonlawyer, have insufficient command of the subject's factual dimensions). Nevertheless, scholastic legal research as reported by such an article should be part of a comprehensive assessment of the literature of forestry law.

The second type of study involving forestry law is descriptive in nature. Such is designed to search and abstract the legal rule system and to present material without extensive legal analysis. Such a study approach could be viewed as a forestry-oriented study *about* law in comparison to a study *in* law. Most frequently, descriptive studies are prepared by a nonlawyer and are directed to an audience of professional foresters. Very often they take a comparative form and are confined to a review of statutory law on several jurisdictions. Such studies attempt to demystify the legal system by extracting only forestry-relevant rules in the hope of providing the professional with organized reference material. A major obstacle to a successful descriptive study stems from the very important role that *process* plays in law. Although descriptive studies are useful for many purposes, the law on any subject cannot be reduced to a set of rules. The full meaning of a statute must be developed so as to apply to specific cases. Analysis of statutory law should include any agency regulations and court decisions which determine the way in which the statute is applied. Where agency actions or court decisions are lacking to guide analysis, the task becomes more difficult but no less important.

Interdisciplinary studies are a third type of forestry-law study. True interdisciplinary research is possible where the disciplines to be integrated are closely related (e.g., biology and chemistry). On this count, law presents a special case, since it does not share a common theoretical basis with any of the disciplines embraced by forestry. Consequently, it might be more appropriate to refer to interdisciplinary studies as project studies. Their distinguishing features are: (1) the research is designed around a site-specific or problem-specific proposal, (2) the theoretical framework of the research is scientific, and (3) the legal issues for study are determined by the project. While this may appear as delegating legal considerations to a secondary or ancillary role, it gives realistic recognition to an instrumental conception of law. There must be a sound basis for a goal-oriented proposition or the selection of an array of alternatives (e.g., a basis in economic, silvicultural, engineering, or managerial efficiency). Questions of legal feasibility can then be effectively formulated. Furthermore, there should be opportunities for feedback and interplay leading to reformulated alternatives in a well organized project. Law, however, should not be a primary determinant of the objectives toward which such alternatives are focused.

The fourth group of forestry-law studies consist of cases. Cases as set forth in law books are appellate court opinions which do not include trial transcripts, exhibits, or a narrative explanation of the factual background of the events in litigation. The court opinion provides only the briefest statement of facts necessary to explain the setting for the legal issues *raised on appeal*. In contrast, cases as developed for instruction in other disciplines are usually devoted to an explanation of the facts, even where legal issues may be involved. For forestry-law research, special efforts should be taken to design case studies which explain both legal issues and factual settings. Some case materials in schools of business or public administration are in such form.

Given the above background and judging the status of current forestry-law information, the following strategic directions would seem appropriate for future research involving forest law:

- Develop improved understanding of legal institutions and process important to forestry.
- Develop an improved legal basis for governmental land use controls.
- Develop improved understanding of environmental regulation as relevant to forestry.
- Acquire enhanced understanding of alternatives to litigation.
- Develop improved understanding of the law of forest ownership and forest products commerce.

Develop Improved Understanding of Legal Institutions and Processes Important to Forestry

Over the last two decades, the institutional environment of forestry has become "legalized" far beyond anything ever anticipated by the forestry profession. Numerous changes have occurred in the system and new constraints have been imposed on management practices. Limitations on the use of established silvicultural practices have followed one after another as litigants succeeded case by case and issue by issue. The response of the forestry profession has been defensive, guarded, and negatively reactive toward courts and lawyers. More useful would be a concerted effort to learn more about how the legal system works.

The new legal environment of forestry results not merely from a quantitative increase in law but more from an increased reliance on administrative agency regulation. Some federal regulatory agencies (e.g., the Environmental Protection Agency) have grown to such proportions that they warrant recognition as the fourth branch of government—even though not mentioned in the Constitution. Over their history, courts have developed reasonably workable approaches to judicial review of agency decisions. Under recent legislation, however, agencies have been delegated sweeping charges and more extensive discretionary authority, all under statutory standards that are less specific. Consequently, the review function of courts frequently becomes an issue. Narrowly viewed, this might be seen as a problem for lawyers concerned with administrative law. However, it poses very important consequences for forestry. Such matters may be difficult for foresters to understand, especially when such professionals tend to focus on end results, not legal technicalities. Research on forestry subjects with explicit treatment of legal process would be useful.

Survey the nature, frequency, and jurisdiction of forestry cases. Research should be undertaken to generate information about the extent of litigation on forest management issues (e.g., over the past 20 years). Focus should be on efforts to generalize the nature, frequency, and jurisdictional occurrence of cases. The research could be divided and expanded in many ways to provide detailed treatment of subjects, including specific attributes of the parties, lawyers, and courts involved. For information of such a nature, the research would have to go far beyond appellate court decisions.

Develop case studies of rule-making functions of agencies administering forestry programs. Selected cases should be developed to demonstrate both the adjudicatory and rule-making functions of administrative agencies. The USDA–Forest Service rule making under the National Forest Management Act is an example. Process requirements, special agency procedures, and public participation could be explicitly assessed. Comparative studies could examine how technical information is processed in management agencies (e.g., the USDA–Forest Service) versus regulatory agencies (e.g., the Environmental Protection Agency).

Examine the effects of judicial decisions on policy making in forest resource agencies. Examination of judicial decision impacts should be undertaken. Such will require researchers trained in public administration—researchers who can develop cases which illustrate the manner in which organizations interpret and respond to judicial decisions. Particularly revealing would be examples which resulted in the adoption of complex strategies, including decisions to appeal and to seek legislative relief.

Describe the delivery system for legal advice about the management of forests. Research describing the delivery system for legal advice about forest management should be organized so as to explain arrangements in terms of private clientele (individuals, small industry, larger corporations), professional associations, administrative agencies, government executives, and courts. Since one might expect to find a "demand" induced development, the research would generate insight on matters such as the appropriate role of government in the provision of legal services. It may also help to answer why legal activities for environmental interests seem to be more highly organized than for forestry.

Develop an Improved Legal Basis for Governmental Land Use Controls

Review of forestry-law literature reveals a history of issues arising from forestry's land-based orientation and from the public-goods dimension of such land. These conditions have set the stage for a complex array of legal questions concerning land use. Private owners of forest land in the U.S. enjoy a highly developed system of land law—one which provides great stability for land acquisition and ownership. Such is a result of Anglo-American legal traditions (dating from Magna Carta time) and of our historic recognition of the value of property law in a capitalistic system. Private land owner rights, however, have never been absolute. Over time, there has been a growth in law that secures for the public, the public interest which often exists in private property. Governments can protect the public interest in forests as either owners and managers of forest lands (e.g., National Forest System) or as regulators of activities occurring on private forest lands (e.g., state forest practice laws).

Legal authority for government intervention to protect the public interest is based on two inherent powers, the power of eminent domain and the police power. Although not set forth in either federal or state constitutions, the scope and definition of these powers have been developed via case law. Furthermore, their legal development has been shaped by constitutional limits imposed by the Fifth

Amendment, namely, "private property" should not "be taken for public use without just compensation." As governments have more frequently employed police power regulation as the means to secure the public interest, courts have been called upon to determine the line between noncompensable regulation and "takings." The rules for testing this issue are not precise. Courts have used at least four theoretical approaches leading to a condition which is difficult for foresters to understand. Consequently, research on the subject is important.

Examine judicial formulas used to decide taking in cases where police power regulations are imposed on forest land uses. Without an exclusive test for determining the limits of police power regulation, courts have had to rely on various theoretical approaches. Such approaches include tests of the degree of physical invasion, the extent to which the restricted use may be judged a nuisance, consideration of the extent to which property values are diminished, and approaches which balance private loss and public benefits. Forest land "taking" cases (as a subset) should be examined to determine the extent to which such approaches have been applied.

Assess forest land valuation procedures as used in eminent domain proceedings. Legal process and economic evaluation techniques employed in physical taking cases should be surveyed with the intent of determining their appropriateness and effectiveness. Such information would be especially useful to forest land owners involved in eminent domain proceedings. Information on appraisal techniques and their treatment in negotiation or litigation would also be useful.

Determine the cost of regulatory limitations on productive uses of forest land. Various land use controls (e.g., wetland preservation laws) should be examined to determine cost transfers to private forest land owners. Such information could be used to influence the design of legislation or amendments, with the intent of providing relief to affected owners.

Assess local government application of police powers to the regulation of private forest management. In recent years, local controls (i.e., timber harvesting ordinances) over private forest management activities have increased significantly in number. Since local ordinances derive authority from state enabling laws, there are many unanswered questions about local government legal authority to regulate such activities. Not only is the legal basis for such ordinances in question, their preemption or limitations related to state forest practices laws is also cause for concern. These issues warrant further research.

Develop Improved Understanding of Environmental Regulation as Relevant to Forestry

Over the past twenty years, there has occurred a dramatic growth in regulation of environmental conditions important to forestry. While points of contact between the environmental movement and forestry have emerged from time to time, there has been less than perfect alignment between the two. Forestry fits squarely within the scope of earlier conservation movements, with a shared concern for natural resources stewardship, scientific management of natural resources, and interest in future availability of forest products. The environmental movement, however, called for a qualitative reexamination of assumptions about growth, development, and industrial externalities. Although forestry has not always been a primary target of environmental regulation, its involvement with extensive land resources has often steered it into the regulatory arena. Forest management activities may be regulated for pollution control, for regulated substances (e.g., pesticides), and for public review procedures established to control changes in natural ecosystems. Indirect effects may result from environmental regulation of the industrial conversion of wood products.

Literature addressing environmental regulation of forestry activities is sketchy and speculative. Such a condition occurs, in part, because many regulatory programs adopt implementation strategies and time tables that place control of forest activities in relatively low priority. For example, the Environmental Protection Agency's action under the 1972 Amendments to the Federal Water Pollution Control Act has focused more on direct discharges and less on nonpoint sources of water pollution. The full impact of the agency's water quality controls on forestry have probably not been fully realized. Also, the subject of environmental regulation has been dominated by consideration of federal law, while state and local law may in fact be more significant for forestry. Experience, for example, with environmental impact assessment is largely in terms of USDA–Forest Service decisions subject to the National Environmental Policy Act (NEPA). State laws modeled after NEPA have more extensive coverage of private sector forestry.

Research concerning environmental regulation will include some subjects over which conflict has been experienced. The central theme, however, should be to develop an objective body of knowledge based on empirical research and to provide better estimates of the scope and dimensions of future problems.

Examine the nature and extent of air pollution controls affecting forestry. At present there is only general knowledge of the scope and effects of air quality programs on forestry. Existing information is based largely on reasoning from the language of the Clean Air Act and subsequent regulations. Industrial point sources are clearly the primary emission targets subject to permit requirements. Even though regulation under the Clean Air Act strives for national uniformity on matters such as new source performance standards, other aspects of control may vary from place to place—in differences in environmental conditions, differences in state implementation plans, and differences in enforcement policy. State or regional research concerning such differences would contribute to an understanding of the actual dimensions of air quality control effects on forest management. Information is also lacking on the effects of regulations on timber harvesting activities (e.g., slash burning). Also not clearly identified are the effects of regulation designed for the "prevention of significant deterioration" in wild areas having air quality exceeding national ambient standards.

Examine the nature and extent of water pollution controls affecting forestry. Pollution control actions under the 1972 amendments to the Federal Water Pollution Control Act had immediate and direct effects on public source dischargers through permit systems implemented in each state. Although the regulation of industrial plants caused changes in the forest products economy, and consequently in forest management, more direct effects on forest practices stem from nonpoint source controls and the regulation of drainage and construction activities under Section

404 of the law. Compliance problems in both such areas are not well documented. Legal requirements may vary from place to place depending on environmental conditions and differences in program implementation. Forestry is made subject to compliance with best management practices under state initiated and Environmental Protection Agency-approved Section 208 plans, only some of which are regulatory.

Examine state environmental protection laws regulating forestry activities. Research should be undertaken to document environmental impact statement requirements as required by state laws. Research could be done in a single state or, for comparative purposes, on a regional basis. The relationships between forest practice laws and environmental quality laws should also be studied.

Develop extended case studies on selected environmental laws relevant to forestry. Regulations addressing pesticides, noise, wildlife, and wildland preservation can often impact forestry activities. The extent of these regulatory impacts on forestry should be assessed. Such may determine a need for further legal research.

Acquire Enhanced Understanding of Alternatives to Litigation

Since the early 1970s, citizens and the organizations which often represent them (e.g., environmental interest groups) have turned to courts to redress perceived wrongs. Relative to legislative or administrative petitions, courts have been viewed as the best avenue for direct action to resolve conflicting interests. However, recent evidence has begun to demonstrate the weaknesses of courts as forums for resolving environmental disputes. Zealous combatants have learned that the judicial decision system is highly structured by long-standing procedures, and that courts are usually circumspect when presented with conflicts produced by changing social values. The predictable result: Losers outnumbered winners because of protracted delays, expense, and the infrequency of clear decisions on the substantive merits of contested issues. The most controversial forestry cases have centered on clear-cutting issues involving the USDA–Forest Service (e.g., Monongahela and Bitterroot National Forests).

Disenchantment with courts has led to emergence of interest in alternative methods of dispute resolution. Hundreds of journal articles have proposed the use of various techniques involving voluntary negotiation and bargaining. Many suggestions completely depart from reliance on an authoritative decision maker. Some suggest creation of special courts, partly in response to alleged inadequacies of generalist courts in dealing with technical information. In fact, the mediation movement has progressed beyond literary treatment to practical application—environmental dispute resolution is an example. Forestry can learn much from environmental dispute resolution, if for no other reason than the similarity between controversial forestry cases and "environmental" cases. Some of the most commonly shared elements are citizen participation in decisions which have public consequences; decisions which are based on technical information and professional judgement; and decisions which involve risk and uncertainty. Without question, alternative dispute resolution is an area which has ripened with experience—to the point where useful, practical research may be possible.

Examine case histories of mediation techniques applied to forest conflict. Most experience with mediation is reported in environmental literature. A survey to identify cases in which forest management practices were at issue would be a useful first step for assessing the utility of the mediation alternative applied to forestry cases.

Identify and assess factors associated with success or failure of mediation as a conflict management approach. Environmental conflict resolved by mediation may have employed certain principles that are relevant to resolution of forest management conflicts. Such principles should be identified and their applicability to forestry conflicts assessed. Numerous environmental cases involving mediation have been reported in the literature. Their implications for forestry have not been thoroughly reviewed.

Assess availability, training, and cost of mediators appropriate to forestry disputes. Dissatisfaction with litigation is often based on doubts about the effectiveness of generalist judges who lack special knowledge of forestry. When mediation is suggested, such concerns become especially acute. Research should determine the availability, training, and cost of mediators suitable for involvement in forestry disputes.

Examine natural resource professional organizations as potential contributors to conflict management. Professional natural resource organizations (e.g., Society of American Foresters, The Wildlife Society, American Fisheries Society, Society for Range Management) might be encouraged to play an important role in developing institutional service arrangements for the mediation of forestry and related natural resource disputes. Implied is an understanding of the mediation needs of the profession and of the manner in which such organizations might organize to answer the needs for mediation. Research focused on professional organizations might contribute to the establishment of a system for certifying forestry mediators.

Develop Enhanced Understanding of the Law of Forest Ownership and Forest Products Commerce

Forestry has long-standing and continuing concerns with laws that enable as well as constrain forest ownership and commerce in forest products. Noteworthy in this respect are two early "forest law" books that deal exclusively with such subjects (Kinney 1917, Falk 1958). Though well intended, these works are outdated. More promising in a contemporary setting would be works that address specific problems and are prepared by well staffed teams of researchers. Such is important because the design of research focused on forest ownership and forest products commerce encounters a number of special problems. In most instances, for example, the research subjects are not referenced to forestry in legal source materials. As an illustration, antitrust law is organized to address business enterprises in general, without specific reference to forest products firms. The research task, therefore, requires planning, design, and staffing at levels which probably transcend the capability of an individual. The subjects are probably more amenable to treatment by the staff of large firms engaged in work on ongoing legal problems. Academic studies may add to general knowledge, but ideally there should be integration with the world of action to a degree not likely to be achieved by research based on published information alone. Also, there are certain subjects which private firms (facing the world of competition and government regulation) would prefer remain confidential. For example, they are unlikely to publish information about expected rates of return on investments or about strategies related to marketing or legal problems. Since such research design problems are also encountered by researchers in schools of business administration, their approach to such research, and their subsequent publication of information on these subjects, may offer models for forestry research in these areas.

Examples of research areas deserving of attention in the area of land ownership and forest products commerce (especially research via extended case studies) include land title and trespass issues, landowner liability, contractual arrangements, product liability, antitrust laws, and labor laws. Since many of these subjects are governed by state law (which often is in varying stages of development), researchers would do well if attention were focused on state conditions.

Summary and Conclusions

Research involving forest law is replete with opportunities as well as challenges. Important to recognize is the multi-faceted nature of legal research and the unique manner in which law research is undertaken. However, legal research and forest resource economics research have—in concert—much to offer the forestry community. As to future research involving law, efforts would prove most fruitful if expended on furthering the community's understanding of legal institutions and processes, legal bases for government land use controls, environmental regulations, alternatives to litigation, and the law of forest ownership and forest products commerce. Accomplishments in such areas would most certainly enhance the efficient application of law to forest resource management.

Bibliography

Brown, G. W., D. Carlson, G. Carter, D. Heckroth, M. Miller and B. Thomas. 1977. Meeting water quality objectives on Oregon's private forest lands through the Oregon forest practices act. Oregon State Department of Forestry. Salem, OR.

Coggins, G. C. 1984. Public rangeland management law: FLPMA and PRIA. In: Developing strategies for rangeland management. National Research Council. Washington, DC.

Coggins, G. C. and C. F. Wilkinson. 1987. Federal public land and resources law. Foundation Press. Mineola, NY.

Cubbage, F. W. and W. C. Siegel. 1984. Forest practice law: Statutory provisions and court decisions. In: Forest resources management: The influence of policy and law by M. Lortie.

Cubbage, F. W., W. C. Siegel and T. K. Haines. 1987. Water quality laws affecting forestry in the eastern United States. In: Proceedings of symposium on monitoring, modeling and mediating water quality. Technical Publication Series TPS 87-2:597-609. American Water Resources Association. Washington, DC.

Demaria, S. L. 1983. The legislative and regulatory environment for forestry enterprises in California. S. J. Hall Lectureship in Industrial Forestry. Department of Forestry and Resource Management. University of California, Berkeley. Berkeley, CA.

Ellefson, P. V. 1984. Forest practice laws as a regulatory force: Organizational and administrative experience in the United States. In: Forestry legislation by F. Schmithusen (ed.). Geneva, Switzerland.

Ellefson, P. V. and R. N. Stone. 1984. U.S. wood-based industry: Industrial organization and performance. Praeger Publishers. New York, NY.

Ellefson, P. V. and F. W. Cubbage. 1980. State forest practice laws and regulations: A review and case study for Minnesota. Bulletin 536-1980. Agricultural Experiment Station. University of Minnesota. St. Paul, MN.

Falk, Harry W. 1958. Timber and forest products law. Berkeley, Howell and North Publishers. New York, NY.

Forest Service. 1983. The principal laws relating to Forest Service activities. Agriculture Handbook No. 453 (revised Sept., 1983). U.S. Department of Agriculture. Washington, DC.

Forest Service. 1985. Supplemental information for the principal laws related to Forest Service activities. Agriculture Handbook No. 453 (revised Sept., 1983). U.S. Department of Agriculture. Washington, DC.

Green, P. F. 1982. Government regulation in the forests: Impacts of the 1973 California Forest Practice Act. Environmental Quality Series No. 36. Institute of Government Affairs and Institute of Ecology. University of California, Davis. Davis, CA.

Griffin, N. and C. Watkins. 1986. Liability for dangerous trees. Quarterly Journal of Forestry 80:23-26.

Henly, R. K. and P. V. Ellefson. 1986. Cost and effectiveness of legal mandates for the practice of forestry on private land: Experiences with state forest practice laws in the United States. Silva Fennica 20:358-365.

Henly, R. K. and P. V. Ellefson. 1986. State forest practice regulation in the U.S.: Administration, cost, and accomplishment. Bulletin AD-SB-3011. Agricultural Experiment Station. University of Minnesota. St. Paul, MN.

Henly, R. K., F. W. Cubbage and W. C. Siegel. 1987. State and local regulation of forest practices: A bibliography. Staff Paper Series No. 61. Department of Forest Resources. University of Minnesota. St. Paul, MN.

Irland, L. C. 1985. Logging and water quality: State regulation in New England. Journal of Soil and Water Conservation 40(1):98-102.

Marcouiller, D. W. and P. V. Ellefson. 1987. Forest land use and management conflicts: A Review and evaluation of approaches for management. Staff Paper No. 65. Department of Forest Resources. University of Minnesota. St. Paul, MN.

Kinney, J. P. 1917. The essentials of American timber law. John Wiley and Sons Publishers. New York, NY.

Mukatis, W. A. and C. F. Sutherland. 1986. Contracts for woodland owners and Christmas tree growers. Extension Circular No. 1192. Extension Service. Oregon State University. Corvallis, OR.

Niesslein, E. and W. Zimmermann. 1986. The impact of environmental legislation of forest conservation. In: Proceedings of Division 4. XVIII World Congress. International Union of Forest Research Organizations. Ljubljana, Yugoslavia.

O'Laughlin, J. and P. V. Ellefson. 1979. Diversified firms entering the wood-based industry by merger. Staff Paper Series No. 7. Department of Forest Resources. University of Minnesota. St. Paul, MN.

O'Laughlin, J. and P. V. Ellefson. 1981. U.S. wood-based industry structure: Part I - Top 40 companies. Forest Products Journal 31(10):55-62.

Sachet, J., S. Keller, A. McCoy, T. Orr, Jr. and N. Wolff. 1980. An assessment of the adequacy of Washington's forest practice rules and regulations in protecting water quality.

Technical Report 208 Forest Practices Assessment DOE 80-7A. Washington Department of Ecology. Olympia, WA.

Rupp, C. W. 1985. Reflections and comments on impact of law and litigation on forest policy. In: Proceedings of 1985 national convention. Society of American Foresters. Bethesda, MD.

Shannon, R. E. 1983. Selected federal public wildlands management laws (Volumes 1 and 2). Forest and Conservation Experiment Station. University of Montana. Missoula, MT.

Shannon, R. E. 1985. The Taylor Grazing Act: A 50-year perspective. Western Wildlands 11(1):2-7.

Siegel, W. C. and F. W. Cubbage. 1984. Environmental protection law and forest management regulation on private lands in the United States. In: Policy analysis for forestry development by A. C. Papastavrou (ed.). Thessaloniki, Greece.

Siegel, W. C. and F. W. Cubbage. 1984. Environmental protection law and forest management regulation on private lands in the United States. In: Forestry legislation by F. S. Schmithusen (ed.). Geneva, Switzerland.

Siegel, W. C., B. M. Kramer and J. D. Mertes. 1984. Forest resource law. In: Forestry handbook by K. F. Wenger (ed.). Society of American Foresters. Bethesda, MD.

Siegel, W. C. and F. W. Cubbage. 1985. Forestry legislative and regulatory trends: Potential impact for the hardwood resource. In: Proceedings 13th Annual Hardwood Symposium of the Hardwood Research Council.

Thompson, A. 1985. Common law, statutes and conservation values: Do they have anything in common? Forestry Chronicle 61:131-134.

Vaux, H. J. 1983. State interventions on private forests in California. In: Governmental interventions, social needs, and the management of U.S. Forests by R. A. Sedjo (ed.) Resources for the Future. Washington, DC.

Wilson, L. W. 1985. The Oregon Forest Practices Act. Journal of Soil and Water Conservation 40:103-104.

21

MANAGEMENT OF FIRE IN FORESTED ENVIRONMENTS

Thomas J. Mills

Reduction of damage due to wildfire has been of interest to forest managers for decades. Such an interest has often been predicated on the notion that the effects of wildfire are universally harmful. In the past, accomplishment of loss reduction objectives were frequently limited by the availability of hand labor and the general remoteness of forested areas (limited access). Such conditions placed a practical limit on fire management investments. Because the effects of fire were overwhelmingly viewed as detrimental (the belief prevailed that all fires must be controlled) and the pool of available fire control resources was limited (total investments were not large), very little attention was directed to the economics of fire management programs—the marginal cost of wildlife control could be greater than the marginal benefit of fire control actions. In addition, the emergency atmosphere surrounding fire management activities contributed to displacement of interest in the economic assessment of fire control actions. Swift and dramatic damages wrought by wildfire, especially the loss of human life, fostered crisis conditions which precipitated immediate action at whatever the cost. Such a crisis-oriented atmosphere has rarely been of concern to decisions about other forest management activities (e.g., silvicultural practices, road construction).

The cost of fire management activities and the benefits resulting from there gradually increased as a concern of forest managers. In the early 1900s, a modest number of economic analyses were undertaken (Headly 1916, Flint 1928, Sparehawk 1925). Unfortunately, such analyses suffered from insufficient economic methodology and quantitative capability. Eventually, however, there evolved the notion of minimizing the sum of fire management costs and fire damages (e.g., the "least cost-plus-loss" notion). When properly applied, the latter is a valid economic efficiency criteria. Unfortunately, seldom was there sufficient supporting methodology or quantitative methods to properly implement the criteria.

Although limited advances were made in understanding the economic efficiency of fire management investments, they were overwhelmed by concern for the stochastic nature of the fire management system. Even if a particular fire management action was economically efficient under "average" conditions, average conditions did not account for the risk presented by more severe fire conditions. Stated another way, there was an implied hypothesis that a fire management program viewed as most efficient under average fire severity conditions was far smaller than the most efficient fire management program required for more severe fire conditions. Inadequacies in methodology and quantitative analysis most certainly inhibited economic efficiency analysis under average fire severity conditions; they were dwarfed, however, by difficulties of joint efficiency and risk analysis of fire management alternatives. Concern with the stochastic dimension of fire management programs was and continues to be extremely important. Economic efficiency information alone is unlikely to have a major influence on fire management decisions unless properly joined with information on risk and probabilities.

The lack of economic efficiency methodology and quantitative analysis capability led the USDA–Forest Service to adopt (in 1934) the "10 A.M." fire management policy (Forest Service 1977, Payne 1982). The strongly suppression-oriented policy stated that fire control actions should be taken to suppress wildfire during the first burning period, or, should that fail, fires should be controlled by 10 A.M. of the next burning period. The assumption underlying the 10 A.M. policy was that the marginal cost of any control action today is less than the marginal damages that will be caused by the wildfire if it continues to burn tomorrow. Supported by the view that wildfire effects were overwhelmingly harmful and by the continued absence of an economic approach for testing such a hypothesis, the 10 A.M. policy persisted until 1978 (Forest Service 1981).

The USDA–Forest Service's 10 A.M. policy was reinforced in 1972 by a similarly oriented policy of controlling all wildfire before it exceeded 10 acres in size (Forest Service 1972). The assumption behind the policy was that the marginal cost of increased initial-attack fire fighting resources was less than the marginal damages of any wildfire that "escaped" to become larger than the 10 acre target. Again, economic capability to test such an assumption was severely lacking; attempts to upgrade such capabilities met with limited success (Gorte and Gorte 1979). One reason for the limited application of economics to fire management was that, unlike silvicultural investments and other profit-oriented concerns, few private companies were directly concerned about the efficiency of fire control activities. Most wildfire control programs are delivered by public agencies (especially federal and state natural resource agencies) and are usually not funded (directly) by the private concerns whose timber resources are protected. Furthermore, since the underlying assumption of most fire control policies rendered economic efficiency irrelevant, public agency research was directed to obtaining a better understanding of fire behavior and fire control technology.

Advances in the application of operations research methods to fire management did, however, occur. Many were not directly applied; they did, however, prove important to latter development of fire management economics (Martell 1982). The Fire Operational Characteristics Using Simulation model (FOCUS) (Bratten et al. 1981), for example, successfully integrated considerable information to simulate fire behavior, dispatch of initial attack crews to fires, and fire containment. The FOCUS model was a valuable tool for the evaluation of alternative initial attack force alternatives, even though it included little information on the economic efficiency of fire management alternatives.

After an almost 50 year submersion, the 1970s saw a renewed interest in the economic efficiency of fire management programs. There was an increased appreciation of wildfire's role in the natural workings of various ecosystems and a recognition that not all effects of wildfire are detrimental (Mutch 1970). Examples include the role wildfire plays in creating a mosaic of habitat types for wildlife and the subtle ways that fire helps perpetuate some preferred commercial timber species. Advances in wildfire control technology also lead to greater concern over program efficiency. New technologies were leading to ever-increasing fire fighting costs. The cost of fire engines, smoke jumpers, helicopters, and bulldozers became much larger than previous costs associated with fire control via hand labor. Also symptomatic of renewed interest in fire management efficiency were questions raised by the Office of Management and Budget (Forest Service 1977) and the U.S. Senate (1978). In response, the USDA–Forest Service revised its fire management policy, discarding the 10 A.M. policy in favor of a policy that included economic efficiency as an objective (Forest Service 1981).

By the 1970s, improved analytical methods, improved computational capacity, greater amounts of data on fire management systems, and advances in economics applied to other areas of natural resource management enabled substantial gains in understanding the economic efficiency dimension of fire management systems. Simard (1976), for example, applied basic concepts of microeconomics to fire management issues, demonstrating that the application of standard economic principles to the fire management question might constitute a difference of "degree" but not of "kind." Advances were also made in the analysis of the economic effects of fire on resource outputs (Mills and Flowers 1986), the economic efficiency of alternative initial attack programs (Schweitzer et al. 1982, Bellinger et al. 1983), the consequences of fuel treatment programs (Barrager et al. 1982), and joint estimates of the expected-value economic efficiency and the probability distribution about that expected value (Mills and Bratten 1988). Although advances have been significant, much remains unknown about the economic efficiency dimension of fire management. Efficiency still plays a relatively minor role in actual fire management program decisions.

Research addressing fire management activities can follow a number of avenues. Each of the latter, however, should demonstrate substantial concern for the development of more effective tools for analysis and evaluation of fire management programs. Important to such development is the application of advanced methods of analysis to a broad range of fire and resource conditions, the intent being to identify systematic patterns in the variables being estimated. Far too often, researchers develop a model and fail to take the next step—determining how a particular variable in question varies under different resource and management conditions. Information about such patterns can have more of an affect on actual natural resource program decisions than models per se. No one is better suited to apply the model and interpret patterns of response than the researcher who built the model.

Strategic research directions involving the economics of fire management would be most significant if they included the following:

- Develop better understanding of net value changes in resource outputs resulting from fire in forested environments.
- Develop enhanced understanding of the efficiency of fire management actions directed at individual fires.
- Develop improved understanding of the economic efficiency of various hazard-reduction treatments and strategies.
- Develop enhanced understanding of the economic efficiency of long-term regional fire management programs.
- Develop improved understanding of risk-efficiency of long-term regional fire management programs.
- Develop risk preferences for the occupance of fire in forested environments.
- Develop information bases needed to support advances in the economics of fire management programs.

Develop Better Understanding of Net Value Changes in Resource Outputs Resulting from Fire in Forested Environments

The notion that fire effects are primarily detrimental to natural resources caused postponement of most research on procedures for estimating fire-caused changes in the present net value of forest outputs. Early research was successful in identifying the major elements to be included in such analyses (Mactavish 1966). Other research efforts, however, contained major flaws in approach (e.g., inclusion of the sunk cost of original stand establishment in the present net value computation). While differing very little from the classical net value computation for silvicultural treatments (with and without the treatment), such concepts require refinement before they can be successfully applied to fire management situations. Specifically, research should focus on

- Improving methods for estimating the change in the present net value of resource outputs with and without wildfire.
- Determining the pattern of change in the present net value as it varies with differing resource conditions, fire characteristics, and resource management objectives.

There is some indication that the net value change (fire management economics term for change in the present net value of resource outputs) is quite sensitive to the completeness of the net value change computation for timber resources (Mills and Flowers 1983). A less complete or simple formulation can yield estimates of net value change that are not only quite different in magnitude from those resulting from a more complete computation but can also be different in sign. That is, a simple formulation of the timber net value change may estimate a modest "loss," while a more complete formulation of net value change will estimate a "gain." For nontimber forest outputs, the net value change due to wildfire appears to be very small and is probably unaffected by the completeness of the net value change formulation. Such is the case for northern Rocky Mountains analyses involving the range resource (Peterson and Flowers 1984), the water resource (Potts et al. 1985) and the recreation resource (Flowers et al. 1984). Further research is needed to determine if the same pattern of net value change among resources exists in other regions of the country.

A particularly important dimension of the problem involves appropriate delineation of the area from which the fire-induced change in resource outputs is measured—the immediate fire site or the management unit within which the fire occurs (Bell et al. 1975, Brown and Boster 1978, Mactavish 1966, McLean 1970, Mills and Flowers 1985, Van Wagner 1979, 1983). Research addressing this aspect of output measurement could greatly affect the net value change due to fire, especially for timber.

Research is also needed on the pattern of net value changes across a broad range of fire management conditions, not only because net-value change estimates are central to any analysis of fire management program alternatives but also because an understanding of net value changes will help focus subsequent research on the broader question of fire management program performance. Such research would provide information similar to currently available information on general patterns of present net value for various silvicultural investments across a broad range of site conditions, treatment types, and management objectives.

Develop Enhanced Understanding of the Efficiency of Fire Management Actions Directed at Individual Fires

Fire management decisions focused on individual fires are made in an environment where time is of essence. Because so little time is available for evaluating management alternatives once a fire has started, there is a need to evaluate the performance of alternative fire management actions under simulated fire conditions. Such implies the availability of analytical tools that can quickly simulate the consequences of proposed management actions. To facilitate the latter, research should focus on

- Improving quantitative models that integrate fire behavior, fire control effectiveness, fire program costs, and net value change dimensions into a computation of the cost-plus-net-value-change of alternative fire management responses to individual fires.
- Applying improved quantitative models to a broad range of initial attack and large fire conditions so as to document the pattern of economic efficiency that might exist among alternative management activities in different fire, resource management, and resource conditions.
- Developing methods that can deliver information on patterns of economic efficiency for different fire management activities focused on individual fires.

Research on initial attacks mounted to control individual fires can build on a fairly rich base of simulation models (e.g., FOCUS model of initial attack (Bratten et al. 1981), recent USDA–Forest Service (1982) model). The primary focus of such research should be to improve estimates of fire management cost and net value change. Relatively little is known about how the efficiency of different types of initial attack actions vary with fire, resource, and management conditions.

Research on the efficiency of alternative fire management actions focused on escaped or large fires is far more challenging. A framework for such analysis is available (Seaver et al. 1983); unfortunately, there is very little modeling capability that can be used by managers facing the prospects of a very large fire. Because of difficulties involved in developing empirical production function information (e.g., size of a large fire as a function of the number and mix of fire control resources applied), large fire problems are far more difficult to address than initial attack questions. The stochastic dimension of changing weather conditions and movement of the fire in response to differing resource conditions (both can be assumed away in initial attack modeling) are complexities that add to problems of large-fire economics research. There are, however, potentially significant payoffs to improved economic efficiency information on strategies for large fire management. Increased mobility and fire management coordination that enhances control of large fires can be very costly. Not uncommon are expenditures of $1 million per day for the suppression of a large fire.

Develop Improved Understanding of the Economic Efficiency of Various Hazard-Reduction Treatments and Strategies

Treatment of forest and range fuels has been a hallmark of fire management for years (e.g., fuel breaks, treatment of slash after harvesting). In the early 1960s, interest in fuel treatment efforts intensified when managers found that exclusion of wildfire in forested areas lead to accumulations of natural fuels. While some analysis of fuel treatment economics has been accomplished, more rigorous analysis is needed. Research should be undertaken to

- Improve methods for estimating the economic efficiency of the treatment of fuels.
- Apply improved methods to a broad range of fuel, fire, and resource conditions to determine major patterns in economic efficiency (if any).

A fundamental question concerning treatment of fuels is whether treatment costs are more than the benefits returned by fuel reduction practices. Slash fuels from timber harvesting, for example, deteriorate naturally (Salazar and Bevins 1984); thus, fuel treatment may only reduce the chances of more severe fire behavior. Reducing such chances is, in turn, important only to the extent that a fire will occur when nontreated fuels are in existence.

Research on the economics of fuel treatments has been undertaken in the northern Rocky Mountains (Wood 1978), the Southwest (Hirsch et al. 1979), and the Pacific Northwest (Barrager et al. 1982). Such research has pointed to very low economic returns to fuel treatment practices. The research, however, has serious flaws or simplifying assumptions (e.g., Wood's assumption that the greatest net value losses would occur in sawtimber stands. In a timber net value change analysis, Mills and Flowers (1985) found that the greatest net value losses occur in poletimber stands in the northern Rocky Mountain region). The relationship between fire hazard reduction objectives and site preparation objectives (e.g., slash treatments) also needs to be researched before conclusive information can be developed on the relative efficiency of fuel treatments.

Develop Enhanced Understanding of the Economic Efficiency of Long-Term Regional Fire Management Programs

The size of an organization charged with fire management responsibilities is fixed in analyses of alternative initial attacks on a single fire. The location and occurrence of the fire is also given. Of concern to managers facing an individual fire is: How many and which of the available initial attack units should be dispatched to a fire? Neither the size of a fire management program nor the number of fires are fixed in analyses of alternative fire management programs for large areas (regions). In such cases, fundamental concern turns to judgments about the most efficient level and mixture of fire management inputs (e.g., initial attack and large fire suppression resources), given the heterogeneous nature of resources within a region and the probability of fire starting at a particular location under certain weather conditions. Addressing such a concern requires careful integration of a number of factors that are physical (fire behavior), biological (fire effects), economic (resource values and costs of fire management actions), statistical (fire occurrence and weather frequencies), organizational (initial attack, fuel treatment, prevention, detection, and large fire suppression) and managerial (fire suppression constraints and resource management objectives) in nature. Failure to address these basic dimensions will lead to answers that may be incorrect in magnitude and sign. Research should focus on

- Improvement of models for evaluating alternative fire management programs for broad regions.
- Application of improved models to determine patterns of economic efficiency that may exist among fire management program alternatives in different regions.

A number of analyses have been made of strategic regional directions for fire management. Schweitzer and others (1982) focused research on six national forests—research that was later expanded upon (Forest Service 1982, Mills and Bratten 1988). Unfortunately, the models developed by such research have numerous shortcomings that could affect economic efficiency estimates. In this respect, Mills and Bratten (1988) list several shortcomings, including simulation of initial attack on fires and calibration of fire behavior distribution for all wildfire in a region. With nationwide fire management expenditures approaching $500 million per year, research on regional fire management strategies could have significant payoffs.

Develop Improved Understanding of
Risk-Efficiency of Long-Term Regional Fire
Management Programs

Estimating the economic efficiency of alternative fire management programs for broad regions is a challenge, but expected value economic efficiency is only one portion of the decision calculus. The prominent stochastic dimension of a fire system continues to be of significant concern. While expected value economic efficiency estimates are important, their augmentation by information about the probability distribution of expected value is likely to be much more important to program decision makers. Research should be focused on

- Improving models that can estimate both the expected value economic efficiency of alternative fire management programs for a region as well as the probability distribution about that expected value.
- Application of improved models to determine how the shape of the probability distribution of economic efficiency varies with different fire management program alternatives and different regions.

As in the 1930s, many managers continue to believe that the most efficient fire management program in the average year is smaller than the most economically efficient program required for more severe fire years. The first credible test of this perspective was by Schweitzer et al. (1982). Studied were alternative levels of initial attack in three years of differing fire severity on six national forests located throughout the nation. In five of the six national forests studied, Schweitzer et al. (1982) found that the most economically efficient level of initial attack in the year of "moderate fire severity" was also the most efficient program level in the "high fire severity" year. In only one of the national forests did an increase in the severity of the fire year lead to a higher, more efficient fire management program.

Mills and Bratten (1988) developed a more rigorous model to treat the probabilistic dimension of fire occurrence and fire weather severity. Applying the model to case study situations in the northern Rocky Mountains, Mills and Bratten (1988) found that how the risk dimension was articulated determined whether a risk-averse decision maker would select a higher fire management program level than a risk-neutral decision maker. Described in traditional fire management form, the answer was the same as that found by Schweitzer et al. (1982)—there was no risk justification for the decision maker to select a higher level fire management program.

Risk-efficiency research is one of the most challenging in the fire management arena. Because the research involves considerable data and specification of several interrelated components of the fire management system, risk-efficiency research should not proceed if research funds are highly constrained. While expensive, such research focuses on the very heart of debates about fire management programs. While certain to be controversial, research of this nature could result in very significant payoffs.

Develop Risk Preferences for the Occurrence of Fire in Forested Environments

The fire management community has long been of the opinion that a risk adverse attitude was acceptable. Some, however, would argue that public agencies should be risk neutral, since public risks are spread among a large number of events (Arrow and Lind 1970). Any single wildfire, for example, is small in relation to the public's total investment in natural resources—no matter how dramatic the fire. Others argue that the public is averse to the risk of fires, and that public decision makers should also be risk averse. Research on risk preferences should focus on

- Adapting to fire management programs, improved risk preference techniques that have been developed in other fields.
- Applying risk preference measurement techniques to determine public risk preferences to fire hazards and associated public decision maker risk preferences.

Research is needed to determine whether risk aversion is justifiable as a fundamental philosophy guiding fire management programs. Are there sufficient irreversible effects of fire that would warrant a risk stance that is different from that warranted by decisions about other public programs? If so, what are the implications of such a stance and how can they best be managed? To focus such research, efforts should be directed to determine the risk preference patterns of well identified segments of the public to various fire situations. For example, Gardner et al. (1985) conducted a survey of 1,646 members of organized conservation groups and found that 58 percent favored the use of prescribed fire—even if an occasional fire escaped control. In a similar study, Cortner et al. (1984) recorded that 72 percent of 1,200 residents surveyed in the Tucson, Arizona metropolitan area were willing to bear the risk of an escaping prescribed fire, given their knowledge of benefits to be gained from such a forestry practice. Research of this nature indicates the public's willingness to assume a more risk-accepting stance than is often reported by fire management specialists. Carefully crafted research could provide valuable insights to the risk dimension of fire management—a dimension that could improve techniques for measuring the probability distribution about expected value performance variables.

Develop Information Bases Needed to Support Advances in the Economics of Fire Management Programs

Research in a number of auxiliary areas must advance before research on the economics of fire management can proceed in a substantial manner. Economic analysis of fire management programs is an integrator of diverse physical, biological, and economic information—an integration process which often leads to measures of economic efficiency. The success of economic analyses is dependent upon the availability of physical and biological information. In this respect, research is needed to

- Improve data on the physical and biological impacts of fires of varying severity on various resource outputs (e.g., timber, recreation, water, wildlife).
- Improve production function information that defines the effectiveness of initial attack efforts and large fire suppression efforts.
- Improve probability distributions that measure fire behavior for the full population of fires to be faced by fire management programs within regions.

Advances in the above areas are especially important for research involving the economic efficiency and risk dimensions of fire management programs applied across broad regions (Mills and Bratten 1988).

Summary and Conclusions

Economics research focused on fire management activities has trailed economic research focused on other forest management practices and programs. In part, this discrepancy exists because methods of economic analysis appropriate to fire management have not been sufficiently developed. Fortunately, significant progress has been made since the mid-1970s. Future research should strategically seek to improve the forestry community's understanding of net value changes in resource outputs resulting from fire, efficiency of fire actions on individual fires, effectiveness of fuel treatment practices, efficiency of long-term regional fire management strategies, preferences for fire risk by the public, and supporting information necessary for economic evaluation of fire management alternatives.

The potential payoff of economic analyses applied to fire management programs is substantial. Advances may well, however, be difficult and expensive, especially the development of techniques for jointly estimating the economic efficiency and risk performance of alternative fire management programs. Regardless of research directions chosen, however, economics research on fire management should proceed with full appreciation of the history of fire management programs. Continuing to exist is a strong belief that fire control is intuitively good and that careful economic analysis of program costs and benefits is not always necessary. Any economics research on fire management issues should proceed with full knowledge of such convictions.

Literature Cited

Arrow, K. J. and R. C. Lind. 1970. Uncertainty and the evaluation of public investments. American Economic Review 60:364-378.

Barrager, S. M., D. Cohan and P. J. Roussopoulos. 1982. The economic value of improved fuels and fire behavior information. Research Paper RM-239. Rocky Mountain Forest and Range Experiment Station. USDA Forest Service. Ft. Collins, CO.

Bell, E. F., R. Fight, and R. Randall. 1975. ACE the two-edged sword. Journal of Forestry 73:642.

Bellinger, M. D., H. F. Kaiser, and H. A. Harrison. 1983. Economic efficiency of fire management on nonfederal forest and range lands. Journal of Forestry 8:373-375, 378.

Bratten, F. W., J. B. Davis, G. T. Flatman, J. W. Keith, S. R. Rapp, and T. G. Storey. 1981. FOCUS: A fire management planning system. General Technical Report PSW-49. Pacific Southwest Forest and Range Experiment Station. USDA Forest Service. Berkeley, CA.

Brown, T. C. and R. S. Boster. 1978. On the economics of timber damage appraisal for forests. Journal of Forestry 76:777-780.

Cortner, H. J., J. J. Malcolm, E. H. Zwolinski, and E. H. Carpenter. 1984. Public support for fire management policies. Journal of Forestry 82:359-361.

Flint, H. R. 1928. Adequate fire control. Journal of Forestry 26:624-638.

Flowers, Patrick J., Henry J. Vaux, Jr., Philip D. Gardner, and Thomas J. Mills. 1984. Changes in recreation values after fire in the northern Rocky Mountains. Research Note PSW-373. Pacific Southwest Forest and Range Experiment Station. USDA Forest Service. Berkeley, CA.

Forest Service. 1972. National fire planning. U.S. Department of Agriculture. Washington, DC.

Forest Service. 1977. Evaluation of fire management activities on the national forest. U.S. Department of Agriculture. Washington, DC.

Forest Service. 1980. National Forest System fire management budget analysis. U.S. Department of Agriculture. Washington, DC.

Forest Service. 1981. Forest Service Manual [FSM] 5130.3: Fire suppression policy. U.S. Department of Agriculture. Washington, DC.

Forest Service. 1982. Forest Service Manual [FSM] 5109.19:Fire management analysis and planning handbook. U.S. Department of Agriculture. Washington, DC.

Gardner, P. D., H. A. Cortner, K. F. Widaman and K. J. Stenberg. 1985. Forest-user attitudes toward alternative fire management policies. Environmental Management 9:303-312.

Gorte, J. K. and R. W. Gorte. 1979. Application of economic techniques to fire management: A status review and evaluation. General Technical Report INT-56. Intermountain Forest and Range Experiment Station. USDA Forest Service. Ogden, UT.

Headly, R. 1916. Fire suppression district 5. Office Report. Rocky Mountain Forest and Range Experiment Station. USDA Forest Service. Ft. Collins, CO.

Hirsch, S. N., G. F. Meyer, and D. L. Radloff. 1979. Choosing an activity fuel treatment for southwest ponderosa pine. General Technical Report RM-67. Rocky Mountain Forest and Range Experiment Station. USDA Forest Service, USDA. Ft. Collins, CO.

Mactavish, J. S. 1966. Appraising fire damage to mature forest stands. Department Publication 1162. Canada Department of Forestry and Rural Development. Ottawa, Ontario, Canada.

Martell, D. L. 1982. A review of operational research studies in forest fire management. Canadian Journal of Forest Research 12:119-140.

McLean, D. L. 1970. Appraisal of damage to immature timber. Information Report FF-X-22, Ser. SB 764-c3. Canada Department of Fisheries and Forestry. Canadian Forest Service. Forest Fire Research Institute. Canadian Fisheries Research Board. Ottawa, Ontario, Canada.

Mills, T. J. and F. W. Bratten. 1988. Economic efficiency and risk character of fire management programs: Northern Rocky Mountains. Unpublished report. USDA Forest Service. Washington, DC.

Mills, T. J. and P. J. Flowers. 1983. Completeness of model specification: An illustration with fire-induced timber net value change. In: Proceedings of seventh conference on fire and forest meteorology. American Meterological Society. Washington, DC.

Mills, T. J. and P. J. Flowers. 1985. Fire-induced changes in net value of timber: Sensitivity to fire size, fire severity, and management objectives. Canadian Journal of Forest Research 15:973-981.

Mills, T. J. and P. J. Flowers. 1986. Wildfire impact on the present net value of timber stands: Illustrations in the northern Rocky Mountains. Forest Science 32:707-724.

Mutch, R. W. 1970. Wildland fires and ecosystems: A hypothesis. Ecology 51(6):1046-1051.

Peterson, D. L. and P. J. Flowers. 1984. Estimating postfire changes in production and value of northern Rocky Mountain-Intermountain rangelands. Research Paper PSW-173. Pacific Southwest Forest and Range Experiment Station. USDA Forest Service. Berkeley, CA.

Potts, D. F., D. L. Peterson and H. R. Zuuring. 1985. Watershed modeling for fire management planning in the northern Rocky Mountains. Research Paper PSW-177. Pacific Southwest Forest and Range Experiment Station. USDA Forest Service. Berkeley, CA.

Pyne, S. J. 1982. Fire in America, a cultural history of wildland and rural fire. Princeton University Press. Princeton, NJ.

Salazar, L. A., and C. D. Bevins. 1984. Fuel models to predict fire behavior in untreated conifer slash. Research Note PSW-370. Pacific Southwest Forest and Range Experiment Station. USDA Forest Service. Berkeley, CA.

Schweitzer, D. L., E. V. Anderson and T. J. Mills. 1982. Economic efficiency of fire management programs at six national forests. Research Paper PSW-157. Pacific Southwest Forest and Range Experiment Station. USDA Forest Service. Berkeley, CA.

Seaver, D. A., P. J. Roussopoulous, and A. N. S. Freeling. 1983. The escaped fire situation: Decision analysis approach. Research Paper RM-244. Rocky Mountain Forest and Range Experiment Station. USDA Forest Service. Ft. Collins, CO.

Simard, J. A. 1976. Wildland fire management: The economics of policy alternatives. Forestry Technical Report 15. Forest Fire Research Institute. Canadian Forestry Service. Department of Environment. Ottawa, Ontario, Canada.

Sparhawk, W. N. 1925. The use of liability ratings in planning forest fire protection. Journal of Agricultural Research 30:693-762.

U.S. Senate. 1978. Report on Department of Interior and related agency appropriations bill. Report 95-1063. Washington, DC.

Van Wagner, C. E. 1979. The economic impact of individual fires on the whole forest. The Forestry Chronicle 55:47-50.

Van Wagner, C. E. 1983. Simulating the effect of forest fire on long-term annual timber supply. Canadian Journal of Forest Research 13:451-457.

Wood, D. B. 1978. Fuel treatment opportunities on the Lolo National Forest: An economic analysis. Research Note INT-272. Intermountain Forest and Range Experiment Station. USDA Forest Service. Ogden, UT.

22

MANAGEMENT OF INSECTS AND DISEASES IN FORESTED ENVIRONMENTS

Lloyd C. Irland

Forest pest management is important to forestry for several reasons. First and most obvious is the extraordinary level of mortality, growth impact, and quality loss imposed each year by forest pests on forest productivity (Jacobi et al. 1981). Mortality alone is similar in magnitude to total industrial wood removals, though much of the pest-induced losses may not have current net economic value. Damage to plantations as well as to mature timber is often severe in local areas. Pest concerns can also limit species selection for planting and limit silvicultural options for stand treatment. Past investments in planting or other treatments increase the willingness of managers to apply pest control treatments to salvage past investments.

Second, costs of responding to forest pests with pesticides or management measures are high and have been rising. Despite the increased application of sophisticated remote sensing technology to pest detection, survey and detection costs are significant in any serious program of integrated management. Uncontrolled pest activity may affect the environment, but spraying and silvicultural treatments can also have adverse effects. Frequently, pest managers recommend costly programs of anticipatory hazard reduction; such programs are often difficult to fund and have uncertain and deferred benefits. Managers are asked to apply diligent efforts to future risks, even though already inundated with immediate crises caused by forest pests. In contrast, when forest pests are on the rampage, managers are called on to propose and manage costly treatment and salvage programs in an atmosphere of urgency.

Finally, pest management actions are often highly controversial, even when they do not involve pesticides. The intensity of such controversy alone would seem a sound argument for better economic and policy analysis applied to the subject. The USDA–Forest Service is now at work on a forest pest management strategy. The preparation of that strategy could benefit much from a better base of economic analysis.

In agriculture, extreme instances of excessive pesticide use, creation of damaging secondary pests, and adverse side effects on wildlife and water quality have led to aggressive research and application efforts in pest management. At its highest development, a system of integrated pest management includes efforts to define minimum acceptable losses and economic treatment thresholds; to carry out advance cultural measures, species selection, and crop rotation techniques to reduce favorable pest habitat conditions; and to avoid inducing pest resistance to pesticides, and still maintain adequate crop yields. Sophisticated systems of integrated pest management have been developed for apples, cotton, and other crops.

Forest pest management has also seen dramatic transitions in approach, technology, and philosophy. There is a higher appreciation of the ecological basis of pest problems and a stronger commitment to analyzing management and anticipatory solutions. In a limited number of instances, such action has risen to a level that could truly be called integrated pest management (Coster and McGregor 1983). There is increased recognition of the importance of survey and detection and of pest complexes as a cause of problems—not just a single pest acting alone. There is widespread recognition that full control is not possible or desirable, and an awareness that there are no mutually exclusive treatment options—there are only different mixes of management and treatment activities. Budget and political pressures have led to more stringent scrutiny of project proposals.

Although pest management is an important and costly function in forest management, attention devoted to the subject by forest economics researchers has not been plentiful (Konrad 1984). A review of the literature quickly reveals the truth of such a statement. As defined here, pest management is principally concerned with insects and diseases; wind, snow, and other damaging agents are excluded. What strategic directions might forest economics research focusing on forest pest management assume? Based on a survey of selected pest management literature, on the author's experiences with and perceptions of forest pest management, and on the reaction of 18 experts on pest management research and application, five strategic research directions are suggested:

- Organize and evaluate biological knowledge important to forest pest management decisions.
- Develop a sound basis for benefit cost analysis of alternative mixes of forest pest treatments.
- Develop improved understanding of the risk of forest ownership and management as influenced by forest pests.
- Develop improved understanding of technological changes affecting forest pest management.
- Develop more effective pest management programs via the application of policy analysis and program evaluation.

Organize and Evaluate Biological Knowledge
Important to Forest Pest Management Decisions

Much of the difficulty in economic analysis of pest management decisions has stemmed from frail biological data and associated projections. The economic analyst needs initial forest conditions and future conditions about tree quality, size distributions, and harvest potential both with and without treatment. Too often, these relationships are poorly defined for making multi-decade projections. Many of the difficulties in economic analyses trace directly to such weaknesses (Stark 1987). Some pests, such as spruce budworm, come and go over long time spans. Since forty or more years may elapse between outbreaks, judging which stands may benefit from a heavy "budworm thinning," which stands will be killed, and which stands may offer the best protection opportunities can be very difficult. While treatment focused on the best timber producing sites may seem most appropriate, there may be practical limitations to such a policy. Also, what may be considered the most effective policy as an outbreak is spreading may not be appropriate once the outbreak is well established throughout the host forest.

Declining biological impact of pests is also in need of understanding. For example, recent gypsy moth outbreaks in Connecticut are having less impact on timber mortality than earlier outbreaks. The pest appears to have habituated itself to the area's environment and perhaps has acquired a suite of predators, thus reducing its impact on the forest. In other examples, pest impacts are often most severe on the marginal sites where trees are already under moisture or other forms of stress. The biological challenge is to accurately identify these situations; the economic and management challenge is to appraise the merit of treating sites having low growth rates.

Uncertainty as to the optimum point to begin spray treatments often exists because of limited biological understanding of forest pests. *Ex post* analyses of spruce budworm control, for example, suggest that intense and early intervention would have been preferable to the policy actually pursued—which was to wait until spraying would make the difference between tree survival and death. The California Medfly episode illustrates the same principle.

Though integrating available knowledge of this kind is not strictly economics, the lack of such integration poses difficulties for economic and policy analyses. The economic and policy analysis community should take a leadership role in working with biologists and control specialists to remedy the situation. Over a period of time, it should be possible to provide syntheses of such information, much like those developed by large programs involving Gypsy moth, spruce budworm, and western budworm (e.g., Hudak and Raske 1982, Montgomery et al. 1983, Amman et al. 1977, Cole and McGregor 1983, Forest Service 1986b, Brookes et al. 1987). A problem, however, is that information of this sort is required in advance of severe outbreaks, a period when spending of funds and time may be difficult to justify.

Specific opportunities for enhancing the integration of knowledge concerning the biology and economics of pest management include the following.

Assess the adequacy of knowledge concerning pest impact and management, and prepare syntheses which integrate biological and economic information. Identify critically important forest pests; determine those for which an inadequate information base exists. Establish a priority ranking for filling information gaps over a five-year period and assign interdisciplinary teams the task of preparing syntheses that integrate economic and biological information. Also, identify loss scenarios with and without treatments for the most likely combinations of site and stand conditions and potential treatments.

Identify and assess management actions addressing forest pests, especially growth and yield implications of forest pest infestations. For example, identify and assess the importance of secondary forest pests, the role and means for advance warning indicators of hazard, and the role of forest management or other conditions in outbreak development. Determine the extent to which existing growth and yield tables account for pest impacts in managed and unmanaged stands.

Develop a system to facilitate access to current pest management information. Develop a systematic program to foster interaction between the economics, management, biological, and pest survey and detection communities. Of special interest should be ensuring the currency of biological, economic, and management information so that quick action can be taken when new field observations, control experiences, or research indicate a need to revise accepted views. Such an approach could be patterned after efforts of research groups currently working on application of artificial intelligence, expert systems, and decision support systems for pest management (Twardus and Brookes 1983, Rauscher 1987). Significant problems facing development of information programs are identification (with more precision) of potential users, description of information requirements, and development of modeling systems that are not excessively complex or that do not make assumptions which current knowledge cannot support. Aggressive involvement by application-oriented economists in the development of information programs is essential in order to ensure that managerially and strategically relevant questions are asked and that displays of results emerge in a form suitable for management use and economic analysis. A number of modeling efforts and syntheses deserve note: Baskerville 1984, Cuff and Baskerville 1982, MacLean 1983, Seymour and Lemin 1987, Sewall Company 1983.

Develop a Sound Basis for Benefit Cost Analysis of Alternative Mixes of Forest Pest Treatments

Forest pest management has often been viewed as simply applying one more management practice in the forest, subject to standard financial analysis. Foresters conducted sanitation cuts, sprayed insecticides, or salvaged stands in hopes of controlling the impact of pests. Today, however, most forest management strategies recognize that pests require a multifaceted program of survey and detection, silvicultural and species plans, direct control measures, planned loss taking in selected areas, salvage, monitoring, and ongoing evaluation and control. The benefit cost analysis of such multifaceted programs is far more complex than analyzing individual management practices. A variety of such work has been done; see for example Bible 1985, Irland 1977, Huff et al. 1984, de Steiguer et al. 1987.

Specific opportunities that would enhance the basis for benefit cost analysis of alternative forest pest treatment mixes include the following.

Evaluate the concept of economic threshold as applied to forest pest management. Developing the basis for effective benefit cost analyses of pest management activities requires readily available, correctly formatted economic information about forest pest control and detection processes. To meet such information needs, further work is needed to determine the extent to which treatment threshold concepts, as used in agriculture, actually apply in forestry. Some suggest that because of forestry's long time horizons and many uncertainties, the treatment threshold concept fails when applied to forestry activities (for a useful discussion in another context, see Walstad and Kutch 1987, pp. 4-8). Also, many important forest insects exhibit only one generation per year. A failure to administer prompt protection cannot be remedied for an entire year, and there may be many ambiguities in interpreting such a failure.

Review and evaluate factors affecting forest pest treatment costs, with a focus on development of guides for project cost estimating. Effective application of benefit-cost analysis to forest pest activities requires an understanding of many variables affecting treatment cost, how such costs change over time, and the stochastic uncertainties in treatment success. For example, a spray treatment or silvicultural hazard reduction treatment may be subject to wide uncertainty as to ultimate effectiveness. Their usefulness may hinge on uncertainties regarding other hazards (e.g., wind, fire) or secondary infestations of bark beetles. The statistical and analytical machinery exists to handle such uncertainties; unfortunately, available biological data is limited.

Develop more effective methods for valuation of losses from forest pests. An often critical issue to forest management is the valuation of losses from forest pests. On the surface, valuing timber losses would appear straightforward, encumbered by few difficulties. Such is not so. Placing a value on timber losses requires assumptions about timber prices, extraction costs, product prices, logging and milling technology, and value of damaged or killed trees (Irland and Runyon 1984, Baumgartner 1987). For example, the development of a waferboard industry in the Rocky Mountain Region gives stumpage value to forest stands where no timber values previously existed. In addition, jobs and local incomes now depend on this resource, a change occurring in but a few short years. But in situations where a species of timber is in surplus supply (for whatever reason), the value of timber attributed to incremental losses resulting from a forest pest may be zero from a social perspective. While such a perspective may be unacceptable to landowners, public land managers, and even to some pest management professionals, it needs to be considered. Current debates over the net residual stumpage value of green timber in much of the Rocky Mountain Region illustrate how difficult such questions are to resolve (LeMaster, Flamm, and Hendee 1987).

Since a control procedure is often an investment in future forest productivity, it will be just as sensitive to assumptions concerning prices, interest rates, and product yields as any intensive management decision. There is still room for economics research that will give managers clearer guidance in developing prudent price assumptions to guide decisions. The concept of a "shadow price" can be useful. Shadow prices (originating in linear programming) are incremental values (mathematically expressed) attributable to marginal changes in an input or output. In terms of valuation, a shadow price is a properly determined social price which is in contrast to a market price that may be biased by taxes, subsidies, externalities, or market imperfections. The point is that the current market price for a stand of timber may be greater or less than its true shadow price. The latter measures real value to society.

Evaluate application of stand level versus forest-wide concepts of forest pest valuation. Timber valuation can be considered from a stand as well as a forest perspective (Brown and Boster 1978, Van Wagner 1979). However, relatively few analyses have compared long-run harvest potential under a range of pest control options for entire forests or regions (Bousfield et al. 1984, Sewall 1983). Such analyses contain many uncertainties; they need to be approached with caution. At the same time, they shed additional perspective on long-term aspects of pest management decisions. Economists and operations researchers have developed a high level of skill and experience with such analyses. It has been determined, for example, that because of allowable cut effects, successful pest management (e.g., fire control) can actually reduce long-term yields by disrupting the transition to a more regulated age class distribution. That is, when there is surplus old growth (which is most often vulnerable to defoliators or bark beetles), the long-run value of preserving particular stands can be zero since there is excess growing stock compared to the amount needed to support the long-run planned harvest level (the surplus is less under the long rotations mandated on public lands). Therefore, the analysis of impact, valuation of loss, and design of treatment response must consider forest-wide and regional effects on volume and prices as well as individual site relationships.

Develop workable methods for appraising nontimber effects of pests and pest management activities. If the valuation of timber losses is fraught with both biological and economic perils, so much more so is the valuation of nontimber effects of pest activity. Unfortunately, only limited economics research has been focused on the importance of pest-induced changes on watershed water yields, on big game habitat, on habitat diversity for birds, and on aesthetic values. Information of this sort can be especially important in wilderness or other noncommercial timber stands. For example, in parts of the Rocky Mountain and Cascade regions (where some timber values may be marginal, some arguing negative), public concern has led to increasing emphasis on control of western budworm for purposes of preserving forest scenery. An environmental impact statement for the Carson National Forest (Forest Service 1985) illustrates the difficulties of performing economic analyses of pest-induced, nontimber forest impacts. Numerous case studies on pest-induced impacts on nontimber output have been undertaken; primary research, however, is costly and not always workable for control project planning. Some basis for extrapolation from case studies, or forming credible shorthand estimates, must be devised. Major payoffs can result from assembling available biological pest information, integrating such information with sound social science concepts, and using such information to guide costly pest management procedures aimed primarily at nontimber benefits.

Evaluate anticipatory hazard reduction treatments for forest pests. Increasingly important are various forms of preventive silviculture which are recommended for both defoliators and bark beetle complexes in major timber types. The biological basis for prescriptions is often less than desirable; there is also a tendency to judge the prescriptions as failures after applying "too little too late." Both in developing the basis for economic analysis and in later program evaluation, anticipatory hazard reduction treatments deserve far more economic analysis. Such treatments are costly and are often applied well in advance of the urgent motivation supplied by "brown needles." Restricting capital flows for long periods of time, they need well grounded economic justification.

Develop useful guidelines for the economic evaluation of salvage projects resulting from forest pests. When decisions are made to take mortality losses rather than apply direct pest control practices, there remains the decision to salvage or not to salvage. In addition to obvious concerns over fire hazard reduction, nontimber values (e.g., watershed and wildlife values), and secondary pests (e.g., bark beetles in down timber), a better understanding of the costs of salvage logging relative to product values would be a useful avenue for research (Baumgartner 1987, Forest Service 1986a, Lowery 1982, Znerold 1985).

Evaluate existing agency guidance on the application of economic analysis of forest pests. There is nothing special about the techniques and formatting of economic analysis applied to forest pests. However, review of public agency guidance (e.g., USDA–Forest Service regional guidance and state practice) concerning the preparation of economic analysis of pest control actions may be useful. Suggested economic procedures should be checked to assure that incomplete or incorrect economic methods have not accidentally found their way into field procedures. A careful student of USDA–Forest Service economic analyses for National Forest System pest control projects may find instances of incredibly high benefit cost ratios and of benefits being attributed to "no-action" alternatives.

Improve methods for evaluating "no-action" options for pest management. There is a major need for more careful economic analysis of the impacts of the "no action" or the "business as usual, take the losses" alternative to forest pest management. Numerous examples of wildly exaggerated costs of "no action" exist; it is a responsibility of research to apply liberal doses of exaggeration control. Extreme claims as to the projected direct and secondary impacts of uncontrolled outbreaks have caused considerable damage to the credibility of public pest control programs. This is indeed unfortunate, given that pest control measures are often essential to intensive forest management. While analysis of tourism impacts of brown foliage and grey tree skeletons is rudimentary, the basic outline of prudent procedure for local impact analysis is no secret to economists. The issue is one of integrating such information more properly into applied analysis (Manning 1982, Milne 1986, Agricultural Enterprises, Inc. 1986, Runyon 1980). In this connection, improved approaches to displaying the complex and basically incommensurable positive and negative impacts of pest activity and control interventions are badly needed. These issues include ethical questions that often arise in pest management decisions. Thoughtful guidance from economists and other social scientists could be most helpful in this respect. Just how to shape this concern into a research agenda, however, is less than clear.

Develop Improved Understanding of the Risk of Forest Ownership and Management as Influenced by Forest Pests

Insects, diseases, and other damaging agents pose a risk for private and public owners of forests. Not only do such pests increase uncertainty regarding projected stand and forest-wide yields, they also increase the variability of annual management costs and thereby have a doubling effect on the variance of net returns to owners. Early foresters recognized this consequence and subsequently advocated forest fire control as a means of reducing variability in net returns, thus improving the incentive for private owners to hold wealth in the form of standing timber. Despite many statements in forestry and conservation literature concerning the high risk of forest landownership due to pests and fire, there exists very little high quality statistical analysis that measures and compares such risks across regions, timber types, and time. Research may well determine that separating out insect and disease risk is too difficult or not useful, in which case a broader view of risks might have to be taken.

Identify and seek agreement on workable statistical measures of risk; empirically apply such measures to major forest pests. A number of specific questions can be raised when thinking about risk—consider two. First, what is the proper statistical measurement of risk? A few case studies have offered risk measures, yet consensus concerning proper methodology is lacking (Anderson et al. 1987). Second, what is the proper organizational posture toward risk? The likelihood of extensive spread of pests has given rise to occasional tendencies to conduct aggressive control operations with little regard for technical feasibility and cost-effectiveness. The DDT Siegfried Line that proposed to contain gypsy moth in New England is but one example. The policy of costly slash burning on clearcut sites in the northern Rockies because of perceived unacceptability of fire hazard is another.

Identify and assess forest landowner postures toward risks imposed by forest pests. An improved basis for designing an organization's posture toward risk should be developed. Such could take advantage of the extensive body of work in statistical decision theory, optimizing under uncertainty, theory of insurance, and portfolio management. A question of concern to analysts of long-term public works investments, and which is highly relevant to forestry, is whether public and private investors should have different postures toward risk.

Research to design shorthand or condensed methods of measuring and expressing risk or relative degree of risk should be undertaken. Such should focus on economical ways of integrating risk into investment analyses of property acquisition and valuations as well as incremental investments in timber growing. Decisions trees are one procedure for doing so (Talerico et al. 1978, Freeling and Seaver 1980). Once a way of integrating risk into investment analysis has been developed, it should be empirically applied to determine if investors systematically overestimate the risk of timberland ownership. Judging by statements of timberland owners and investment professionals, a consensus on this point does not presently exist.

Identify and evaluate relationships between forest management activities and changes in the level of risk imposed by forest pests. One of the most common pest management prescriptions involves more intensive treatments during the life of a stand—to improve vigor of growth, to modify species composition, to otherwise modify habitat in ways unfavorable to a pest, or to remove high risk or infected individuals (sanitation cutting) before they infect others. But these treatments bring a risk of root and stem damage to residual trees and possibly introduce slash and other sources of hazard. The road needed for access may increase fire risk by increasing hunting and recreational use. Successful treatments, by increasing stand value, may also increase values at risk to some other pest or mischance.

There is a deeper issue to consider. Some experts have argued, for example, that western budworm hazard has been increased by fire prevention and cutting practices which have increased the abundance of vulnerable species (Gara et al. 1984, Williams 1980). Elsewhere in the West, fire exclusion, low cut levels, and long rotations have eliminated patchwork mosaics of vegetation, replacing them with extensive, uniform, single-species, over-mature stands which are excellent bark beetle habitat. In fact, the principal ecological role of the beetle may be to maintain vigorous nutrient cycling by eliminating stagnated old stands. Some observers have argued that past cutting practices increased the abundance of fir in spruce-fir forests, increasing both susceptibility and vulnerability to spruce budworm. There may be legitimate debate about the validity of these arguments, but they suggest that past actions or inactions may have helped create present and future pest problems (Leuschner 1986). So pest management economics must involve a searching, ecologically based critique of present management practice in order to be fully effective. Some writers go beyond this and argue that many impacts of tree-damaging insects in the forest are ecologically beneficial, and that such benign impacts need to be considered as well.

Clarify long- and short-run questions of risk and assess the value of diversification as a means of addressing risk imposed by forest pests. A different perspective may be needed for short-term versus long-term risks imposed by forest pests. Certainly the issues raised previously above concerning valuation are relevant. If biological losses below a certain threshold are not economically significant, such should be reflected in a reduced assessment of risk over the long term. And in a forest-wide context, the benefits of diversification across species and sites may mean that stand-level measures of risk need to be fitted into a wider context to have meaning for forest management decisions. This is related, of course, to well-seasoned discussions of forest fire insurance: Why is such insurance not commercially available and should government create a market in such contingent claims?

A practical question before some organizations is the extent to which diversification by timber types and regions will reduce overall risk in a managed portfolio of timberland (Irland and Howard 1987). Once better measures of actual risk are in hand, such questions can be more effectively answered. To date, the few studies on diversification concern the role of timberland in diversified portfolios from the viewpoint of price risk.

Assess the manner in which private firms make forest pest management decisions. Most discussions of risk imposed by forest pests assume governmental decision making, especially as related to public lands. Yet private owners also make pest management decisions. Well developed case studies of how private concerns evaluate risks and decide treatment investments would be of special value.

Develop Improved Understanding of Technological Changes Affecting Forest Pest Management

Technological change has been rapid in forest pest management, often exceeding administrative ability to absorb and effectively implement. Yet there exist chronic technical problems that have been resistant to solution by sustained research programs. In insect control, changes in pesticide availability and aircraft types have regularly placed managers and researchers back to earlier points on new learning curves. At the same time, there exist promising technologies in the form of new unregistered pesticides, new detection and remote sensing methods, new concepts in application, and new outbreak and impact forecasting systems. Unfortunately, many of these new technologies languish for lack of resources to expand their development.

Some factors that hinder development of pest management technology are the high cost of individual trials, the long time lags before treatment results can be known with confidence, the frequent availability of only one generation per year of target insects, and the numerous weather, application, and other factors that can upset test plans or complicate interpretation of test results. The many costs and delays involved in registering improved insecticides are subject of frequent complaint by forest managers and the pesticide industry (General Accounting Office 1986). Annual research funding cycles and the difficulty of initiating research in advance of need—and maintaining it once the brown needles have faded—are additional obstacles. The sheer number of pests and diseases which could present future problems tends to fragment research and development efforts. These difficulties are all the more reason why effective economic analysis should be brought to bear on technological change in pest control.

Undertake and assess case studies of technical innovations important to the management of forest pests. Focused case studies of the invention and adoption process in pest management should be carried out with the intent of identifying common features that could be learned from successful and unsuccessful pest control efforts. Case studies will require application of public administration principles as well as techniques employed to evaluate research effectiveness (Coster and Lachance 1986).

Evaluate the effectiveness of economic incentives available to suppliers of pest management technologies. Private firms are responsible for developing numerous pest management technologies, including survey and detection equipment (e.g., pheromone traps), pesticides (B.t.), application methods (nozzles), and timber harvesting equipment used for partial cuts. These private sector suppliers arguably provide the bulk of society's investment in improved technology. Whether firms invest or not has a significant effect on the cost and success of forest pest control efforts. For example, extensive private investments in improving B.t. efficacy were essential in making the product useful for controlling spruce budworm (Irland and Rumpf 1987). For another, federal investments made to develop Zectran, a highly effective and environmentally safe insecticide, were largely wasted for many years because a private sector producer was unavailable.

As with many new technologies, the forestry market for insecticides is often small and erratic, and is often prey to the vagaries of government budget decisions, political outcries and litigation, and natural conditions affecting pest outbreaks. Since the market for forestry insecticides is but a tiny fraction of the total pesticide market, incentives would appear to be very modest for a large chemical company to risk capital on forestry applications—even for currently registered farm pesticides.

Yet, improved generations of safer and more target-specific pesticides are needed to backstop effective protection programs. For many years, pesticide regulatory programs have been virtually paralyzed in the Environmental Protection Agency. Furthermore, laws and procedures regulating pesticides contain requirements that may unnecessarily constrain progress while accomplishing limited improvement in health or environmental safety. For example, some insecticides are sprayed in swamps to control mosquitoes. But the identical material cannot be sprayed in the same swamp to control a forest pest without a time-consuming and costly registration process.

Evaluate the economics of survey and detection programs. Incentives for developing costly survey technologies (e.g., remote sensing) have been dominated thus far by subsidized "demonstration projects." These have demonstrated technical virtuosity but not operational and low-cost attractiveness. At the other end of the spectrum, research is needed to design means of more effectively utilizing, for purposes of insect and disease detection, persons already working regularly in the forest areas. Cost-effectiveness analysis of survey and detection needs to be better integrated with effective appraisal of new technologies.

Develop More Effective Pest Management Programs via the Application of Policy Analysis and Program Evaluation

A recurring theme in discussions about pest management programs is the need for more intense policy analysis and evaluation. What forms should pest management policies take to support improved management decisions and swifter development and adoption of new, safer, and more effective technologies? How effective have previous pest management efforts been, and what lessons might be learned from these experiences? Unfortunately, almost no systematic economics work is currently being focused on such questions. Considering how controversial and how important pest management activities can be, such a void cannot continue. Some method must be developed for funding and carrying out truly independent policy analysis and program evaluation. Opportunities for applying policy and program analysis to pest management programs include the following.

Evaluate the efficiency and effectiveness of pest management programs involving cost-sharing. What have been the incentive effects of past cost-sharing policies focused on pest management? Have federal cost shares for spraying encouraged excessive reliance on spraying as a management tool? Have subsidies for sanitation and salvage treatments had the intended effects? Have they encountered unanticipated difficulties? Has suitable on-the-ground follow-through been accomplished in these cases?

Identify and evaluate the effectiveness of institutional arrangements used to carry out pest management programs. What have been the most effective institutional arrangements for carrying out survey and detection programs and for implementing pest control programs? Between state and federal agencies, there exist a significant range of ways to organize such functions. For example, would private pest management cooperatives be appropriate in certain circumstances, or would local pest control districts be more applicable (similar to mosquito control districts)?

Evaluate management responses to recent major outbreaks of forest pests. Comprehensive evaluations of technical and policy responses to recent pest outbreaks should be carried out. These case studies should be designed to maximize independence of view on the part of researchers. There are few fields in forestry that could benefit more from truly independent program evaluation. An internal USDA–Forest Service evaluation of agency programs and priorities is now underway (Forest Pest Management Economic Analysis).

Review and evaluate state pest management policies and the application of National Environmental Policy Act (NEPA) procedures. Given the importance of state agencies in forest pest management efforts, an evaluation of state capabilities and policy issues should be undertaken. In a related vein, studies of the administrative and coordination problems involved in pest management programs, including NEPA compliance, would be most useful (Irland 1983). While NEPA procedures are well institutionalized at the federal level, there is unevenness in review of pest management activities by federal agencies, and some states have only limited experience in environmental impact statement processes concerning pest management.

Evaluate (ex post) benefit-cost analyses of pest management programs and activities. A critical review of past benefit cost analyses and *ex post* evaluation of the resulting projects would be productive.

Summary and Conclusions

The research agenda in forest pest management is vast. The need for better managerial decision support requires that economists take the lead in assembling meaningful analyses of the biological facts, the costs and impacts of treatments, and statistical analysis of risks. In addition, extensive work on the economics of technological change and on program and policy evaluation is needed. Setting priorities for such an agenda is difficult.

A comprehensive and rigid view of pest management research priorities would be less than useful. Knowledge varies by pest and by region; emerging pest problems differ as well. In all likelihood, there will never be sufficient professional talent nor adequate financial resources to work on all of today's pest management problems while also getting a head start on tomorrow's problems. Realistically, there probably never will be more than a few forest resource economists sufficiently familiar with the biology of forest pests and the nature of appropriate control technologies. As such, resource economics as a field applied to forest pest management is likely to progress very slowly. One way to foster the involvement of forest resource economists in pest management would be to involve them more closely in ongoing pest management projects at earlier stages. Another would be to employ them in evaluation and case studies designed to learn the most valuable lessons from past experience. Still another means would be to focus their attention on generic questions such as rigorous methods of risk evaluation and loss valuation and on immediate practical technology advances for controlling major pests. If progress can be made on such an initial agenda, a more complete effort could be made to broaden the work to address the many other research directions identified above. We could then make significant progress in helping to move forest pest management away from hasty responses to the latest brown needles.

Literature Cited

Agricultural Enterprises, Inc. 1986. Impact assessment of mountain pine beetle and western spruce budworm. Contract no. 53-82x-0-166.

Anderson, W. C. et al. 1987. Assessing risk to plantation investments from insect attacks. Research Paper SO-231. Southern Forest Experiment Station. USDA Forest Service. New Orleans, LA.

Amman, G. D. et al. 1977. Guidelines for reducing losses of lodgepole pine beetle in unmanaged stands in the Rocky Mountains. General Technical Report IN-36. Intermountain Forest and Range Experiment Station. USDA Forest Service. Ogden, UT.

Baskerville, G. 1984. A critique and commentary on the 1983 supply-demand analysis for the spruce-fir forest of Maine. Faculty of Forestry. University of New Brunswick. Fredericton, N.B., Canada.

Baumgartner, D. C. 1987. Salvaging fire-damaged timber in Michigan. Northern Journal of Applied Forestry. 4(3):149-157.

Bible, T. D. 1985. Economic considerations for spruce budworm management in North America. In: Recent advances in spruce budworm research. CANUSA spruce budworms research symposium.

Bousfield, W. E. et al. 1984. Economics of Douglas-fir tussock moth control. Forest Pest Management. USDA Forest Service. Washington, DC.

Brookes, M. H. et al. 1987. Western spruce budworm and forest-management planning. Technical Bulletin No. 1696. Forest Service and Cooperative State Research Service. U.S. Department of Agriculture. Washington, DC.

Brown, T. C. and R. S. Boster. 1978. On the economics of timber damage appraisal for public forests. Journal of Forestry 76(12):777-780.

Cole, W. E. and M. D. McGregor. 1983. Estimating the rate and amount of tree loss from mountain pine beetle infestations. Research Paper INT-318. Intermountain Forest and Range Experiment Station. USDA Forest Service. Ogden, UT.

Coster, J. E. 1984. Concepts of integrated forest insect management. In: Proceedings of integrated forest pest management symposium by S. J. Branham and G. D. Hertel (eds.). University of Georgia. Athens, GA.

Coster, J. E. and D. Lachance. 1986. Evaluation of the Canada-United States spruce budworms program: Organizational and administrative effectiveness. Proceedings of conference on evaluation and planning of forestry research by D. P. Burns (ed.). Northeastern Forest Experiment Station. USDA Forest Service. Broomall, PA.

Cuff, W. and G. Baskerville. 1982. Ecological modelling and management of spruce budworm infested fir-spruce forests of New Brunswick, Canada. In: Analysis of ecological systems: State of the art in ecological modelling. Elsevier Scientific Publishing Co. New York, NY.

de Steiguer, J. E. et al. 1987. Optimal level of expenditure to control the southern pine beetle. Research Paper SE-263. Southeastern Forest Experiment Station. USDA–Forest Service. Asheville, NC.

Forest Service. 1986a. Highlights, southwide suppression program for the southern pine beetle. Southern Region. U. S. Department of Agriculture. Atlanta, GA.

Forest Service. 1986b. Western spruce budworm in the northern region: 1985 Situation analysis. Report 86-12. U. S. Department of Agriculture. Washington, DC.

Forest Service. 1985. Final environmental impact statement: Western spruce budworm management program for portions of the Carson National Forest. U. S. Department of Agriculture. Washington, DC.

Freeling, A. S. and D. A. Seaver. 1980. Decision analysis in Forest Service planning: Treatment of mountain pine beetle. Technical Report 80-8. USDA Forest Service. Washington, DC.

Gara, R. I. et al. 1984. Influence of fires, fungi, and mountain pine beetles on development of a lodgepole pine forest in south-central Oregon. In: Lodgepole pine: The species and its management by D. M. Baumgartner et al. (eds.). Washington State University. Pullman, WA.

General Accounting Office. 1986. Pesticides. EPA's formidable task to assess and regulate their risks. GAO/RCED 86-125. U. S. Congress. Washington, DC.

Huff, D. J. et al. 1984. The economics of spruce budworm outbreaks in the Lake States: An overview. The Great Lakes Entomologist 17(4): 239-253.

Hudak, J. and A. G. Raske, eds. 1982. Review of the spruce budworm outbreak in Newfoundland. Report N-X-205. CFS submission to the Royal Commission on Forest Protection and Management. Newfoundland Forest Research Centre. St. Johns, Newfoundland, Canada.

Irland, L. C. 1977. Notes on the economics of spruce budworm control. Technical Note No. 67. School of Forest Resources. University of Maine. Orono, ME.

Irland, L. C. 1983. Improving the EIS process: A case study of spruce budworm control. In: Improving Impact Assessment by S. L. Hart et al. (eds.) Westview Press. Boulder, CO.

Irland, L. C. and K. L. Runyon. 1984. Economics of spruce budworm management strategy. Handbook No. 620. Spruce Budworm Handbook. USDA Forest Service. Washington, DC.

Irland, L. C. and T. H. Rumpf. 1987. Cost trends for *Bacillus thuringiensis* in the Maine spruce budworm control program. Bulletin of the Entomological Society of America. Summer: 86-90.

Irland, L. C. and T. E. Howard. 1987. Innovative forms of timberland ownership: What are the driving forces? In: A clear look at timberland investment. Forest Products Research Society. Madison, WI.

Jacobi, W. R. et al. 1981. Multisource inventories: Procedures for assessing the damage caused by insects and diseases. Research Paper 221. Southeastern Forest Experiment Station. USDA Forest Service. Asheville, NC.

Konrad, Gary D. 1984. Insect pest control: Economic theoretical models, costs and benefits (a review of literature). Research Note No. 39. Department of Forestry. North Carolina State University. Raleigh, NC.

LeMaster, D. C., B. Flamm and J. Hendee 1987. Below-cost timber sales. Conference proceedings. The Wilderness Society. Washington, DC.

Leuschner, W. A. 1986. Effects of below-cost timber sales on forest protection. Proceedings of below-cost timber sales conference by D. C. LeMaster et al. (eds.). The Wilderness Society. Washington, DC.

Leuschner, W. A., and James M. Ferguson. 1987. Assessing the economic impact of atmospheric deposition in Eastern forests. Working Paper No. 48. Southeastern Center for Forest Economics Research. Research Triangle, NC.

Lowery, D. P. 1982. The dead softwood timber resource and its utilization in the west. General Technical Report INT-125. Intermountain Forest and Range Experiment Station. USDA Forest Service. Ogden, UT.

MacLean, D. A. 1983. Effects of spruce budworm outbreaks on the productivity and stability of balsam fir forests. Maritime Forest Research Centre. Canadian Forestry Service. Department of the Environment. Fredericton, N.B., Canada.

Manning, G. H. 1982. Impact of the mountain pine beetle on the economy of British Columbia. Proceedings of joint Canada/USA workshop on the mountain pine beetle related problems in western North America. Environment Canada. Canadian Forestry Service. Ottawa, Ontario, Canada.

Milne, G. R. 1986. Economic impact of the spruce budworm outbreak in Newfoundland. Information Report N-X-237. Newfoundland Forestry Centre. St. Johns, Newfoundland, Canada.

Montgomery, B. et al. 1983. The spruce budworm manual for the Lake States. Technical Manual 82-6. Michigan Cooperative Forest Pest Management Program. Michigan State University. E. Lansing, MI.

Rauscher, H. M. 1987. Expert systems for natural resources management. The Compiler (5)5:1-41.

Runyon, K. L. 1980. Economic impact of spruce budworm: An eastern Canada perspective. Proceedings of economics working group session. Eastern spruce budworm research work conference. College of Forest Resources. University of Maine. Orono, ME.

Seymour, R. S. and R. C. Lemin, Jr. 1987. Adapting Forman for timber supply analysis in Maine. College of Forest Resources. University of Maine. Orono, ME.

Stark, R. W. 1987. Impacts of forest insects and diseases: Significance and measurement. CRC Critical Reviews in Plant Science 5(2):161-203.

Sewall, J. W., Company. 1983. Spruce-fir wood supply/demand analysis. Forest Service. Maine Department of Conservation. Augusta, ME.

Talerico, R. L. et al. 1978. Pest control decisions by decision-tree analysis. Journal of Forestry 76(1):16-19.

Twardus, D. B. and M. H. Brookes. 1983. A decision-support system for managing western spruce budworm: Report from a workshop. Pacific Northwest Forest and Range Experiment Station. USDA Forest Service. Portland, OR.

Van Wagner, C. E. 1979. The economic impact of individual fires on the whole forest. The Forestry Chronicle. pp. 47-51.

Walstad, J. D. and P. Kuch (ed.). 1987. Forest vegetation management for conifer production. John Wiley Publishers. New York, NY.

Williams, J. T. 1980. Silvicultural and fire management implications from a timber type evaluation of tussock moth outbreak areas. Proceedings of 6th conference on fire and forest meteorology. Society of American Foresters. Washington, DC.

Znerold, M. 1985. Management of lodgepole pine under catastrophic conditions (Deschutes National Forest). Proceedings of national silvicultural workshop. Timber Management. USDA Forest Service. Washington, DC.

23

STRUCTURE AND PERFORMANCE OF NONINDUSTRIAL PRIVATE FORESTS

William B. Kurtz

Nonindustrial private forests (NIPFs) make up some 58 percent of the nation's commercial timberland (Birch et al. 1982). In addition to producing nearly half the nation's roundwood output, they are an especially significant source of nontimber benefits, including recreation, wildlife, watershed and scenic beauty. Unfortunately, NIPFs have long been considered a problem by many resource managers and policy makers. They cite as causes of the NIPF dilemma the heterogeneous nature of the forests in question and the diverse character, motives, and objectives of the landowners. More fundamental, however, are the often divergent interests of society and individual NIPF owners relative to the appropriate type and magnitude of forest outputs to be produced by NIPFs. As stated by Quinney (1964), "There are some striking differences between a stereotype social viewpoint of the small forest owner and the owner's actual identity and own perception of his role."

The problems ascribed to NIPFs are many. Most commonly identified are low levels of investment in forest management practices, generally low levels of response to public programs aimed at encouraging forest management, and lack of participation and cooperation with private initiatives designed to promote the production of forest benefits. These difficulties are compounded by the fact that some, maybe many, NIPF owners view their forests as consumer goods yielding psychological benefits. Such is in contrast to the view that owners of NIPFs are producers of goods that have tangible dollar values.

From a policy perspective, there are many issues concerning NIPFs. Especially noteworthy is the lack of forest management practices applied to hardwood timber stands (Ellefson 1984) and the decline in reforestation of cut-over softwood stands in the South (Fecso et al. 1982). Since much of the nation's hardwood resource is in NIPF ownership, timely application of forest management practices (including harvest) would seem of paramount importance if a sustained flow of quality hardwood timber products is deemed socially desirable. In the South, inadequate reforestation of privately held forest lands threatens to seriously disrupt future supplies of timber for the southern wood-based industry. Technical and financial assistance to NIPF owners may play an important role in reducing these adverse effects.

Past Research Concerning Nonindustrial Private Forests

A very significant proportion of past NIPF research has been descriptive in nature, namely, relating fundamental landowner characteristics (e.g., age, income) to landowner forestry objectives, management activities, and size of forestry holdings. Such research has been a valuable contribution in that it provided useful information about the clients (NIPFs) served by agency service foresters, forestry consultants, and various public policy makers. Likewise, a series of descriptive studies (undertaken in conjunction with forest survey) focusing on Northeastern U.S. have provided especially useful information for estimating timber availability from NIPF lands (Kingsley and Birch 1977, Birch 1986). The information enabled a thorough examination of past owner harvesting activities and future harvesting intentions.

Considerable research has been focused on timber as an output of NIPFs. Such research has been aimed primarily at determining timber availability, especially as might be determined by owner intentions and past harvesting activities. Research by Schallau (1965) in northern Lower Michigan examined relationships between the economic availability of timber and tract size, owner residence, and owner turnover rate. Schallau's research suggested that timber availability might be restricted by absentee ownership and fragmented land ownership. Working with USDA–Forest Service forest survey information, Birch (1986) utilized a refined descriptive process to determine owner potential for timber management and harvest. Unfortunately, such an approach failed to yield definite estimates of available timber. Alig (1986) added to the stock of methods for determining timber availability from privately owned forest land. Such was accomplished via development of an econometric model which applied classical land rent theory to factors that influence forest acreage changes in the Southeast. Owner objectives and responsiveness to market forces were not explicitly recognized by the model. Binkley (1981) expanded microeconomic research focused on NIPFs by developing an owner behavior model which presumes landowner timber harvest decisions are made with the intent of maximizing a utility function that recognizes consumption of both timber and nontimber forest outputs. In testing the adequacy of the model on a sample of New Hampshire NIPF owners, empirical results were found to be in agreement with the model's expectations. Further research by Brooks (1985) led to development of reforestation behavior equations (southeast and south central regions of U.S.) for use in an inventory simulation model (Southern Pine Age-class Timber Simulator). The intent was to project softwood timber inventories on privately owned forest land. The information was used to modify the Timber Assessment Market Model (TAMM) and ultimately to determine the effects of alternative levels of cost-share payments on acres harvested and planted.

A number of research efforts have been undertaken with the express purpose of guiding policy development and program decision making. The products of such research have proven useful to state forestry personnel charged with the development of public programs designed to encourage management of NIPFs. In this regard, Kurtz and Lewis (1981) devised a decision model based on attitudes (motivations, objectives, and constraints) that lead to adoption of a forest management strategy. The model was expanded and refined so as to provide more of a systems perspective to the decision processes of NIPF owners (Kurtz et al. 1983). To further understand factors influencing timber production decisions of NIPF owners, Young and Reichenbach (1987) applied the Theory of Reasoned Action. Their intent was to identify attitudinal and subjective components of owners' decisions. Owners having no intention of producing timber within a specified period (ten years) were identified; a framework was developed to change their beliefs and thus their intentions.

Research concerning NIPFs has also focused on management and harvest investment decisions. Various policy instruments are available to motivate NIPF owners to invest in forestry practices and harvesting activities (Worrell and Irland 1975). Generally, however, investment in NIPFs is direct investment by individuals, completely financed from their own resources and generally of a short-term nature (investment strategy has been suggested as a basis for categorizing NIPF owners) (Yoho 1985). Like most other investment opportunities, an investment in non-industrial forestry is expected to yield a suitable return over time. Mills and Hoover (1982) evaluated risk and diversification aspects of forest land investments utilizing portfolio theory. A continuing and most perplexing problem, however, is whether public cost-share payments actually induce private investment in reforestation or whether they simply replace capital that would have been invested without government involvement. The problem has been addressed with an aggregate investment model that examined reforestation investment behavior as a function of real personal income, real sawtimber stumpage price expectations, real interest rate expectations, and total real dollars for cost-sharing assistance (de Steiguer 1983). Application of the model has lead to the conclusion that changes in investment levels are related to changes in real personal income and real interest rates.

Evaluation of NIPF and Related Programs

Programs designed to encourage timber production from NIPF lands in a direct fashion can take the form of technical assistance, education, cost-sharing, regulatory mandates, or some combination thereof. A complex relationship exists between these mechanisms, especially tax privileges from forestry investments and cost sharing of forestry practices (Royer and Moulton 1987). The effects of this latter combination are often difficult to evaluate because of the long-term nature of forestry investments and the indirect consequences of tax privileges often associated with investment in forestry practices.

A number of program evaluations addressing financial efficiency have been undertaken. For example, the overall economic performance of the 1974 Forestry Incentives Program (FIP) was evaluated by Mills and Cain (1979). The evaluation computed financial returns to the government and to NIPF owners and projected increases in timber yields resulting from program participation. Risbrudt and Ellefson (1983) evaluated the 1979 FIP in a similar fashion; program efficiency was found to have increased. Using the Risbrudt and Ellefson (1983) data set, Dicks et al. (1983) computed direct tax payments resulting from timber produced because of landowner participation in the 1979 FIP. Income, severance, and yield taxes were found to return $0.96 to federal and state treasuries for every FIP dollar expended over the life of the timber rotations. Based on landowner investment behavior, Tikkanen (1983) developed econometric models for evaluating the effectiveness of programs designed to encourage forestry investments. The models estimate the level of subsidization needed to meet public policy goals.

Evaluations have also been made of the effectiveness of programs focused on NIPFs. Assessments of program success have been measured in terms of planted acreage retained over a period of years as well as condition of existing stands. Such evaluations provide information on which judgments about a program's ability to increase future timber supplies can be based. Reported rates of retention have ranged from a low of 35 percent for 40-year-old Civilian Conservation Corps's plantations in northern Mississippi (Williston and Dell 1974) to over 96 percent for seven year old Forestry Incentives Program reforestation in the North (Ellefson and Risbrudt 1987). Retention rates for Soil Bank Program plantations in the South have been reported as high as 86 percent (Alig et al. 1980). Kurtz et al. (1980) reported that 95 percent of the tracts planted with conifers under the Agricultural Conservation Program (ACP) in five eastern states had been retained to some degree 10 to 15 years following planting. Alig et al. (1980) has identified further relationships between retention rates and plantation acreage.

As review of the above problem areas indicates, research focused on the nation's nonindustrial private forests has been significant. Over the years, much has been accomplished; more needs to be done. Future economics research focused on NIPF lands and their owners should be framed so as to facilitate owner interest in maximizing satisfaction from their forestry endeavors while at the same time encouraging production of a socially desirable mix of consumptive and nonconsumptive goods and services. Forest economics research oriented to NIPF ownerships would be most effective if oriented around the following strategic areas:

- Acquire better understanding of the rationale for ownership of nonindustrial private forest land.
- Develop more effective complementary relationships between public and private programs focused on NIPFs.
- Design more effective programmatic and institutional mechanisms for encouraging management of NIPFs.
- Develop more efficient broad-based resource policies and programs influencing the use and management of NIPFs.
- Develop better understanding of factors influencing NIPF owner decisions to harvest and market timber.

Acquire Better Understanding of the
Rationale for Ownership of Nonindustrial
Private Forest Land

The private landowner's decision to own forest land is generally driven by an economic motive to maximize returns to present and planned expenditures. However, the economic rationale for holding forest land is often confounded by a variety of circumstances, including why the land was obtained, the manner in which it was obtained, and the role played by a particular land parcel vis-à-vis the remainder of an owner's property. Needed is research that will provide a better understanding of the reasons for and conditions under which NIPF land is purchased and held. Such knowledge would enable present and potential NIPF owners to make the most efficient use of investment capital. Specific research opportunities that would enhance the development of such knowledge include the following.

Characterize NIPF owners according to investment strategy. The investment strategy of an individual owner is reflected in the manner in which a given property is managed and ultimately disposed of. A number of strategies comprised of similar combinations of management-disposal (or holding) intentions could be identified as being relevant for NIPF owners. Evaluating the land investment with respect to other investments in an owner's portfolio and linking that to the owner's characteristics provides an indication of investment stability relative to future timber supplies. By developing a model describing the influence of various economic factors on owner behavior, owner response to both market and nonmarket forces might be anticipated.

Characterize the physical nature of forest land involved in land transactions. Important characteristics of forest land that affect suitability for investment are stocking (e.g., species, volume, age, condition), productive potential (site index), location, access, and tract size. These characteristics, which undoubtedly should be reflected in transaction price, must be interpreted in terms of the owner's investment strategy and intentions for the land. While physical-biological information is available from the USDA Forest Service (e.g., forest survey or similar data), it has not been effectively linked back to the land transaction and to owner intentions. By segmenting the characteristics of forest land with respect to various products, their respective contributions to land price can be estimated.

Assess capitalization of timber management investments in land transactions. Although interested in forest land because of potential tax benefits and stability in value, many investors are reluctant to invest because finances must be tied up for unusually long periods of time. A salient question can be raised regarding the capitalization of timber management investments in forest land transactions: If a NIPF owner invests in timber stand improvements, can the cost of the investments plus an adequate rate of interest be captured upon sale of the land prior to final harvest? Although such transactions can easily be documented for forestry plantations, obtaining such information for natural stands is more difficult. Sales of NIPF land need to be examined to determine whether the value of past improvements in the forest situated on the land are in fact recognized by both buyers and sellers.

Evaluate the extent to which sold timbered tracts are disaggregated. Once sold, forest land is frequently divided into smaller tracts; the predominant use changing from timber production to uses such as homesites, recreation, wildlife, or unmanaged (natural). The long-term implications of such changes in land use include reductions in the timberland base and subsequent declines in timber supply. Short-term implications include changes in county tax bases and possibly a one-time sale of timber if the forest land is cleared. In areas where a substantial amount of disaggregation is occurring, research should be undertaken to determine the effect of such disaggregation on timber supply and timber markets.

Identify factors influencing NIPF owners to purchase forest land. Forest land is purchased by individuals for a variety of reasons. Seldom, however, are such reasons recorded immediately after a transaction has taken place. Gathering such information for recently purchased NIPF parcels and relating it to owner characteristics, forest land characteristics, and investment intentions would provide forestry professionals with an information base from which to target the forestry information needs of new landowners.

Evaluate effects of taxes on the purchase and ownership of forest land. Taxes can play a significant role in land purchase decisions and in the management strategies followed by many forest land purchasers. To what extent do tax provisions influence such transactions? How do provisions of federal and state tax codes interact to influence purchase and management strategies of owners? Research concentrating on the tax aspects of NIPF owners' decisions could yield information that enables development of tax policies and tax rates that are appropriate to private and public forestry interests.

Assess institutional arrangements for encouraging stability in the ownership of NIPFs. A commonly held belief is that ownership of NIPFs is continuously changing (a problem that precludes completion of management and harvest plans) because owner objectives keep shifting. Because of the inherent difficulty in tracking changes in owners' goals as land parcels change ownership, little information is available to substantiate such assertions. Ownership changes should be tracked and potential shifts in timber availability from NIPF lands established. Of equal importance is why such changes occur—and what might be done institutionally to lessen their supposed adverse impacts. What institutional mechanisms can be developed that will encourage more stability of ownership patterns? Institutional factors with the potential to influence NIPF owner decisions should be cataloged and their relationship to owner decisions examined. Research could be undertaken to determine owner awareness and use of such arrangements and to assess their affect on owner decisions to sell or retain ownership of forest land.

Develop More Effective Complementary Relationships Between Public and Private Programs Focused on NIPFs

A complex relationship exists between the NIPF owner, the wood-based industry, consulting foresters, and various government organizations in terms of efforts to provide a stable supply of forest benefits from NIPF lands. Because of the sovereignty of NIPF owners relative to the maximization of their unique utility function, such organizations generally have only an indirect influence over NIPF owner decisions. Examining existing relationships could provide the necessary background for their improvement and provide information useful to the design of new, more effective organizational relationships. Opportunities for research include the following.

Evaluate the use of increment contracts to increase intensity of timber management activities. Attention has been directed to the use of increment contracts for increasing the production of timber from NIPFs (Zinn and Miller 1984). Increment contracts are long-term timber management contracts whereby landowners receive regular payments based on the average annual growth of their timber. Such contracts have been used in the South (especially in competitive timber markets) to help ensure a stable supply of stumpage for timber processors. Their potential for increasing the availability of timber from NIPF lands in other regions should be examined. Additional research might be concentrated on determining the manner in which individual NIPF owners utilize the contract payments received.

Assess opportunities for timber management contracts between NIPF owners and the wood-based industry. Various forms of contractual arrangements are possible between NIPF owners and the wood-based industry. Especially useful would be information about the types and characteristics of contracts currently used by industry to ensure sustained access to timber from NIPF lands. Such information could be used to link the characteristics of firms using contracts with the characteristics of NIPF owners agreeing to such contracts. Linkages could be assessed in light of NIPF investment strategies, land use goals, personal and resource characteristics.

Assess characteristics of agency forestry professionals assisting NIPF owners. State forestry agency professionals often have a distinctly different role from private consulting foresters; state forestry personnel are frequently called upon to provide management information that does not directly lead to increased timber production. In order to assure NIPF owner receptiveness to technical forestry information, a correspondence must be established between the beliefs and approaches of service foresters and the management beliefs and goals of the forest landowner. The beliefs and attitudes of agency service foresters should be examined relative to management of NIPFs for various consumptive and nonconsumptive products. Knowledge of beliefs and attitudes could be useful for development of training programs designed to help service foresters become more sensitive to NIPF owner needs and to the role public service foresters play in the NIPF owner management decision process.

Evaluate means of facilitating the establishment of private consulting services. NIPF timber production activities might be substantially increased if private consulting services were readily available to NIPF owners. To embark on such a profession requires the expenditure of substantial capital and the assumption of significant risk. This is especially so in states with well-developed state-agency service forestry programs. The economic feasibility of assisting qualified individuals, particularly minorities and women, to establish a consulting business needs to be determined for areas where forestry assistance is needed by owners of NIPFs. Subsidies (e.g., salary supplements, start-up cost assistance), with anticipated payback from fee earnings, should be examined in the framework of benefit-cost analysis.

Evaluate the effectiveness of forestry advice provided by private consultants and public service foresters, especially in terms of subsequent forestry accomplishments on NIPF lands. In many instances, NIPF owners have the choice of requesting technical advice and assistance from a state agency forester at no fee (or a nominal fee) or from a private consulting forester at a market-determined fee. Questions have been raised as to differences in the quality of service rendered by the respective groups (Cubbage 1983). A comparison should be made of the physical differences in forestry accomplishments resulting from the advice offered by the two groups.

Design Effective Programmatic and Institutional Mechanisms for Encouraging Management of NIPFs

Decisions to harvest timber or to apply forest management practices are influenced by many factors. Various mechanisms, both direct and indirect, can be used to influence such decisions (e.g., forest cooperatives, landowners associations, program targeting). Importantly, the mechanism selected should be the most effective in terms of technical and economic ability to achieve desired results. A broad array of research directions could be undertaken in this respect, the products of which could be most helpful to program planners and to industrial and agency decision makers. Research opportunities include the following.

Evaluate the economic and administrative feasibility of establishing and maintaining management and marketing cooperatives. Cooperatives have been established to provide marketing, processing, and management services to NIPF owners. Yet the history of forestry-oriented cooperatives does not speak of much success. Assuming the existence of sufficient interest and a sufficient NIPF resource base in a given area, encouraging development of a cooperative might prove beneficial to NIPF owners and the markets they serve. Cooperatives currently in operation should be studied to determine strengths and weaknesses, operating problems, and institutional support requirements. Likewise, the records of failed cooperatives should be examined to see if reasons for lack of success can be inferred.

Assess implications of size of NIPF as affecting owner objectives and management activities. Size of forest land holding, particularly the size of individual tracts, has a significant economic influence on the operability of timber stands. This relationship exists because of the fixed costs associated with management and harvesting activities on a given tract. Also related to size of holding is the use to which an owner puts a forested tract—a decision which can directly affect the tract's availability for timber production. On a regional basis, fixed and variable costs should be identified and modeled for several different-sized NIPF ownerships. Once accomplished, representative ownerships could be developed for the different-size classes based on owner management strategies and ownership characteristics. The overall intent would be to test the potential of alternative policy and institutional scenarios on timber availability.

Evaluate management of NIPFs for consumptive and nonconsumptive forest uses. Owners' objectives for NIPF holdings typically include consumptive as well as nonconsumptive uses. In some situations, these uses can be complementary or supplementary in their relationship; in other cases the relationships are competitive. For areas where multiple uses of NIPFs are prevalent, management strategies for various combinations of forest land uses need to be described and the outputs of various products and services need to be quantified. Trade-offs between potential earnings and opportunity costs associated with nonrevenue generating uses should be identified and quantified.

Evaluate labor requirements for mixed enterprises (timber/agricultural) involving NIPFs. Management of NIPF lands that are held in combination with other enterprises (e.g., agriculture) may be facilitated by the use of underutilized labor. A form of complementariness may exist between some enterprises in terms of labor use. Research could be undertaken to identify enterprise combinations in various regions and the labor requirements associated with each. Understanding these relationships, particularly in terms of labor requirements, could be beneficial to industry and to private and public service foresters. The information could help define limits to the application of existing labor supplies and the degree to which supplemental labor is available for increasing timber management activities.

Identify NIPF owners most receptive to the application of forest management practices. Certain owners of NIPFs are more prone to undertaking forest management activities than others. The reasons for a positive attitude toward the application of management activities are both internal and external to the NIPF owner's decision process. Questions regarding reasons for undertaking practices could be asked of NIPF landowners. Such information could be related to the characteristics, management strategies, and intentions of different owners.

Evaluate means of targeting programs to receptive NIPF landowners. Public and private programs designed to encourage management of NIPFs can be targeted to any number of owner groups. The intent of such efforts would be to heighten awareness of management opportunities and appropriate forest management techniques for capturing such opportunities. Identifying potential target groups and designing appropriate messages for their use can facilitate the economic efficiency with which programs are administered. Equally important is assessment and evaluation of the results of a targeting effort. Long-term research which tracks the accomplishments of targeted programs should be undertaken.

Develop educational programs to acquaint NIPF owners with available techniques of forest management. Learning processes require students to be attuned to subject matter both in terms of receptiveness and ability to comprehend the material. With respect to forestry, some NIPF owners have difficulty with the subject material because they are unfamiliar with the more fundamental aspects of forest management. Many educational programs, attempting to achieve desired results in a short time with minimum effort, include material that is beyond the ability of recipient owners to comprehend. A reverse process needs to be undertaken. Desired educational outcomes (e.g., application of forest management practices) should first be identified, then the educational approach should be designed in light of existing levels of owner knowledge. A survey of several target groups of NIPF owners should be taken to determine their understanding of fundamental forest management concepts. Information gathered could then be utilized by extension and agency educational specialists to design educational materials and approaches commensurate with a landowner's level of understanding.

Assess economic and administrative feasibility of NIPF landowner associations as means of encouraging forest management practices. Landowner associations (usually organized on a regional basis) are often composed of individuals who are sincerely interested in improving the productivity of their forest lands. Encouraging membership in such organizations, no matter how informal, is a significant means by which landowners can share forestry experiences. Research on the characteristics and management objectives of NIPF owner groups could provide information on how to make such groups operate more effectively and could clarify reasons why individuals elect to join a particular group(s).

Evaluate the effectiveness of paraprofessionals in assisting management decisions of NIPF owners. Use of paraprofessionals appears to be a cost-effective means by which well-trained individuals (full or part-time) can provide technical assistance to NIPF owners. Paraprofessionals, under the direction of experienced professional foresters, can assist in the implementation of management plans that call for timber harvesting and the application of forest practices. The costs of training and supporting a cadre of paraprofessionals in a given geographic area should be assessed as should their effectiveness in encouraging management of NIPFs.

Develop More Effective Broad-Based Resource Policies and Programs Influencing the Use and Management of NIPFs

A wide spectrum of public and private policies and programs are applicable to management problems of NIPFs. Some focus specifically on owners of NIPFs, while other are not unique to such owners. Many programs focus only on increasing and stabilizing timber supplies, yet many explicitly recognize nontimber uses of NIPFs (e.g., state forest practice laws). Especially useful would be research focused on this broad array of public and private programs. The intent of this research would be to provide information that will enable such programs to be more effective in encouraging management of NIPFs. Research opportunities include the following.

Assess the relationship of policies guiding national forest system timber sales to the management strategies of NIPF owners. National forest system timber sale policies can have a significant effect on timber prices received by NIPF owners. Price effects can be beneficial or detrimental, depending on the degree of competition within a given market. In areas where competition is suspected, research could be instituted to determine the proportion of stumpage coming from the respective sectors and to estimate the influence of national forest system timber sale policies on stumpage prices received by NIPF owners. An understanding of the degree of price competition between national forest timber and timber available on NIPF lands could be useful for the formulation of more effective timber sale policies in areas where a significant degree of market competition exists.

Assess the characteristics of agricultural enterprises enrolling land in the Conservation Reserve Program. The Conservation Reserve Program (CRP) provision of the Food Security Act of 1985 is designed to reduce erosion from presently cropped erodible soils by promoting establishment of permanent vegetative cover. Tree planting on qualified soils has been undertaken by many farmers, although the majority of acres enrolled have been planted to pasture species. To provide an accurate assessment of the program's effectiveness, and to provide valuable information for near term program adjustments, farmers' decisions with respect to choice of cover species should be documented. A comparison should be made of CRP participants based on their decisions to plant pasture or to plant trees. Various owner and resource characteristics need to be inventoried and, subsequently, related to management strategies and land use intentions over different planning horizons.

Evaluate the efficiency of Conservation Reserve Program tree planting activities. Participants in the Conservation Reserve Program (CRP) receive various cost-share incentives to cover establishment costs and receive an annual payment for ten years for withdrawing land from crop production. In addition, the Program stipulates that a substantial amount of technical assistance be provided each participant in terms of soil qualification determination, conservation plan development, and adherence to cover specifications. The various costs associated with each aspect of the Program should be quantified for different planting regimes. Following a sample of established plantings, the anticipated product flows could be projected and revenues estimated.

Assess retention, condition, and efficiency aspects of Forestry Incentives Program tree plantings and timber stand improvement practices. Many of the products of the Forestry Incentives Program have been in existence for nearly 15 years. The intermediate condition and retention of plantings and timber stand improvement (TSI) practices installed with FIP investments should be evaluated. Many NIPF owners have had the opportunity to incur treatment costs or gain revenues from material removed prior to stand maturity. Research could be initiated to determine the stocking and condition of planted and treated stands cost-shared under the FIP at least ten years ago. In addition, costs and revenues for practices applied after the initial cost-shared practice should be evaluated to determine the accuracy of existing stand management regimes used to project financial returns.

Assess the retention and condition of Agricultural Conservation Program conifer plantings. Conifer plantations that were cost-shared under the Agricultural Conservation Program (ACP) were examined 10 or more years ago by Kurtz et al. (1980). Such plantings are now more than 20 years old. Significant information would be provided through charting the stability of retention and the condition of plantations examined earlier. Changes in plantation characteristics could be related to changes in management strategies and ownership objectives.

Measure induced investment in timber production from government subsidy programs. The purpose of many government subsidy programs is to stimulate or induce capital investment which would not otherwise have been invested by the private sector. In the context of estimating net benefits, it is important that an accurate measurement be made of the amount of net investment induced by the opportunity to participate in a subsidy program. Methods need to be developed for ascertaining the amount and subsequent impact of the net investment induced by the different forms of programs. Several efforts should be undertaken to refine direct research methods (e.g., questionnaires) and aggregate investment models. Research should also be undertaken to explore the potential of using contingent valuation as a means of estimating the level of induced investment as a form of bid to receive the subsidy.

Identify and measure state-level costs and benefits of cooperative forest management programs. The Cooperative Forest Management (CFM) Program (administered by the USDA–Forest Service, State and Private Forestry) provides technical information to state forestry agencies. The intent is to assist states in their role of supporting NIPF owners in various aspects of resource management and protection. This information flow is an indirect subsidy to NIPF owners that, in effect, induces them to make investments in forestry activities. Research should be conducted to determine who participates in the various components within the CFM structure, and what the nature is of the efficiency and effectiveness of such involvement.

Assess the economic and physical effects of the Renewable Resources Extension Act. The Renewable Resources Extension Act (RREA) is designed to support state Cooperative Extension Service departments in their efforts to develop and deliver natural resource educational programs focused on the general public. RREA funding has been used in a number of different ways to support a variety of program emphases. Research should be undertaken to identify and value the increments of production attributable to increased program funding made available by the Act. Such information could be used by the USDA–Cooperative Extension Service to evaluate the effectiveness of the program and to compare its efficiency with other Cooperative Extension Service programs.

Compare the economic efficiency and social equity aspects of public incentive programs important to NIPFs. Comprehensive judgments about public incentive programs must consider overall economic efficiency as well as the equitable distribution of benefits and costs (Ellefson and Wheatcraft 1983). The different incentive programs aimed at encouraging application of forest management practices on NIPF lands obviously have different efficiency and equity implications. Models depicting the benefits, costs, products, and economic impacts among recipients associated with the various incentive or educational programs should be constructed.

Analyze, in a comparative sense, programs focused on NIPFs in foreign countries. Much can be learned from foreign country experiences with public and private forestry programs focused on NIPFs. Foreign countries in which unique political and cultural constraints coupled with great reliance on the timber-based economy have developed innovative policies for dealing with NIPF owners (Metz 1986). A thorough review of foreign NIPF literature should be instituted with the intent of yielding insight into foreign NIPF problems and the policies and institutions developed to deal with them.

Evaluate the economic and administrative efficiency of state regulatory mechanisms requiring forestry investments by NIPF owners. State regulation of practices undertaken by NIPF owners has increased dramatically since the late 1960s. Taking the form of legal mandates to invest, such regulations often require NIPF owners (by force of law and fear of penalty) to invest in forest practices (e.g., reforestation standards) that are designed to achieve specific forestry or environmental objectives (Henly and Ellefson 1987). Most regulations are imposed by state governments via state forest practice laws. Unclear is the effectiveness of such laws and how they can be used to complement programs which provide NIPF owners with fiscal and technical assistance. Further review of the costs and benefits of regulatory programs addressing NIPFs is needed.

Identify and evaluate financial incentives for timber management activities of NIPF owners. Financial incentives are a unique means of encouraging the application of timber management practices for several reasons: They immediately result in a cost reduction, the owner may have some choice regarding tax liability, and the payment constitutes a private transfer of capital for exclusive use. Because of such characteristics, financial payments are generally viewed by landowners in a manner different from other forms of assistance. Research should be initiated to determine the manner in which such subsidy payments relate to the financial strategies adopted by NIPF owners. In the context of benefit-cost analysis, their overall economic efficiency should be measured and attempts made to isolate their singular influence on decisions to participate in cost-sharing programs and adopt practices.

Develop a Better Understanding of Factors Influencing NIPF Owner Decisions to Harvest and Market Timber

The supply of timber from NIPFs is basically a function of the supply of land used for timber production and the availability of timber from such land. Predominant factors influencing these variables are owner objectives, alternative investment opportunities, timber markets, and public incentives. Most efforts directed toward stabilizing or increasing timber supply from NIPFs must relate to such factors. A number of research avenues addressing constraints to timber production by NIPF owners are open, including the following.

Evaluate impacts of NIPF land market transactions on timber supplies. As various tracts of forest land are exchanged in forest land markets, the supply of timber from such tracts changes according to the intentions of the new owners. Econometric models of forest land markets need to be developed. Models should accurately identify and describe factors related to land transactions, specify the amount of land involved and its characteristics, and define and specify the effects of land transaction factors on land prices and available timber. Price elasticities of demand and supply for forest land should be estimated to describe the causal relationships between land transaction prices and the various land market factors.

Evaluate the effects of land use changes on timber supplied by NIPFs. Shifts in the direction of a timber supply curve, and subsequent effects on timber supply, are directly related to the land management intentions of the owner. Research should be undertaken to track the movement of forest land into and out of timber use, the characteristics of the tracts involved in such changes, and the factors underlying land use intentions of owners as specifically related to land use changes. In addition, changes in the supply of timber from the tracts should be projected, with special attention addressed to the timing of such shifts.

Define and assess factors affecting the availability of timber from NIPFs. A NIPF owner's decision to make timber available for harvest is a function of timber prices and the owner's land-use intentions (present and future)—all conditioned by necessity to sell. The decision is dynamic in nature, possibly more so than any other decision made by the owner. As such, the task of ascertaining in advance the nature of the owner's decision is difficult. However, efforts should be made to do so; they should be concentrated on identifying and prioritizing factors that impact such decisions and should determine, in a dynamic sense, the relationship of such factors to various social and market conditions. By doing so, a probability model of NIPF owner decisions might be formulated—one which reflects current and potential timber availability.

Evaluate the effects of ownership change on the availability of timber from NIPFs. Change in land ownership is a specific factor which has significant influence over timber available from NIPFs. In some instances, a change in ownership will lead to land clearing at which time timber will be marketed prior to maturity— or not marketed at all. In other cases, land clearing will not occur, yet the owner's intention for the forest will change—timber may or may not be available for harvest. Needed is a more thorough understanding of the types of land transactions that occur and the characteristics and intentions of land purchasers. An understanding of these relationships is necessary to determine changes in the availability for harvest of timbered tracts as ownership changes take place.

Identify and evaluate factors influencing timber harvesting and timber selling behavior of NIPF owners. Timber harvesting and selling by NIPF owners is unique in that the decision in many, if not most, instances represents an event made only once or twice in a person's lifetime. Many factors influence such a decision. Research is needed to provide a better understanding of the role such conditions play in the decision process of NIPF owners. Comparative studies need to be made of factors influencing the decisions of owners recently harvesting timber versus owners possessing marketable timber but are unwilling to harvest and sell.

Assess decisionmaking processes employed by NIPF owners. The forestry decision processes followed by NIPF owners are important to production of various forest outputs. However, it has particular significance to timber supply since the preponderance of the nation's commercial timberland is held by owners of NIPFs— owners who have a variety of forest management objectives. The role of numerous factors causing NIPF owners to take a particular action at a given time within the life of a timber stand needs to be better understood. A wealth of secondary data exists, much of which is still in original questionnaire form. A thorough review of such data, coupled with surveys of NIPF owners in selected regions, should be undertaken with the intent of developing quantitative linkages between owners' decisions and the quantity of timber made available for harvest over time.

Summary and Conclusions

Controlling nearly two-thirds of the nation's commercial timberland, nonindustrial private forests are an especially important segment of the forestry community and those dependent upon it. Nonindustrial private forests are often viewed as a social problem because their owners have land use objectives other than timber production. The problem can be cast as a divergence of the interests of society at large and the interests of individual landowners. Various public and private programs have been designed to encourage owners of NIPFs to manage their forests more aggressively. Most such programs have focused on timber production; most have met with only limited success.

Future research focused on NIPFs should be planned to encourage a better understanding of the rationale for ownership of NIPF land, the complex relations existing between public and private programs focused on NIPFs, the programmatic and institutional mechanisms available for encouraging management of NIPFs, the broad-based resource policies often affecting NIPFs, and of factors influencing NIPF owner decisions to harvest and market timber. Forest resource economics research within such areas should lead to a more thorough understanding of problems associated with NIPF ownership; this would enable landowners and society in general to capture many of the benefits represented by owners of nonindustrial private forests.

Literature Cited

Alig, R. J. 1986. Econometric analysis of the factors influencing forest acreage trends in the southeast. Forest Science 32(1):119-134.

Alig, R. J., T. J. Mills and R. L. Shackelford. 1980. Most soil bank plantings in the south have been retained; some need follow-up treatments. Southern Journal of Applied Forestry 4(1):60-64.

Binkley, C. S. 1981. Timber supply from private nonindustrial forests. Bulletin No. 92. School of Forestry and Environmental Studies. Yale University. New Haven, CT.

Birch, T. W. 1986. Forest land owners of Maine, 1982. Resource Bulletin NE-90. Northeastern Forest Experiment Station. USDA Forest Service. Broomall, PA.

Birch, T. W., D. G. Lewis and H. F. Kaiser. 1982. The private forestland owners of the United States. Resource Bulletin WO-1. USDA–Forest Service. Washington, DC.

Brooks, D. J. 1985. Public policy and long-term timber supply in the south. Forest Science 31(2):342-357.

Cubbage, F. W. 1983. Measuring the physical effects of technical advice from service foresters. In: Nonindustrial private forests: A review of economic and policy studies by J. P. Royer and C. D. Risbrudt (eds.). School of Forestry and Environmental Studies. Duke University. Durham, NC.

de Steiguer, J. E. 1983. The influence of incentive programs on nonindustrial private forest investment. In: Nonindustrial private forests: A review of economic and policy studies by J. P. Royer and C. D. Risbrudt (eds.). School of Forestry and Environmental Studies. Duke University. Durham, NC.

Dicks, M. R., W. B. Kurtz, D. E. Ervin, R. J. McHugh and G. A. Myles. 1983. Impacts of the 1979 Forestry Incentives Program on state and federal taxes. In: Nonindustrial private forests: A review of economic and policy studies by J. P. Royer and C. D. Risbrudt (eds.). School of Forestry and Environmental Studies. Duke University. Durham, NC.

Ellefson, P.V. and A.M. Wheatcraft. 1983. Equity and the Evaluation of Forestry Incentives Programs Funds. In: Nonindustrial private forests: A review of economic and policy studies by J.P Royer and C.D. Risbrudt (eds.). School of Forestry and Environmental Studies. Duke University. Durham, NC.

Ellefson, P. V. 1984. Encouraging growth and use of eastern hardwoods: Public and private programs for the task. Forest Products Journal 34(10):18-23.

Ellefson, P.V. and C. D. Risbrudt. 1987. Forestry Incentive Program investments in the North: Retention rates for acres treated in 1974. Northern Journal of Applied Forestry 4(3):133-135.

Fecso, R. S., H. F. Kaiser, J. P. Royer and M. Weidenhamer. 1982. Management practices and reforestation decisions for harvested southern woodlands. Staff Report AGES821230. Statistical Reporting Service. U.S. Department of Agriculture. Washington, DC.

Henly, R. A. and P. V. Ellefson. 1986. State forest practice regulation in the U.S.: Administration, cost, and accomplishment. Bulletin AD-SB-3011. Agricultural Experiment Station. University of Minnesota. St. Paul, MN.

Kingsley, N. P. and T. W. Birch. 1977. The forest land owners of New Hampshire and Vermont. Resource Bulletin NE-51. Northeastern Forest Experiment Station. USDA Forest Service. Upper Darby, PA.

Kurtz, W. B., R. J. Alig and T. J. Mills. 1980. Retention and condition of Agricultural Conservation Program conifer plantings. Journal of Forestry 78(5):273-276.

Kurtz, W. B. and B. J. Lewis. 1981. Decision-making framework for nonindustrial private forest owners: An application in the Missouri Ozarks. Journal of Forestry 79(5):285-288.

Kurtz, W. B., T. D. Marty and C. B. Trokey. 1983. Motivating the nonindustrial private forest landowner. In: Proceedings of 1983 convention. Society of American Foresters. Bethesda, MD.

Metz, A. 1986. Influence of forest owners as an interest group in achieving the forest policy goals in Finland: The programme "Forestry 2000." Silva Fennica 20(4):286-291.

Mills, T. J. and D. Cain. 1979. Financial efficiency of the 1974 Forestry Incentives Program. Journal of Forestry 77(10):661-666.

Mills, W. L., Jr. and W. L. Hoover. 1982. Investment in forest land: Aspects of risk and diversification. Land Economics 58(1):33-51.

Quinney, D. N. 1964. Small private forest landownership in the United States: Individual and social perception. Natural Resources Journal 3(3):379-393.

Risbrudt, C. D. and P. V. Ellefson. 1983. An economic evaluation of the 1979 Forestry Incentives program. Bulletin 550. Agricultural Experiment Station. University of Minnesota. St. Paul, MN.

Royer, J. P. and R. J. Moulton. 1987. Reforestation incentives. Journal of Forestry 85(8):45-47.

Schallau, C. H. 1965. Fragmentation, absentee ownership, and turnover of forest land in northern lower Michigan. Research Paper LS-17. Lake States Forest Experiment Station. St. Paul, MN.

Tikkanen, I. 1983. Effectiveness of forest policy programs: A policy analysis perspective. In: Nonindustrial private forests: A review of economic and policy studies by J. P. Royer and C. D. Risbrudt (eds.). School of Forestry and Environmental Studies. Duke University. Durham, NC.

Williston, H. L. and T. R. Dell. 1974. Growth of pine plantations in north Mississippi. Research Paper SO-94. Southern Forest Experiment Station. USDA Forest Service. New Orleans, LA.

Worrell, A. C. and L. C. Irland. 1975. Alternative means of motivating investment in private forestry. Journal of Forestry 73(4):206-209.

Yoho, J. G. 1985. Continuing investments in forestry: Private investment strategies. In: Investments in forestry by R. A. Sedjo (ed.). Westview Press. Boulder, CO.

Young, R. A. and M. R. Reichenbach. 1987. Factors influencing the timber harvest intentions of nonindustrial private forest owners. Forest Science 33(2):381-393.

Zinn, G. W. and G. W. Miller. 1984. Increment contracts: Southern experience and potential use in the Appalachians. Journal of Forestry 82(12):747-749.

24

PRODUCTION AND VALUATION OF WATER FROM FORESTED WATERSHEDS

K. William Easter

Accounting for over 60 percent of all U.S. stream flow, forested watersheds are an especially important source of the nation's supply of usable water (Megahan et al. 1981). Because of their location in areas of significant precipitation, they annually yield 13 more inches of runoff than originates from nonforested lands (4 inches) located throughout the nation (Forest Service 1986). Only recently, however, has the application of economics to the management of forested watersheds become of significant interest to researchers (Easter et al. 1986; Gregersen et al. 1987). The bulk of past watershed management research has involved hydrologic studies of the physical relationships between management practices/hydrologic projects and subsequent flows of water (e.g., relationship between streamflows and vegetation management). For example, large increases in water yield have been found to occur after removal of all forested vegetation (Lull 1970; Hoover 1944; Rothacher 1970), while selective tree harvesting apparently has little or no effect on water yield. To obtain noticeable effects on water yield, tree cutting intensity must reach a point where approximately 20 percent of forest and related vegetation is removed (Hibbert 1967). Even though research of such a nature has been conducted, it has not been widespread nor has it fostered widespread consensus on the amount and quality of water that is likely to result from the application of various forest management practices. To further compound the problem, existing production function linkages are generally not in a form that can be directly used in economic analysis (Gregersen et al. 1987, Ellefson and Miles 1985). In such an environment, it is not surprising to find even less agreement on the economic feasibility of forest management practices undertaken to augment water yield (Bowes et al. 1984).

The lack of well-functioning markets for the allocation of water from forested watersheds has also had an adverse effect on research concerning the economics of watershed management and water use. Water in various quantities and qualities is typically not sold in competitive markets. Thus, alternative approaches (e.g., shadow prices) must be used to determine the value of water and associated goods and services. Of further concern for analysis and research is that water is generally an input to the production of other goods and services. Researchers must look to the products of water (e.g., agricultural crops, electric power, fish, transportation, recreation activities, improved health) in order to determine the benefits of increased water production. In some cases markets exist for these secondary products; very often they are lacking.

Economics research focused on the production and valuation of water from forested watersheds is an especially important topical area. Such research should concentrate on the following strategic directions:

- Develop more effective means of valuing and planning the use and management of forests for the production of water.
- Develop improved understanding of means by which water from forested watersheds can be allocated and reallocated.
- Develop improved understanding of political and administrative structures important to the production of water from forested watersheds.
- Acquire improved understanding of the economics of hydrologic projects and construction activities germane to forested watersheds.
- Develop improved understanding of local and regional economic and social impacts of water produced by forested watersheds.
- Acquire improved understanding of the economics of managing forested watersheds to improve water quality.

Develop More Effective Means of Valuing and Planning the Use and Management of Forests for the Production of Water

Develop more effective means of measuring the benefits and costs of producing water from forested watersheds. Management of forested watersheds for the production of water assumes that the benefits and costs of doing so can be readily identified, easily measured, and appropriately valued. Such is not always the case. For example, one of the most difficult valuation problems involves valuation of instream water uses—especially difficult when many instream uses do not preclude subsequent downstream water use. For example, water used for recreation or power production can later be used for irrigation purposes. Although there can be considerable return flows from irrigation, the reverse is generally not true. Thus, an important economic question concerns the valuation of water flows that result in joint products. Daubert and Young (1981) have addressed the issue in terms of recreational demand for instream flows in the Cache la Poudre River (in Northern Colorado). Using a contingent valuation approach to estimating the shadow price of instream flows for fishing, they found that during low flow periods (late in summer), the value of instream flows for fishing exceeded those for irrigation.

Peak flows or floods also pose valuation problems. Watersheds are often managed to reduce downstream flooding—an objective that presumes benefits can be measured in terms of damages prevented. An important knowledge gap is the lack of information about differences in the damage susceptibility of various types of units (e.g., homes, commercial buildings). For example, what is the relationship between the value of a building's contents and the value of the structure per se, and how does this relationship change over time (Ramirez et al. 1988)?

Many countries, including the U.S., have a long history of undertaking projects to increase water supplies—projects that have not always passed the test of economic efficiency. Recently, however, arguments have been made that water supplied by such projects has reduced or prevented conflicts between various users of water. If conflict reduction is to be viewed as a benefit of a water project, researchers need to be able to estimate the value of conflict reduction. What value should be placed on new water supplies that reduce conflict and possible subsequent legal action among water users?

Develop more effective means of accounting for the distribution of water and water related benefits and costs. Water rights have historically been a key variable in determining the distribution of benefits produced by a water project (Saliba et al. 1987). If programs to increase the supply of water from forested watersheds are selected, decision makers will find that they must address issues concerning how rights to use the additional water are to be distributed. Under what conditions can water rights be exercised, and how can such rights be transferred between users and uses (e.g., agricultural vs. municipal)? The transfer question is of particular concern to downstream users. If the latter are to be accommodated, water rights might well specify that a change in stream diversions is agreeable as long as increases in consumption do not occur (i.e., increased return flows are just as large as the increase in diversion).

Economic and legal barriers to participation in water markets are also deserving of research. Are there certain restrictions in existing water markets that prevent potential users from purchasing water? If so, will such barriers apply to new supplies coming from forested watersheds? What distributional impacts do such barriers pose? Of related concern is who pays for water projects (incidence of costs). In many cases, the direct user of water pays only a small portion of the total cost of supplying the water. Is this the case for water produced in forested watersheds? Such is likely to be true if it is difficult to identify the beneficiaries of water projects or if the beneficiaries are well-organized politically (e.g., farmers).

Determine appropriate discount rates and opportunity costs. Selection of an appropriate discount rate for evaluating water projects is a widely debated subject in the benefit-cost literature. Its importance in long-term project planning is obvious—a high discount rate suggests that little weight is given to projects that produce long-term benefits. An important area of research is determination of the impact of various discount rates on project selection. Did the higher interest rates of the late 1970s and early 1980s cause a shift away from long-term projects? Are some water users and watershed projects favored by high or low discount rates? Given the international nature of today's capital markets, research focused on the opportunity cost of capital has become especially important.

Develop more effective system-wide watershed planning techniques. During the past two decades, significant improvements have been made in planning techniques (e.g., the watershed planning framework developed by Hufschmidt [Easter et al. 1986]). However, current tendencies are to err on the side of being all-inclusive at the expense of being able to identify the inputs and outputs that are critical to success or failure of a project. Inclinations are to include everything that might be effected and thus forego the flexibility required to focus planning activities. Too much data can be just as bad as too little information. "To provide timely information for policy-making and program development, diagnostic methodologies are needed for rapid assessment of the biophysical and socioeconomic conditions of a watershed and to formulate and evaluate possible courses of action. Such rapid appraisal methodologies must be adaptable to limited data and analytical resources" (Easter and Hufschmidt 1985). Also needed are planning procedures that more effectively involve local participants. Such is particularly important where groups believe they are being ignored in the decision making process. "Studies are required to formulate and evaluate alternative approaches for involving local participants in planning, particularly disadvantaged groups and women" (Easter and Hufschmidt 1985).

Develop Improved Understanding of Means by Which Water from Forested Watersheds Can Be Allocated and Reallocated

Many times water problems are not really physical supply problems but are problems caused by inadequate procedures for allocating or reallocating water among users. In states which are almost always concerned about limited water supplies (e.g., Western States), elaborate rules for allocating water among users and uses have been developed. Evaluation of such rules could lead to important information useful to other states and regions interested in more effective means of allocating water.

Assess regional economic interdependence on water produced from forested watersheds. Water reallocation is an especially complicated activity, primarily because of the interdependence of many water users and the importance of water to some regional economies (e.g., power production, food processing, input to a regional economy via agriculture). In some regions, constraints on economic development may require reallocating limited supplies of water to highly valued users. Development of water multipliers can significantly aid in the making of such reallocation decisions; they can be used to estimate the impact of changes in final sector demand on the demand for water. For example, Ching (1981) found that an increase in final demand for agriculture's alfalfa sector had a much greater impact on water demand than any other sector. Similar research is needed in other water-scarce regions. Once accomplished, such research can define the economic impacts of water reallocation to sectors with higher value uses and can suggest what has to be given up in terms of employment and income.

Development of input-output tables and water multipliers can provide policy makers with information about the impacts of water transfers from one region to another. In most cases, this information is not available and decision makers are forced to make uninformed judgments about such impacts. Without information concerning the impacts of water transfers, decision makers tend to "play it safe" by strongly opposing water transfers from their regions—they are concerned that such transfers will foreclose future opportunities for regional economic development. Research, however, may demonstrate that water is not a key constraint to economic development.

Evaluate pricing and charging systems for water produced from forested watersheds. As a policy tool for allocating water, pricing of water has received only limited attention. Some argue that for most users the elasticity of demand is so low that the price of water would have to become extremely high in order to induce any significant water reallocation during periods of water scarcity. Others argue that water pricing policies exclude external costs and benefits that frequently occur to downstream users. And finally, some argue that access to water is a basic right of all citizens; water pricing forces low-income persons to bear the brunt of adjustments when water is scarce. Diverse arguments of this sort highlight the need for research concerning existing water markets and the prices that result therefrom. Do market determined water prices serve as an effective tool for water allocation? Are they appropriate measures of the value of water? To what degree are there positive or negative externalities from water use that are not captured by water markets? And what is the elasticity of demand for different uses of water allocated by market systems? Can "fair" markets be established for water, given existing water distribution systems (transaction costs may be too high)? And how can sale or trade of water throughout a water system or between systems be facilitated?

If the value of water is low, development of complex systems for levying water changes may not be efficient. As the value of water rises, however, more elaborate methods of charging for water tend to emerge. Research concerning alternative methods of charging for water is needed. What are the transaction costs involved in establishing alternative systems of water changes? What rules should be established and how will such rules be policed?

A number of difficult but researchable issues surface when any system of water charges is about to be implemented. If markets are not used, what should be the basis for water charges? Should charges be based on the opportunity cost of water, the cost of supplying the water, or the consumer's ability or willingness to pay? Also a major issue concerning water charges is the manner in which they are to be applied. A number of alternatives exist, including a direct charge based on the volume of water used, a flat charge per unit of time (usually a month or quarter), or a charge collected as part of a general land or property tax. For irrigation water, the charge might be based on the amount of area irrigated. Charges per volume of water can be increasing, constant, or decreasing as water use increases. Research is needed to determine how these alternatives affect water use and the nature of the transaction costs involved in using different means of charging.

A third and last concern over charges and pricing systems used to allocate water involves distributional consequences. If water charges are increased, what impacts might be expected on different income classes and different regions of the country? For example, higher water fees may have a large impact on eastern water users since they generally have not had to pay the full cost of developing water supplies. Higher water charges may also be an incentive for some people and firms to conserve water. More difficult to predict is how changes in water pricing policies might affect income and employment levels and the migration of people from one region to another.

Evaluate alternative means of allocating water produced from forested watersheds. In many cases, markets for allocating water have not been developed. Where such is the case, what means might be employed to effectively allocate water produced from forested watersheds? In the West, for example, the traditional water allocation rule has been "first in use is first in right to use the water." An often used alternative rule is the riparian doctrine whereby water rights are based on a landowner's location along a stream or other body of water. Neither of these rules can be considered efficient unless the implied water rights are readily tradeable.

An alternative system worthy of testing and research involves the use of "tradeable permits." The latter could be part of an appropriative or riparian rights system. Before a tradeable permit system can be used, however, a number of important questions need to be addressed. First, what are the economic consequences of a "time of use" (appropriative rule) permit system versus a system based on priority types of use (municipal vs. agriculture)? Second, what is the likely impact of permit trading on downstream users? Third, what restrictions might be placed on trades and what would be the economic impacts of such restrictions? Finally, how would uncertainty concerning future water demands and the health of regional economies and heavy water-using industries influence the trading of permits?

Develop Improved Understanding of Political and Administrative Structures Important to the Production of Water from Forested Watersheds

Effective political and administrative structures are critical to the implementation of watershed management programs. Without effective administrative structures, program success is highly unlikely. Similarly, well developed political systems must be available in order to facilitate public involvement in program development and public expression of interest in investing in watershed management programs.

Evaluate comprehensive organizational mechanisms for management of flood plains and forested watersheds. Watershed management projects may involve overlapping responsibilities of various agencies operating within a particular watershed and associated flood plain. In contrast, some situations may involve agencies that have few linkages between a watershed and downstream flood plain management responsibilities. Needed is research to determine the extent to which such linkages are lacking and the means by which they can be encouraged or enhanced. Under what circumstances is it best to implement watershed-based projects with centrally based authorities in contrast to implementation via local administrative units (if they exist) (Easter and Hufschmidt 1985)?

Multiple ownership patterns within forested watersheds can also pose difficulties for the implementation of watershed management programs. For example, removal of forested vegetation by a single owner of a small tract of forest land in a forested watershed may have little adverse consequences for downstream water users.

However, if all forest owners in a forested watershed remove forest vegetation, serious downstream damages can result. Thus it is important to research the likely cumulative effect of separate land use decisions within a watershed and develop institutional mechanisms for addressing such effects. "In some cases, a form of private ownership will likely be appropriate, but in others, community ownership may work best. Alternative property rights need to be tried and evaluated under different socioeconomic conditions" (Easter and Hufschmidt 1985).

Assess conflicting governmental interests in the production of water from forested watersheds. In addition to agencies having overlapping responsibilities within a watershed or flood plain, agencies may also have different management objectives for a particular watershed. Do these different objectives precipitate conflict and inefficiency between agencies? Can administrative structures be designed to prevent such inefficiencies? Does land trading and/or sales and purchases of land among agencies lead to more effective management of watersheds? Who gains and who loses from such sales or transfers (Edwards 1988)?

Evaluate alternative mechanisms for guiding public involvement in watershed program development. Effective management of conflicting public interests focused on forested watersheds located within public forests is frequently of concern to resource managers. How can public agencies more effectively manage conflicting public interests so as to better shape watershed management policies? Has the growth in environmental interest influenced water production policies? How do the influences of various interests change over time and between different political administrations? What institutional or organizational changes might be made to help resolve conflicts among interest groups? Resolving these questions could help public resource agencies do a better job of managing resources as well as responding to different public concerns. Recent research suggests that increased local partic-ipation in resource use decisions can help resolve many land use and land management conflicts. An important research question is: How have various levels of government organized to accommodate local public participation? What can be learned from these efforts? What means have or could be used to accommodate those not directly affected by a government action or project but have an interest in the project's outcome? What incentives are necessary to obtain informed and broad-based local participation (not necessarily support) in watershed program activities?

Evaluate impacts of changing public water policy on the administration of forested watersheds. In response to environmental concerns and increasing demands for recreational opportunities, public water policy has undergone substantial change over the past two decades. It has led to significant shifts in emphasis among various uses of water. For example, as a user of water, water-based recreation has improved its priority relative to water used for irrigation and power production purposes. There has also been an important shift away from constructing new irrigation projects to interest in more efficient water management within existing systems. What do these shifts suggest for management of forested watersheds and the allocation of water flowing from them? The production of quality water and recreational services will likely become more important in the future. What changes in watershed management strategies will be required to accommodate such trends?

More fundamentally, will the trend toward increased water-based recreation continue? As the U.S. population ages, recreational demands could change—how might such age-driven changes influence future demands for water-based recreation? Similarly, if income is closely correlated with age, will per capita income and its distribution become of growing importance as a dynamic variable influencing water allocations to recreational uses?

Assess alternative organizational and institutional arrangements for water use and allocation. As changes occur in demands for water, institutional and organizational arrangements will have to adjust in order to meet such changes. A first step in proposing new arrangements is a clear understanding of the institutional settings within which forested watersheds are currently managed. Who has the rights to resource flows from forested watersheds? How are decisions concerning water rights made? If the rights to water use are changed, how will such changes affect water flows? In many areas, substantial differences in interests and backgrounds exist between those in the upper portions of a watershed and those in a watershed's lower portion. How do these differences influence watershed management? How might linkages be improved between upland and lowland groups so that watershed management can be improved? If linkages and relationships already exist, what are they and how can they be strengthened?

Evaluate interwatershed and interresource planning structures and techniques. Over the years, various attempts have been made to develop planning structures at the interwatershed level. In the 1960s and 1970s, for example, a widely used approach to water resource planning was river basin planning. Although it was discontinued in the 1980s, little research has been conducted to ascertain the effectiveness of such broad-scale planning efforts. If they were not effective, what were the problems? Has the abandonment of river basin planning adversely affected water resource management efforts? Also used as a means of improving watershed management has been the establishment of special watershed districts. To what extent have they been effective? How might they be organized and managed to be more effective in the future?

Regional resource models have also been used as planning tools to help identify trade-offs between different resource outputs (e.g., water, timber, wildlife, and range) in a region. In order to improve the usefulness of these models, additional research is needed to quantify the feedbacks among various outputs. For example, Joyce et al. (1986) found that "feedbacks from other resources on timber, or the management of other resources, often mitigates the impacts of timber on forage and wildlife." If resource models are to be developed for other regions, extensive resource inventory data will be required. For the resource models to fit together, they must be developed from a common land-base definition (Joyce et al. 1986).

Acquire Improved Understanding of the Economics of Hydrologic Projects and Construction Activities Germane to Forested Watersheds

Evaluate the economics of establishing and designing reservoirs in forested areas. Decisions concerning the construction and management of reservoirs in forested areas should be predicated on an evaluation of a wide range of alternatives for accomplishing the same objective (e.g., flood control). For example, could vegetation management within a forested watershed provide the same water management benefits as a reservoir, but at a lower cost? In terms of environmental impacts, what are the trade-offs between reservoirs and alternative management actions, including vegetation management? How might upper watershed resource use change after construction of a reservoir and how might such a change alter sediment loads reaching a reservoir? Does reservoir design need to be modified to accommodate possible high sediment loads? Given a reservoir is to be constructed, what returns might be expected from alternative watershed management programs designed to reduce sedimentation (Easter et al. 1986)? What economic trade-offs exist when a reservoir is designed small enough so as to deter significant environmental damages? Is the scale of a reservoir critical in determining environmental impacts? If not, what are the critical determinants?

Assess vegetative management alternatives to traditional water supply projects. Management of forest vegetation offers very positive opportunities for improving downstream water supplies. However, large scale studies of vegetative management in forested watersheds are needed—especially studies with an economic focus. They are especially needed to determine the off-site and downstream economic impacts of different forest management practices in watersheds drained by third and larger order streams (Harr 1980).

> Although the practice of manipulating vegetation for the purpose of augmenting water flows has been extensively researched by hydrologists going back a half century, there is no consensus in its economic feasibility . . . The role of vegetation cover in determining stream flow is highly complex and not yet fully understood, even by forest hydrologists and ecologists . . . It is one thing to establish with reasonable assurance that water augmentation practices can increase the yield from a treated watershed. It is another to establish that the hydrologic achievement is necessarily beneficial. There are two considerations that must be taken into account. The first relates to the timing of the increment in yield in relation to peak flows, particularly flow peaks. The second is the valuation of an increment in yield known to occur at a time other than during a time of flood flows (Bowes et al. 1984).

In many watersheds, there will be the added concern of water quality. The economic value of the water increment will depend not only on the timing but also on the quality of the water produced.

A related concern is joint product relationships in vegetation management. For example, Bowes et al. (1984) found that opportunities for watershed management changed significantly when water and timber were considered as joint products. Additional research should be employed to estimate the benefits of managing for joint products in a forested watershed—research which presumes the availability of joint production functions (O'Connell and Brown 1972). This is essential to the development of better estimates of the multiple economic benefits that can result from managing vegetation in forested watersheds (Canham 1986).

Evaluate economic implications of road designs in forested watersheds. "It is not the temporary elimination of trees in a timber harvest operation that precipitates watershed damage. Severe environmental damage to watersheds is usually the result of improper techniques used in the removal of the harvested trees" (Klock 1976). Poorly designed roads are often major contributors to soil erosion and downstream damages, particularly in steep forested watersheds (Sullivan 1985). "Research has shown that stricter prescriptions for logging roads governing intensity, drainage and maintenance are essential to good water quality" (Melbourne and Metropolitan Board of Works 1980). In most cases, road building is likely to be the only major construction project in a forested watershed. Thus, the economics of road design and construction deserve special attention by research.

What are the costs of different forest roads designed so as to minimize environmental damages? What are the downstream impacts of different road design and densities (Kochenderfer et al. 1984)? Is road design most important, or is road density the critical factor that determines the magnitude of downstream damages? For areas with slopes greater than 25 percent, some argue that timber should not be harvested when water production is of major concern (Melbourne and Metropolitan Board of Works 1980). Would these conclusions change if alternatives to roads were used to transport harvested timber from a forested watershed (Klock 1976)? What criteria should be used to select the best mode of transporting timber while at the same time protecting the environment? Clearly, slope and soil erosion potential must be considered along with transportation costs. Should potential downstream damages also be an important part of the criteria?

Assess economic trade-offs involved in managing forested watersheds so as to reduce floods. One of the key questions involved in managing forested watersheds is how to reduce floods. In some high rainfall areas, flood prevention can be a major objective of management. What type of projects and management practices are most effective in preventing floods downstream? The more general question is: What are the benefits of flood prevention projects and practices and what are their opportunity costs? These benefits and costs include resources used to provide the flood control practice as well as any outputs foregone because of the projects or practices. For example, the water level in a reservoir may be kept low to guard against high rainfalls which might cause flooding. If heavy rains do not arrive and the water level stays low, there will likely be economic losses due to reduced recreation, irrigation, and electric power production.

Develop Improved Understanding of Local and Regional Economic and Social Impacts of Water Produced by Forested Watersheds

Assess employment and income impacts of managing forests for the production of water. If additional water supplies from forested watersheds enables a region's economy to expand, the local and regional income and employment impacts of such supplies can be traced and estimated with input-output models. Information of this sort can be used to value many different outputs from forested watersheds. Canham (1986) argues that regional economic impacts and actual expenditures should be used as two measures of the value of timber and recreation products from public forest areas (concern should be with impacts on the local community as well as with the economic efficiency with which forest resources are managed). Where water is the primary concern of managers, water multipliers can be estimated.

> Water multipliers, when used correctly, provide one way of assessing the impact of alternative economic development strategies on total water use. By embodying the operation of a regional economy (input/output model) and the use of water by sectors in the economy (direct water coefficient) the water multipliers form a crucial link between an economic development strategy and a scarce natural resource. Thus, water multiplier analysis is a convenient and logical way to relate total water use in a regional economy and a particular economic development alternative (Ching 1981).

From a forested watershed perspective, considerably more research is needed to determine how water multipliers vary by sector and by type of region.

Evaluate regional differences in the economic impact of managing forested watersheds. In studies of different regional economies, one of the more interesting research issues concerns the ability of local economies to quickly adjust to economic change. What community characteristics facilitate adjustment to seasonal or yearly fluctuations in water supplied by forested watersheds? How do differences in end use of water effect the economic impacts of changes in water supply? Are broad-based economic impacts of water more important than impacts arising from relatively few, high-valued water uses?

Acquire Improved Understanding of the Economics of Managing Forested Watersheds to Improve Water Quality

The quality of water flowing from the nation's forests has been of increasing concern since the 1960s. In the 1980s, ground water quality also assumed a prominent position—a position which it is likely to retain for the foreseeable future. Since so much of the nation's water comes from forested watersheds, it is not surprising that concern over the production of quality water is of major interest to forest managers.

Assess the economics of treatment facilities in forested watersheds. In more developed forested watersheds, treatment facilities are required to cope with water pollutants caused by various sorts of developments. Major research questions associated with such facilities focus on the relative cost and effectiveness of alternative technologies. Do, for example, costs and associated project effectiveness vary with scale of facility and the type of pollutants encountered? On a more narrow focus, are forested watersheds economically efficient when used as nutrient filters, particularly if agricultural lands are present in the watershed (Lowrance et al. 1984)?

Evaluate the economics of managing nonpoint source water pollutants originating in forested watersheds. Water pollutants from forested watersheds can result from changes in natural ecosystems (including changes in sediment rates, water temperature, and chemical transport) or from management practices which introduce undesirable substances (including bacteriological pollutants, pesticides, fertilizers, and fire retardants) (Megahan and King 1985). Although considerable research has been focused on the first type of nonpoint source pollution, additional research is needed on management-induced pollutants in forested watersheds. For example, what is the economic relationship between a forestry practice (e.g., road construction) and various levels of pollutants in water flowing from a forested watershed? Such research is confounded by "sketchy production function information which links timber harvesting practices to specific levels of water pollutants generated or curbed" (Ellefson and Miles 1985).

For the forest economics researcher interested in watershed management analyses, production information is critical to project and economic analyses. Irland (1985) argues that top priority should be given to additional field research on the effectiveness and cost of sediment control practices, on the impacts of sedimentation on fisheries, and on cost and effectiveness of low-impact logging machinery and systems. Also cited as important is the need to validate models that predict water quality and streambed and stream biota responses to different forest practices (Ice and Whittemore 1983). Once such information is available, watershed planning and management models need to be developed and, subsequently, estimates provided of both economic and pollution impacts of watershed management practices.

> Watershed planning models are applied to specific water quality problems and evaluate impacts of management practices, subsidies and taxes on pollution and farm income . . . However, the economic components of these models are better developed than components for prediction of pollution control . . . This is due largely to the limited availability of tested chemical transport models . . . It is important for researchers, practitioners and policy-makers to realize, . . . that only a modest beginning has been made. The control of point-source discharges of waste waters is based on more than a hundred years of research and testing, and a continued investment in nonpoint-source models will be necessary to establish a comparable level of technology (Haith 1982).

Additional research is also needed to determine the demand for reduced downstream damages that are caused by nonpoint source pollutants. Watershed investments have often focused only on physical concerns (supply side) (e.g., funds for watershed protection concentrated in areas causing the largest amount of downstream physical damage). The perspective needs to be reversed by first asking where the demand for improved water quality is the greatest—then designing programs to meet that demand. By doing so, one may logically find that demand for water quality varies considerably across the nation and across regions. People in certain areas may demand much higher water quality than persons in other areas. Focusing concern on the demand for quality water could result in a healthy shift in emphasis from exclusive supply-side considerations to a combination of demand and supply concerns. It would also mean that uniformly applied water quality standards are not appropriate.

Research concerning riparian forests in a water quality context is also needed. "Riparian zones are particularly important because of their high productivity, habitat value, and effects on stream properties" (Megahan et al. 1981). Research should focus on understanding these areas and developing cost-effective practices (e.g., buffer zones) required to produce higher quality water from them.

Research should also be directed to determining the effects of soil erosion on the productivity of forest and range lands, especially the ultimate economic effects. These relationships are poorly understood (Gregersen et al. 1987). Although Klock (1982) provides a bioassay approach to evaluating the impacts of soil erosion on forestry productivity, the approach falls short of that needed for economic analysis of the impacts of erosion on soil productivity. Research similar to that done in the agricultural sector (Wen and Easter 1987) should be carried out in forestry.

Evaluate economic implications of managing ground water pollutants in forested areas. Forested watersheds make an important contribution to ground water recharge and, in the process, they may become potential sources of ground water pollution. What forestry practices are cost-effective means of enhancing ground water recharge? What type of pollutants are most likely to pollute ground water within forested watersheds? What would be the economic consequences of drastically curtailing (eliminating) such pollutants from forested watersheds? Research is needed to determine the most economically efficient or best management practices for reducing surface water pollutants and their possible contribution to ground water pollution (Lynch et al. 1985). How might the practices causing ground water pollution be modified to reduce ground water pollution? What are the costs of modifying such practices to landowners, timber harvesters, and final consumers of wood products?

Research is also needed to estimate benefits of preventing ground water pollution and how the demand for such benefits might change in the future. How will different levels and types of ground water pollutants affect future uses? A higher level of pollutants may be acceptable for certain uses of water (e.g., irrigation, certain commercial uses) but not for other uses (e.g., human consumption). Needed is a clearer understanding of how demands for ground water are likely to change in the years ahead. Once understood, more accurate economic analyses can then be made of benefits accruing from additional ground water protection measures.

Summary and Conclusions

Economic and policy research addressing the production and valuation of water from forested watersheds is deserving of special consideration given the importance of forested watersheds as a source of the nation's residential and commercial water supplies. This is especially true as water quality and ground water supplies assume more prominent positions on the natural resource agendas of local, state, and national governments. Future research investments would be most useful if focused on developing more effective means of valuing and planning the use and management of forests for water production purposes, improving methods for allocating and reallocating water from forested watersheds, enhancing political and administrative structures for managing forests for water production purposes, improving the economic understanding of hydrologic projects in forested areas, acquiring better understanding of local and regional economic impacts of forests managed for water production purposes, and enhancing knowledge of economically effective means of managing forests to improve water quality. Economics and policy research guided in strategic directions of this sort would go far toward improving the efficiency with which forested watersheds meet regional and national demands for even greater quantities of quality water.

Literature Cited

Bowes, M. D., J. V. Krutilla, and P. B. Sherman. 1984. Forest management for increased timber and water yields. Water Resources Research 20(6):655-663.

Canham, H. O. 1986. Comparable valuation of timber and recreation for forest planning. Journal of Environmental Management 23:335-339.

Ching, C. T. K. 1981. Water multipliers: Regional impact analysis. Water Resources Bulletin 17(3):454-457.

Daubert, J. T. and R. A. Young. 1981. Recreation demands for maintaining instream flows: A contingent valuation approach. American Journal of Agricultural Economics 63(4):666-676.

Easter, K. W. and M. M. Hufschmidt. 1985. Integrated watershed management research for developing countries. Workshop Report. East-West Center. Honolulu, HI.

Easter, K. W., J. A. Dixon and M. M. Hufschmidt. 1986. Watershed resource management: An integrated framework with studies from Asia and the Pacific. Westview Press. Boulder, CO.

Edwards, J. D. 1988. A look at proposed institutional changes for the management of public lands. Paper presented at American Agricultural Economics Association Meeting. July.

Ellefson, P. V. and P. D. Miles. 1985. Protecting water quality in the Midwest: Impact on timber harvesting costs. Northern Journal of Applied Forestry 2(2):57-61.

Forest Service. 1986. Final environmental impact statement (1985-2030 resources planning act program). FS 403. U.S. Department of Agriculture. Washington, DC.

Gregersen, H. M., K. N. Brooks, J. A. Dixon, and L. S. Hamilton. 1987. Guidelines for economic appraisal of watershed management projects. Conservation Guide 16. Food and Agriculture Organization. United Nations. Rome, Italy.

Haith, D. A. 1982. Models for analyzing agricultural nonpoint source pollution. Research Report RR-82-17. International Institute for Applied Systems Analysis.

Harr, R. D. 1980. Scheduling timber harvest to protect watershed values. Proceedings of interior west watershed management symposium. Pacific Northwest Forest and Range Experiment Station. USDA Forest Service. Portland, OR.

Hibbert, A. R. 1967. Forest treatment effects on water yield. In: Forest hydrology by W. E. Sopper and H. W. Lull (eds.). Pergamon Press. New York, NY.

Hoover, M. D. 1944. Effect of removal of forest vegetation upon water yields. American Geophysical Union Transactions, 25:969-975.

Ice, G. G. and R. C. Whittemore. 1983. Review of model use in evaluating nonpoint source loads from forest management activities. In: Proceedings of Stormwater and Water Quality Model User Group Meeting. University of Florida. Gainesville, FL.

Irland, L. C. 1985. Logging and water quality: State regulation in New England. Journal of Soil and Water Conservation 40(1):98-102.

Joyce, L. A., T. W. Hoekstra and R. J. Alig. 1986. Profile: Regional multi-resource models in a national framework. Environmental Management 10(6):761-771.

Klock, G. O. 1976. Estimating two indirect logging costs caused by accelerated erosion. General Technical Report PNW-44. Pacific Northwest Forest and Range Experiment Station. USDA Forest Service. Portland, OR.

Klock, G. O. 1982. Some soil erosion effects on forest soil productivity. In: Determinants of soil loss tolerance. ASA Special Publication 45:53-66. Soil Science Society of America. Madison, WI.

Kochenderfer, J. N., G. W. Wendel and H. C. Smith. 1984. Cost of and soil loss on 'minimum-standard' forest truck roads constructed in the Central Appalachians. Research

Paper NE-544. Northeastern Forest Experiment Station. USDA Forest Service. Broomall, PA.

Lowrance, R., R. Todd, J. Fail, Jr., O. Hendrickson, Jr., R. Leonard and L. Asmussen. 1984. Riparian forests as nutrient filters in agricultural watersheds. Bioscience 34(6):374-377.

Lull, H. W. 1970. Management possibilities for water yield increases. Proceedings of joint FAO/USSR international symposium on influences and watershed management.

Megahan, W. F., D. H. Boelter, S. P. Gessel, J. W. Hornbeck and J. R. Meiman. 1981. Forest land: Conservation needs, technology and policy alternatives. In: Soil and water resource: Research priorities for the nation. Soil Science Society of America. Madison, WI.

Megahan, W. F. and P. N. King. 1985. Identification of critical areas on forest lands for control of nonpoint sources of pollution. Environmental Management 9(1):7-17.

Melbourne and Metropolitan Board of Works. 1980. Water supply catchment hydrology. Research Report No. MMBW-W-0012. Melbourne, Australia.

O'Connell, P. F. and H. E. Brown. 1972. Use of production functions to evaluate multiple use treatments on forested watersheds. Water Resources Research 8(5):1188-1198.

Ramirez, J., V. L. Adamowicz, K. W. Easter and T. Graham-Tomasi. 1988. Ex-post analysis of flood control: Benefit-cost analysis and the value of information. Water Resources Research 24(8):1397-1405.

Rothacher, J. 1970. Increases in water yield following clear-cut logging in the PNW. Water Resources Research 6(2):653-658.

Saliba, B. C., D. B. Bush and W. E. Martin. 1987. Water marketing in the Southwest: Can market prices be used to evaluate water supply augmentation projects? General Technical Report RM-144. Rocky Mountain Forest and Range Experiment Station. USDA Forest Service. Ft. Collins, CO.

Sullivan, K. 1985. Long-term patterns of water quality in a managed watershed in Oregon: Suspended sediment. Water Resources Bulletin, 21(6):977-987.

Wen, F. H. and K. W. Easter. 1988. Soil erosion and the loss in productivity: An example of the Terril soil series in Minnesota. Bulletin 577-1987. Agricultural Experiment Station. University of Minnesota. St. Paul, MN.

Bibliography

Anderson, H. W., M. D. Hoover, and K. G. Reinhart. 1976. Forests and water: Effects of forest management on floods, sedimentation and water supply. General Technical Report PSW-18. Pacific Southwest Forest and Range Experiment Station. USDA Forest Service. Berkeley, CA.

Archey, W. E. and J. C. Mawson. 1984. Municipal watershed management: A unique opportunity in Massachusetts. Journal of the New England Water Works Association 98(2):138-147.

Brown, T. C. 1981. Tradeoff analysis in local land management planning. General Technical Report RM-82. Rocky Mountain Forest and Range Experiment Station. USDA Forest Service. Ft. Collins, CO.

Brown, T. C. 1982. Monetary valuation of timber, forage and water yields from public forest lands. General Technical Report RM-95. Rocky Mountain Forest and Range Experiment Station. USDA Forest Service. Ft. Collins, CO.

Coats, R. N. and T. O. Miller. 1981. Cumulative silvicultural impacts on watersheds: A hydrologic and regulatory dilemma. Environmental Management 5(2):147-160.

Douglass, J. E. and W. T. Swank. 1972. Streamflow modification through management of eastern forests. Research Paper SE-94. Southeastern Forest Experiment Station. USDA Forest Service. Asheville, NC.

Forest Service. 1979. An assessment of the forest and range land situation in the United States (review draft). U.S. Department of Agriculture. Washington, DC.

Harris, T. R. and C. T. K. Ching. 1983. Economic resource multipliers for regional impact analysis. Water Resources Bulletin 19(2):205-210.

Hickman, C. A. and B. D. Jackson. 1979. Economic impacts of controlling soil loss from silvicultural activities in east Texas. Forest Science 25(4):627-640.

Klemperer, W. D. 1979. On the theory of optimal forest harvesting regulations. Journal of Environmental Management 9(1):1-13.

Krutilla, J. V., M. D. Bowes and P. Sherman. 1983. Watershed management for joint production of water and timber: A provisional assessment. Water Resources Bulletin 19(3):403-414.

Lee, M. T. and K. L. Guntermann. 1976. A procedure for estimating off-site sediment damage costs and an empirical test. Water Resources Bulletin 12(3):561-575.

Lynch, J. A., E. S. Corbett and K. Mussallem. 1985. Best management practices for controlling nonpoint-source pollution on forested watersheds. Journal of Soil and Water Conservation 40(1):164-167.

Miller, W. L. and H. E. Everett. 1975. The economic impact of controlling nonpoint pollution in hardwood forestland. American Journal of Agricultural Economics 57(4):576-583.

Napier, T. L., K. W. Easter, D. Scott and R. Supalla. 1983. Water resources research: Problems and potentials for agriculture and rural communities. Soil Conservation Society of America. Ankeny, IA.

Ponce, S. L. and J. R. Meiman. 1983. Water yield augmentation through forest and range management: Issues for the future. Water Resources Bulletin 19(3):415-419.

Shelby, R. 1984. Estimating monetary values for use permits on western rivers. Journal of Forestry 82(2):107-109.

Sopper, W. E. 1971. Watershed management: Water supply augmentation by watershed management in wildland areas. Report to the National Water Commission. National Technical Information Service. Springfield, VA.

Worley, D. P. and J. H. Patric. 1971. Economic evaluation of some watershed management alternatives on forest land in West Virginia. Water Resources Research 7(4):812-818.

PART THREE

Future Research Program Directions

25

CHALLENGES AND AGENDAS FOR FOREST RESOURCE ECONOMICS AND POLICY RESEARCH IN THE COMING DECADE

H. Fred Kaiser, Richard L. Porterfield,
and Paul V. Ellefson

Forest economics and policy researchers have a rich tradition of providing a variety of important information for decisions about the use and management of forests and the goods and services flowing therefrom. As with any research enterprise, however, there is substantial merit in periodically reviewing the nature of ongoing research activities and in appraising research challenges that may loom on the horizon. As stated by noted economist T. W. Schultz (1964): "A particular profession can become obsolete. We, too, are subject to these risks. Thus it should be salutary, now and then, to remove our workday blinders and look at our approach, the problems on our research agenda, the tools we use, and the way we are organized."

Offered in the spirit of furthering constructive discussion among researchers and others interested in the products of forest economics and policy research, what follows is an agenda of strategic research directions for the field and an enumeration of challenges which must be faced if the suggested research agenda is to be effectively accomplished.

An Agenda: Research Emphasis for the Future

Strategic directions for forest economics and policy research in the next decade will be influenced by a variety of social and economic settings which will arise from domestic as well as international environs. Such settings will ultimately determine research priorities and will have a significant influence on the magnitude of research investments made by public and private organizations. Of the many economic and political conditions capable of compelling change in the forestry community, some are likely to be especially significant to the research agenda of forest economic and policy sciences. The following developments—and associated research directions—are apt to be of special significance. Detailed descriptions of strategic directions (including example research topics) can be found elsewhere in this volume.

Forestry will experience growing linkages to the world economy. U.S. interests in the world economy and the influence of the world economy on U.S. economic activity have posed significant challenges to U.S. forestry. There is ample reason to believe that such challenges will increase in the decades ahead. In such a context, research should develop an improved understanding of

- International trade in forest and related products
- Social and economic growth of developing nations

Forest resource policy making and program development will increase in complexity. The processes by which public and private forest resource policies are developed and implemented have grown increasingly complex in recent years. State and federal governments have installed a plethora of complex procedures that are designed to encourage due process. There is ample reason to believe that such processes will become increasingly complex as the demand for products and services produced by forests grows in intensity. In such a context, research should develop an improved understanding of

- Policy development and program administration
- Forest resources law and legal processes
- Institutional arrangements directing use and management of forests
- Taxation of forest products and forest resources

Forest-based industries will continue significant restructuring within domestic and worldwide economies. The nation's forest-based industries have experienced significant change in terms of economic structure, conduct, and performance. The industries in question range from traditional wood-based enterprises to forest-based firms actively pursuing interests in tourism and recreation. Changes in the structure of such industries will undoubtedly continue in the years ahead. In such a context, research should develop an improved understanding of

- Economic structure and performance of forest-based industries

Rural needs for social and economic development will increase as a focus of forestry interests. Rural economies located within forested areas often experience significant economic and social depression. If properly focused and effectively organized, the forest and related resources located in and near such communities can offer significant opportunity for achieving economic and social vitality. In such a context, research should develop an improved understanding of

- Community and regional economic growth and development

Scientific and technical developments will increasingly influence resource use and productivity. Scientific advances can be of special significance to the efficient use and management of forests and the products therefrom. Properly developed and willingly adopted, new technologies can offer considerable opportunity for gains in productivity. In such a context, research should develop improved understanding of

- Development, dissemination, and adoption of new technology

Management and production processes will grow increasingly complex. Management and production decisions in forestry are multiple in number and complex in character. There is substantial reason to believe that the complexity of such decisions will grow significantly in the future and that such complexity will heighten the level of risk and uncertainty to be faced by decision makers. Therefore, research should develop an improved understanding of

- Wood fiber production
- Timber harvesting
- Production and valuation of forest and wildland recreation
- Management of fire in forested environments
- Management of insects and diseases in forested environments
- Production and valuation of water from forested watersheds
- Distribution and marketing of forest resource products
- Structure and performance of nonindustrial private forests

Information requirements and information management will increase as a concern of forestry interests. Faced with complex forest management decisions imbedded with significant risk and uncertainty, the forestry community has made significant investments in information-gathering processes and activities. There is convincing evidence, however, that information requirements will dramatically increase in the future as will the task of managing such information. In such a context, research should develop an improved understanding of

- Resource assessment, information management, and communications technology
- Forecasting demand and supply of forest resources, products, and services

Forest use and management decisions will increasingly reflect environmental quality interests. Demand for quality forested and related environments has increased tremendously over the past 20 to 30 years. There is significant evidence that the general public and selected interests therein will continue to have high expectations for quality forested environments. In such a context, research should develop improved understanding of

- Forestry sector environmental effects

The emerging settings and strategic directions previously described paint a challenging future for forest economics and policy research. Considerable research within the identified strategic directions is already underway; some organizations have plans to move in such directions. Some identified directions represent research areas yet to be developed. The identified strategic directions have not been prioritized (subjects that were omitted were obviously judged to be of lesser priority). All are sufficiently important to be worthy of attention by the community of researchers talented in the fields of economics and policy. A specific researcher's efforts should be guided by the economic and social importance of the strategic direction, the probability of carrying out successful research within a particular direction, the likelihood that research results will have a significant impact, and the intellectual interests of the researcher.

Challenges in Performing the Research Job

The community of professionals which composes the nation's intellectual capital in the area of forest economics and policy research represents significant resources for addressing many of the issues and strategic research directions previously presented. Although the ability to do so is optimistically bright, there are a number of significant challenges looming on the horizon. If met in a forthright manner, professionals applying economic and policy sciences to forestry and related issues may well be in a position to make even greater contributions to efficient use and management of forest environments. The following should be considered.

Recruiting and educating scientific talent in forest resource economics. Researchers claiming expertise in forest economics and policy sciences number between 500 and 600 individuals nationwide. The effectiveness of this talented pool of individuals is dependent in large measure on their intellect, their formal education, their professional experiences, and the sustained flow of new talent into the research community. There is ample reason to believe, however, that future availability of new professional talent will be less than satisfactory. As forest economists of the 1960s and 1970s approach retirement in the 1990s, the number of forest economists available to take their places may be woefully insufficient—a serious shortage in economic expertise may occur. To address this potential problem, public and private agencies should make concerted efforts to encourage talented individuals to seek an education in forest economics and policy and, where possible, should facilitate the formal educational interests (including graduate study) of interested individuals. Concerted efforts should be made to inform prospective students about opportunities in forest economics and policy research; to provide financial support for graduate education in forest economics and policy sciences; and to facilitate continuing educational opportunities for researchers currently practicing in the field.

Maintaining a healthy creative relationship between theory and applied research activities. Forest resource economics and policy sciences are often the applied expressions of a fundamental body of economic, social, and policy theories. The counterpart to the field's theoretical roots is the application of economic and policy sciences to real-world forest resource problems. Choosing an appropriate balance between theory and application is important. Excessive emphasis on theoretical research can lead to an inordinate concern over theory building and the development of new methodologies. The product of such an emphasis can be a focus on trivial problems and a decline in communication between researchers and forest managers, policy officials and important constituency groups. In contrast, excessive focus on practical problems can lead to research activities that suffer from the lack of directed inquiry which only theory can provide.

Future growth and relevance of forest economics and policy research implies the maintenance of a healthy creative tension between theory and application. Forest economics and policy research should strive to maintain close linkages with fundamental disciplines such as economics, political science, law, mathematics, statistics, and related supporting areas. Intellectual isolation from parent or core disciplines can easily lead to a demise in the flow of intellectual capital on which an applied field is founded. Conversely, the forest economics and policy community should strive to strengthen its reputation for addressing economic and related problems that are of practical significance to managers and users of forests and forest resources.

Establishing and maintaining linkages with other applied fields. Forest economics and policy sciences are often looked to as among the "integrating sciences." As such they have an extremely important role to play in addressing highly complex forestry and closely related issues. To do so, however, requires establishment and maintenance of strong linkages with a variety of applied fields—both within (e.g., silviculture, forest protection, mensuration) and outside the forestry community (e.g., engineering, biometrics, public affairs, agricultural economics). In concert, such fields have much to say about complex contemporary forest resource problems. Reaching out to other disciplines and serving in an integrative and facilitative role can and should be one of the major strengths of forest economics and policy research. Even when unable to provide definitive answers to complex forestry problems, forest economists can serve to redefine important issues so that effective solutions can ultimately be developed.

Monitoring of forest economics research priorities, investment levels, and intellectual capital. Knowledge about the nature and extent of forest economics research—and trends therein—is sometimes an elusive commodity. Frustration often arises because so little effort is focused on systematically collecting evidence that accurately portrays the nature of the field and its trends. Admittedly, there are a number of circumstances that do not facilitate monitoring efforts (e.g., the large number of organizations involved in forest resource economics research, the large number of researchers actively engaged in the subject, and the broad scope of economic and related problems being researched). Such problems aside, there is a need to appraise in a timely and systematic fashion the status of forest resource economics research. Who is doing what? How much is being invested? What techniques and approaches are being employed? What is the academic training of researchers? And what changes in problem orientation are occurring? Such information is essential for establishing program priorities, monitoring program progress, and determining need for additional intellectual capital. The USDA–Cooperative State Research Service gathers considerable information about agricultural and forestry research (e.g., CRIS). Such can be an important starting point for annual compilations about the status of the field.

Identifying high priority research needs and planning the allocation of resources required to address such needs. Only three times over the past 50 years have forest economics and policy researchers in the U.S. comprehensively reviewed the status of the field and attempted to systematically specify potential research directions (Social Science Research Council 1936, Duerr and Vaux 1953, and Clawson 1977). The diversity of the subject material and the pluralism of the research has not always facilitated more frequent assessments. However, since research problems and research methods are increasingly prone to change, a comprehensive review of the field should be undertaken at least once per decade. To be useful and credible, reviews should be more than "laundry lists" or amalgams of platitudinous wishes. Reviews should be keyed to well-thought-out assessments of the major directions forestry is likely to follow in the future; the principal economic and policy problems which are likely to arise from such directions; the ability of the forest economics community to address such problems; and the additional financial and human resources required to undertake appropriate economic and related research. Comprehensive reviews of the field should strive to avoid a self-serving tone; they should be rigorous in content and professional in presentation. One approach might be for a professional organization to periodically "commission" a group of scholars to undertake reviews of the field.

Articulating and communicating the mission of forest resource economics to broader institutional and political bases. Research involving forest economics and related policy sciences has much to contribute to the resolution of forestry and related issues. Such may—or may not—be appreciated by clients traditionally served by the field. Needed, however, are greater efforts to communicate the capabilities of the field to a broader range of constituents. Organized special interest groups and newly established environmental agencies are examples of organizations that may not be fully aware of the contributions that forest economics and related sciences can make to decision-making. In addition, groups that are afflicted with economic and social problems and are less than well organized to express concern about them are deserving of the field's attention (e.g., institutional arrangements for increasing Native American business opportunities in forest products) as are the often beleaguered fiscal, organizational, and managerial capacities of rural governments operating in forested areas. Although possibly viewed as self-serving, such clients can also become important bases of support for furthering the capabilities of forest economics and related research.

Achieving high levels of sustained funding commensurate with the value of the research services produced. Funding of forest economics and related research has experienced substantial declines in real terms in the last decade. In some cases, the declines have been dramatic (e.g., market development and efficient marketing of timber). At the same time, pressure for greater amounts and varieties of goods and services from the nation's forests are mounting—a condition begging for the attention of an applied allocation science such as forest economics. How can high levels of sustained funding be achieved? Obviously, no single funding initiative will suffice. Each institution must develop plans and strategies adopted to unique circumstances, including clientele, research priorities, and availability of research financing in general. Nevertheless, there are some common multi-institutional, discipline-wide activities that warrant consideration. For example, there is need for a more effective means of annually monitoring research funding and allocation on a regional and national basis. Lacking reliable and timely information makes judgments about and reactions to supposed trends in funding nearly impossible. Attainment of high levels of sustained funding would also be facilitated by more clearly defined discipline-wide research priorities—by specification of the products to be realized from investment in such priorities and by identification of the decision makers to be served by such research. Funding of forest economics and related research would also benefit from a more purposeful stance toward the need for financial support. Unless the forest economics community is prepared to exert influence over those responsible for allocating research funding (e.g., Congress, state legislatures, university administrations, public research agencies), the development of research plans and priorities may be of little avail—however elegant in formulation and exposition. Politicizing forest economics and policy researchers is not advocated; however, indirect and properly applied influence is most certainly suggested. Funding of forest economics and policy research would also benefit from efforts to seek funding more creatively. Traditional funding sources should not be neglected (e.g., agency funding, agricultural experiment stations, McIntire-Stennis funding), yet consideration should be given to competitive grants, private corporate funding, forest product check-off systems, private-public cooperative efforts, establishment of charitable funds or foundations, and dedicated positions (chairs) in universities.

High levels of sustained funding of forest economics and related research is an obvious necessity if the benefits of such research are to be fully realized. Significant to the field is the growing tendency away from general research project support toward dependence on problem-specific contracts and policy evaluations undertaken as though they were research activities. Such may be appealing in terms of serving the short-term needs of decision-makers, but it is often at the expense of building long-term supplies of knowledge required to address economic and policy problems of the future. Meeting the challenge of building a healthy stock of forest economics knowledge will require the field to innovate institutionally, to establish directions and priorities, and to actively plan—even coordinate—research activities. Such implies that researchers aligned with forest economics and policy sciences should continue to view themselves as collectively providing services for the betterment of forestry and the public at large. In this sense, collective action in the development of appropriate funding strategies is appropriate.

Transferring the technologies produced by forest economics research to a variety of clients. The products of investments in forest economics and related research are substantial in number, great in variety, and significant in potential impact. To be useful, they must be placed in the hands of appropriate decision-makers in a timely fashion. The forest economics community should facilitate existing technology transfer programs (e.g., extension services) and mechanisms (e.g., periodicals, conferences) and be creative in its efforts to develop new means of conveying the products of research to interested users (e.g., electronic imagery).

Achieving effective coordination and cooperation among research institutions involved in forest resource economics research. Forest economics and related research is undertaken within a wide variety of public and private organizations. This institutional landscape is characterized by a high degree of decentralization. It tends to encourage independence, intellectual freedom and responsiveness to local problems while precluding singular, authoritarian dogma in theory and methods. Decentralization, however, tends to complicate communication and coordination, often fosters duplication of efforts, and can inhibit concerted discipline-wide initiatives. The institutional landscape of forest economics is also extremely diverse and pluralistic in mission, goals, methods, and funding sources. Surely there are strengths to pluralism. But such strengths can also make difficult the attainment of a clear focus on problems and the establishment of discipline-wide strategies to address them. The forest economics research community should continue to explore a variety of institutional mechanisms that would help facilitate more efficient research in decentralized, pluralistic environments—yet preserve the benefits of such environments. Centers or consortiums with special knowledge and expertise (e.g., international trade, demand forecasting, policy analysis) and regional economic research projects supported by national and local research organizations are a few examples. As the United States moves toward more involvement in a global community, new international research arrangements should also be explored (e.g., expansion of links with the International Union of Forestry Research Organizations). An expanded role for existing regional forest economics groups (Midwest Forest Economists, Southern Forest Economics Workers, Northeast Forest Economists, Western Forest Economists) should be examined, especially as they can facilitate research planning, priority selection, and research funding. Whatever the chosen mechanism, forest economics researchers should be constantly in search of new institutional arrangements that strengthen abilities to accomplish and coordinate research activities.

Encouraging multi-disciplinary research. Major forestry and related issues can seldom be addressed in an effective manner by a single discipline or applied field. The ecological, economic, and social climate from which such issues flow is often far too complex to permit the application of narrowly disciplined research. Forest economics researchers have a history of being able to develop useful and innovative ways of encouraging research across disciplinary boundaries. They should continue to actively participate in and where possible facilitate research involving a wide range of disciplines.

Evaluating the contributions of forest economics and policy research. In recent years, policy makers, administrators, and constituency groups have increasingly recognized the need to improve the planning, management, and evaluation of research. How can emerging issues be anticipated so that research programs can respond in a timely fashion? How might research programs and projects that have outlived their usefulness be identified and subsequently redirected? And what returns can be expected from additional investments in forestry research? Issues of this nature cannot always be precisely answered—research involves an inherent high level of uncertainty. Regardless, informed judgments about the productivity and relevance of research programs must be made. Faith that the "invisible hand" of the scientific marketplace will guide researchers to the most important problems cannot always be assumed.

Investments in biological, physical, and mechanical research have often resulted in new technologies that can be clearly identified as new products or new production processes. In the case of forest economics and policy research, the information generated frequently results in improved management decisions or in new or improved policies and institutions. That such information has value is attested to by the frequency with which managers seek such information as they attempt to adjust to changing economic and political conditions. Unfortunately, systematic assessment of the effectiveness and efficiency of investments in forest economics and policy research is nearly nonexistent. This latter condition may exist for good reason. For example, linking quantitatively the products of forest economics research to institutional change (e.g., new laws, revised procedures, changed organizational structure) can be extremely difficult. Any particular change may be the result of multiple research investments, each of which contributed pieces of information that eventually added to a useful total. However, the efficiency and effectiveness of investments in forest economics and policy research is a natural concern of persons and organizations responsible for allocating scarce financial and intellectual resources. As forestry's reservoir of talent concerning allocation questions, the forest economics community has a special responsibility to assess its own research house. Anderson and Evered (1986, p. 799) state the responsibility quite well:

> The scientific community has a major task ahead of it. It must become more aware of the demands of society—namely, that its activities should be relevant and directed to meeting perceived needs. The researchers must not only be accountable but they must also assume the burden of collecting, evaluating, and communicating information on their activities which will justify continued public support.

The case for continued support of research activities must be based on hard evidence whenever possible—not on anecdote or tradition. Sophisticated planning and evaluation will never eliminate risk and uncertainty from research endeavors, but it can improve the allocation of scarce research dollars and help direct research programs into socially profitable areas. Unfortunately, too few attempts have been made to do so in the field of forest economics and policy research.

Conclusion

Forest economics and policy research has made substantial progress in supplying the type and quality of information necessary for effective decisions concerning the use and management of forests. At no time in the past has such research had so far reaching an application to forestry and related problems. This application has increasingly involved multi-disciplinary research activities and a growing emphasis on economic problems arising from increased use of forests for the production of noncommodity outputs including water, recreation, and environmental amenities. The need for information supplied by forest economics research will grow at a healthy pace in the years ahead. Researchers will be challenged to fully utilize their unique economics and policy training, to select relevant and researchable problems for investigation, and to direct research findings to audiences that are in need of such information. Forest economics and policy research should be well-positioned to meet such challenges.

Literature Cited

Anderson, J. and D.C. Evered. 1986. Why do research? The Lancet Vol. II: 799-802.

Clawson, M. C. 1977. Research in forest economics and forest policy. Resources for the Future. Washington, DC.

Duerr, W. A. and H. J. Vaux. 1953. Research in the economics of forestry. Charles Lathrop Pack Forestry Foundation. Washington, DC.

Schultz, T. W. 1964. Changing relevance of agricultural economics. Journal of Farm Economics 46:1004-1014.

Social Science Research Council. 1936. A Survey of Research in Forest Economics. Bulletin No. 24. Subcommittee on Scope and Status of Research in Forest Economics. Committee on Social and Economic Research in Agriculture. New York, NY.

ABOUT THE EDITOR
AND CONTRIBUTORS

David N. Bengston is research economist with the USDA–Forest Service's forestry research evaluation work unit at the North Central Forest Experiment Station (St. Paul, MN). His research interests include economics of technical and institutional change, valuation of nonmarket forest outputs, and planning, management, and evaluation of forestry research.

Thomas C. Brown is economist with the USDA–Forest Service's Rocky Mountain Forest and Range Experiment Station (Fort Collins, CO). His research interests include water economics and management, scenic quality assessment, and valuation of nonmarket goods and services.

Daniel E. Chappelle is professor in the Department of Forestry and Department of Resource Development, Michigan State University (East Lansing, MI). He is past economist with the USDA–Forest Service's Pacific Northwest Forest and Range Experiment Station.

Frederick W. Cubbage is associate professor in the School of Forest Resources, University of Georgia (Athens, GA). He is past associate economist with the USDA–Forest Service's Southern Forest Experiment Station and service forester with the state of Kentucky. His research interests include forestry law, timber harvesting economics, and technical forestry assistance programs.

Benjamin V. Dall is professor of environmental law and policy, College of Environmental Science and Forestry, State University of New York (Syracuse, NY). He is a former self-employed attorney, professor at the University of Nevada's Desert Research Institute, and chair of the Department of Managerial Science and Policy at the State University of New York (Syracuse).

Gilbert P. Dempsey is research economist with the USDA–Forest Service's Northeastern Forest Experiment Station (Princeton, WV). His research interests include forest industry development, international trade, industrial resource use, and industrial productivity.

J. E. de Steiguer is leader of the USDA–Forest Service's pest impact assessment technology work unit at the Southeastern Forest Experiment Station and an adjunct professor of the forestry faculties at Duke University and North Carolina State University (Research Triangle Park, NC). His research interests include nonindustrial private forestry, forest damage from insects and air pollutants, and timber supply and demand.

K. William Easter is professor of resource economics in the Department of Agricultural and Applied Economics, College of Agriculture, University of Minnesota (St. Paul, MN). He is past agricultural economist, USDA–Economics Research Service (Washington, DC), leader of an India agricultural development program for the Ford Foundation, and leader of a University of MN–Colorado State University–USAID project on water management in developing countries. He is editor of *Watershed Resource Management: An Integrated Framework with Studies from Asia and the Pacific.*

Paul V. Ellefson is professor of forest economics and policy in the Department of Forest Resources, College of Natural Resources, University of Minnesota (St. Paul, MN). He is past director of resource policy programs for the Society of American Foresters, resource economist for the Michigan Department of Natural Resources, research forester for the USDA–Forest Service Northeastern Forest Experiment Station, and forester on the Mendocino National Forest (California). He is a fellow of the Society of American Foresters and is author of *U.S. Wood-Based Industry: Industrial Organization and Performance* and *Forest Resource Policy: Process, Participants and Programs.*

Hans M. Gregersen is professor of forest and resource economics in the Departments of Forest Resources and Agricultural and Applied Economics, University of Minnesota (St. Paul, MN). He is past forestry officer for the United Nation's Food and Agriculture Organization, research forester for the USDA–Forest Service and archaeological surveyor for the Tikal Project in Central America. He is author of numerous books and articles addressing natural resource economics and development.

Perry R. Hagenstein is president of Resources Issues, Inc. (Wayland, MA) and president of the Institute for Forest Analysis, Planning and Policy (Phoenix, AZ). He is past executive director of the New England Natural Resources Center, senior policy analyst for the U.S. Public Land Law Review Commission and economist for the USDA–Forest Service's Northeastern Forest Experiment Station. He is a member of the Board of Minerals and Energy Resources of the National Academy of Sciences and past president of the American Forestry Association.

Thomas E. Hamilton is associate deputy chief for research, USDA–Forest Service (Washington, DC). For the USDA–Forest Service, he is past director of the Resources Program and Assessment Staff (Washington, DC), assistant director of the Northeastern Forest Experiment Station, and research economist for the Pacific Northwest Forest and Range Experiment Station. He is responsible for organizing and implementing USDA–Forest Service research programs conducted in cooperation with private organizations, states, universities, and other federal agencies.

Lloyd C. Irland is president of The Irland Group and faculty associate at the College of Forest Resources, University of Maine (Bangor, ME). He is past state economist for the State of Maine, Director of Maine's Bureau of Public Lands,

Coordinator of the Maine Department of Conservation's Spruce Budworm Program, professor at the Yale School of Forestry and Environmental Studies, and economist with the Chicago Board of Trade. He is author of *Wilderness Economics and Policy* and *Wildlands and Woodlots: The Story of New England's Forests.*

H. R. Josephson is director of the Department of Forestry and Conservation's Distinguished Visitor Program, University of California (Berkeley, CA). He is past director of the USDA–Forest Service's Division of Forest Economics and Marketing Research and researcher at the Pacific Southwest (California) Forest and Range Experiment Station.

H. Fred Kaiser is director of the USDA–Forest Service's Forest Resources Economics Research Staff (Washington, DC) and is coordinator of IUFRO Division IV. He is past director of the Policy Analysis Staff (Washington, DC), staff economist for State and Private Forestry (Washington, DC), staff economist for Programs and Legislation (Washington, DC), researcher for the Forest and Range Task Force (Lincoln, NE), and researcher at the Southern Forest Experiment Station. He is responsible for planning and coordinating USDA–Forest Service research programs related to forest resource economics.

W. David Klemperer is associate professor of forest economics in the Department of Forestry, Virginia Polytechnic Institute and State University (Blacksburg, VA). He is a former self-employed forestry consultant and staff of Associated Forest Industries, Oregon State University and Washington State University Extension Service. His research interests include valuation, taxation and investment analysis.

William B. Kurtz is professor of forest economics and agricultural economics at the University of Missouri (Columbia, MO). He is past professor of natural resources management at California Polytechnic State University and range conservationist with the USDA–Soil Conservation Service. Research interests include economics of hardwood management and decision making of nonindustrial private forest landowners.

Jan G. Laarman is associate professor of forest economics and policy in the Department of Forestry at North Carolina State University (Raleigh, NC). He is principal investigator for a multi-year USAID forestry development project in Ecuador and Central America and a Fulbright researcher of economic development issues in Costa Rica. His research interests include forestry development, labor economics, forest products trade, and technology transfer in forestry.

Allen L. Lundgren is director of the Forestry for Sustainable Development Program and an adjunct professor in the Department of Forest Resources, College of Natural Resources, University of Minnesota (St. Paul, MN). He is past principal economist with the USDA–Forest Service's North Central Forest Experiment Station, visiting professor at the University of Arizona's School of Renewable Natural Resources, and research fellow at the East-West Center's Environment and Policy Center. His research interests include economics of forest management and evaluation of forestry research.

William G. Luppold is leader of the USDA–Forest Service's economic research unit at the Northeastern Forest Experiment Station (Princeton, WV).

James R. Lyons is staff assistant for the Committee on Agriculture of the U.S. House of Representatives (Washington, DC). He is past director of Resource Policy Programs for the Society of American Foresters and program analyst for the USDA–Fish and Wildlife Service (Washington, DC).

Thomas J. Mills is director of the USDA–Forest Service's Resource Program and Assessment Staff (Washington, DC). For the USDA–Forest Service, he is past director of the Policy Analysis Staff (Washington, DC), staff assistant to the Deputy Chief for Research, group leader of economics research with the Forest Resources Economics Research Staff, and leader of the fire management planning and economics research unit with the Pacific Southwest Forest and Range Experiment Station. He is responsible for coordinating activities devoted to the implementation of the Forest and Rangeland Renewable Resources Planning Act.

Jay O'Laughlin is associate professor in the Department of Forest Science, Texas A&M University (College Station, TX). He has experience in management and administration in the public and private sectors. His research interests include structural changes in wood-based industrial economies.

George L. Peterson is leader of the USDA–Forest Service's valuation of wildland resource benefits and economics and quantitative ecology work unit at the Rocky Mountain Forest and Range Experiment Station (Fort Collins, CO). He is past professor of engineering at the University of California and professor of civil engineering and chair of the Council for Urban and Regional Planning at Northwestern University. He is author of *Valuation of Wildland Resource Benefits*. His research interests include regional planning, outdoor recreation, and natural resource economics.

Richard L. Porterfield is director of Administrative and Participative Management for Champion International Corporation (Stamford, CT). For Champion International Corporation, he was past general manager of the Alabama Timberlands Region, planning manager of Southeastern Operations, manager of Timberlands Raw Material (Stamford, CT), and director of Timberlands R&D (Stamford, CT). He is also past professor of forestry and forest economics at Mississippi State University and at the University of Arkansas at Monticello.

Clark Row is a consulting forest economist (Phoenix, AZ). He is past resource economist with the USDA–Forest Service (Washington, DC). His research interests include forest products markets and timber supplies and trade.

Richard A. Skok is dean of the College of Natural Resources, University of Minnesota (St. Paul, MN) and associate director of the Minnesota Agricultural Experiment Station. He is North American Representative to the International Union of Forest Research Organizations. He is past president of the Association of State College and University Forestry Research Organizations and past member of the Joint Council on Food and Agricultural Sciences, U.S. Department of Agriculture.

Carl H. Stoltenberg is dean of the College of Forestry at Oregon State University (Corvallis, OR). He is past forest economist at the University of Minnesota, Duke University, and Iowa State University, and economics researcher with the USDA–Forest Service in Pennsylvania and Washington, DC. He is chair of the Oregon Board of Forestry and past president of the Society of American Foresters.

Larry W. Tombaugh is dean of the College of Forest Resources, North Carolina State University (Raleigh, NC). He is past chair of Michigan State University's Department of Forestry, assistant director of the National Science Foundation, and principal economist with the USDA–Forest Service. He is a fellow of the Society of American Foresters.

J. Michael Vasievich is leader of the USDA–Forest Service's regional economics work unit of the North Central Forest Experiment Station (East Lansing, MI) and adjunct faculty member of the Department of Forestry at Michigan State University. He is past research forest economist with the USDA–Forest Service's Southeastern Forest Experiment Station and faculty member of the Department of Forestry and Environmental Studies at Duke University. His research interests include quantitative analysis of forest production problems.

Henry J. Vaux is professor emeritus at the University of California (Berkeley, CA). He is past chair of the California Board of Forestry and dean of the University of California's School of Forestry (Berkeley). He is coeditor of *Research in the Economics of Forestry* and a fellow of the Society of American Foresters and the American Association for the Advancement of Science.

Thomas R. Waggener is director of the Center for International Trade in Forest Products and professor of forest economics and policy in the College of Forest Resources, University of Washington (Seattle, WA). He is past staff economist for the U.S. Public Land Law Review Commission. His research interests include forest economics, forest policy, and international trade in forest products.

Henry H. Webster is chief (state forester) of the Division of Forest Management in the Michigan Department of Natural Resources (Lansing, MI). He is past chair of the Department of Forestry, Iowa State University. He serves as chair of the State of Michigan's Interagency Task Force that leads and focuses Governor James Blanchard's target industry program for forest resources and industries.

INDEX